Data Handling and Analysis

HISTOPATHOLOGY

CLINICAL IMMUNOLOGY

CYTOPATHOLOGY

BIOMEDICAL SCIENCE PRACTICE
EXPERIMENTAL & PROFESSIONAL SKILLS

TRANSFUSION & TRANSPLANTATION SCIENCE

HAEMATOLOGY

MEDICAL MICROBIOLOGY

CELL STRUCTURE & FUNCTION

CLINICAL BIOCHEMISTRY

fundamentals OF
biomedical science

Fundamentals of Biomedical Science

Data Handling and Analysis

Second edition

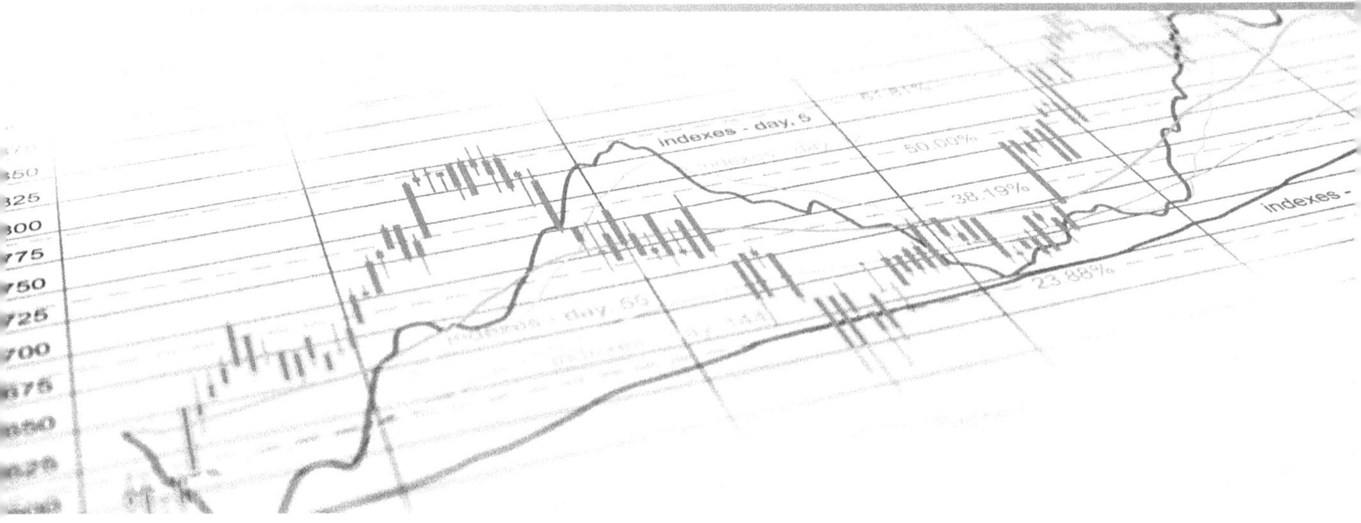

Dr Andrew D. Blann
PhD FRCPath FRCP(Ed) FIBMS CSci
Institute of Biomedical Science
Coldbath Square, London

Great Clarendon Street, Oxford, OX2 6DP,
United Kingdom

Oxford University Press is a department of the University of Oxford.
It furthers the University's objective of excellence in research, scholarship,
and education by publishing worldwide. Oxford is a registered trade mark of
Oxford University Press in the UK and in certain other countries

First edition 2014

Published in the United States of America by Oxford University Press
198 Madison Avenue, New York, NY 10016, United States of America

British Library Cataloguing in Publication Data
Data available

Library of Congress Control Number: 2018950299

ISBN 978-0-19-881221-0

Contents

An introduction to the Fundamentals of Biomedical Science series

Biomedical scientists form the foundation of modern healthcare, from cancer screening to diagnosing HIV, from blood transfusion for surgery to infection control. Without biomedical scientists, the diagnosis of disease, the evaluation of the effectiveness of treatment, and research into the causes of and cures for disease would not be possible. However, the path to becoming a biomedical scientist is a challenging one: trainees must not only assimilate knowledge from a range of disciplines, but must understand—and demonstrate—how to apply this knowledge in a practical, hands-on environment.

The *Fundamentals of Biomedical Science* series is written to reflect the challenges of biomedical science education and training today. It blends essential basic science with insights into laboratory practice to show how an understanding of the biology of disease is coupled to the analytical approaches that lead to diagnosis. Produced in collaboration with the Institute of Biomedical Science, the series provides coverage of the full range of disciplines to which a biomedical scientist may be exposed.

Lecturer support materials

The online resources to accompany *Data Handling and Analysis* feature figures from the book in electronic format, for use in lecture slides. To register as an adopter, visit www.oup.com/uk/blann2e and follow the on-screen instructions.

Thanks to Doug Altman for inspiration

1

Information in biomedical science

Learning objectives

After studying this chapter, you should be able to ...

- Appreciate the variety of different types of information in biomedical science
- Explain the bases of different types of data, e.g. from blood tests and from tissues
- Describe key features of qualitative information
- Understand the fundamentals of quantitative information
- Recall the importance of the mean and standard deviation and of the median and interquartile range
- Outline the importance of the numerical basis of the reference range
- Recall that scaling methods, such as the Likert scale, permit the quantitative analysis of qualitative information

Biomedical science is dominated by information, almost all of which refers to the particular health status (or ill-health, as the case may be) of an individual or groups of individuals. This chapter introduces the different types of information that exist and describes the ways in which this information is presented and used. It gives a broad overview of the issues surrounding information handling and statistical analysis as applied to biomedical science, and explores the different ways in which information can be described and analysed, and the issues that relate to best practice when it is being handled.

1.1 Introduction

Many laboratory scientists are directly involved in the generation of numerical or descriptive information derived directly from human body fluids and tissues. For haematologists, this information may be the number of red and white blood cells in a sample of venous blood, while for biochemists it may be the concentration of glucose in that blood. Microbiologists may collect information that tells them which particular microbes are present in a sample of sputum from a patient with a lung problem, and histologists may study lung tissues from that patient; the information that they handle may determine the presence or absence of a cancer. This information may not just be collected at a single time point—in some cases we are interested in how an index (such as level of cholesterol in the serum) changes as someone undergoes a change in their diet and lifestyle, or is established on a therapy to reduce levels of this lipid.

Information handled by a laboratory scientist is typically generated in one of two ways: either obtained directly by the scientist (for example, using microscopy, as in many microbiology, histology, and cytology analyses) or indirectly, through the use of some kind of automated analyser (as used for most tests in haematology, immunology, and biochemistry). Immunologists and haematologists often use both analysers and microscopy. Scientists in training in Higher Education will be unlikely to generate clinical data such as these, but biochemistry and cell biology experiments often provide the opportunity to develop analytical and statistical skills.

Whatever the means by which the information is collected, however, its value and manner of interpretation must be understood. We must appreciate the limitations of the techniques and what steps we can take to ensure that these techniques are delivering the correct data. Failure to grasp these concepts may lead to error and possibly loss of life.

1.2 Types of information and how to describe them

Let us first consider why information is needed. The understanding and treatment of many human diseases and conditions rely heavily on the generation and interpretation of precise and scientifically generated numerical **data**. But what is data? The *Oxford English Dictionary* defines it as *facts and statistics collected together for reference or analysis.* Similarly, **statistics** is defined as *the practice or science of collecting and analysing numerical data in large quantities, especially for the purpose of inferring proportions in a whole from those in a representative sample.* However, as scientists we need to further refine these definitions, because only through this can the rigour of scientific principles be applied. Hence data is effectively numerical information, and we use statistics to interpret this data—to make sense of and extract meaning from it.

Data
information expressed in a numerical form

Statistics
the formal comparative analysis of data

For example, a person with diabetes needs to monitor and control their blood glucose levels, and also ensure their blood pressure is to target. They need to collect the information (by measuring the level of glucose in the blood and blood pressure) and interpret this as data. The purpose of this is to determine whether the numbers are too high, too low (and so may need to be altered), or are within acceptable limits. In another setting, such as the study of large numbers of people, we may want to determine those factors (such as high blood pressure, obesity, and smoking) that are shown statistically to increase the risk of cancer, stroke, or heart attack in an entire population, and so which are also likely to influence the health of an individual.

Often, a laboratory scientist handles more than just a single value (for example, a single blood glucose measurement), but needs to compare sets of data (that is, several measurements from different groups). The analysis of data in this way starts to draw upon reasonably sophisticated numerical and statistical tools. Quite often, these tools seem quite intimidating at first glance; the aim of this chapter, however, is to help you to understand how and why these tools are used so that you can make these methods work for you. The practical use of these statistical tools will be developed in Chapters 6 to 8.

The nature of information

In biomedical science, information generally comes in one of two types:

- **Quantitative**: where the information is assigned a directly measurable, numerical value, and so can be described as data. Examples of this are height, weight, age, red blood cell count, temperature, serum potassium, or perhaps the proportion or percentage of patients with a particular problem (such as an infection or with diabetes), or the number of people with a particular ABO blood group. In most cases, the value of the particular data is defined not by an individual but by an objective observation or scale, itself often derived from a machine such as a biochemistry autoanalyser.

Quantitative
information in a numerical form, as data

As laboratory scientists we are called upon to help clinical practitioners confirm or refute an initial diagnosis, and similarly to report on the success (or not) of a particular treatment. This often calls for an assessment of levels of a certain molecule or cell in the blood, and, in so doing, determining whether such levels are different from those for an individual derived from a population assumed to be healthy. The latter is described as the **reference range**. We shall return to this towards the end of the chapter.

Reference range
a set of numbers with which a particular item or set of data is compared

- **Qualitative** information, on the other hand, is formed from words (often descriptive terms such as good, reasonable, occasional, frequent, moderate, and severe). These items of information are generally obtained directly from people being investigated, not merely patients with a particular health problem, often through the use of a questionnaire. A major issue associated with word information of this type is that the value or importance of the word itself is defined by the individual, not by the scientist, and therefore has a high degree of subjectivity.

Qualitative
information in the form of words

In spite of the value of these two types of information, some information lies somewhere between the two. It is also possible for some types of qualitative information to be converted into numerical forms (data) for rigorous analysis—that is, for it to be changed from a qualitative to a quantitative form. A weakness with qualitative information is its subjective nature.

Consider the concept of pain: what one person may consider to be moderate, to another person can be severe. This gives a degree of uncertainty. However, a study may compare the proportion of people (perhaps 25 from 50) who use the word 'moderate' to describe their pain, with the proportion (possibly 10 from 50) that use the word 'severe', or even 'excruciating'. Although the assessment of pain is being evaluated in a qualitative way (i.e. the 'quality' of the pain), it can also be assessed in a quantitative manner, with the number of people involved.

SELF-CHECK 1.1

Which of the following are qualitative and which are quantitative? The age of a group of people, how tired they are, how much noise they are making, how far they have travelled today, the type of entertainment they would choose, and the degree of hunger they have.

Categorical data
data that can be allocated to one of a small number of discrete categories

Continuous data
data that can (within reason) take any particular number within a scale

Titre
a semi-quantitative method for estimating how much of a substance, such as an antibody, is present

1.3 Quantitative data

One of the challenges of medicine is that often quite useful information is provided by the patient, such as the degree of pain in the chest or abdomen. This form of qualitative information is difficult to use in a wider sense as it cannot easily be converted into a number that can be transformed into an item of quantitative data. Broadly speaking, we can classify almost all data in biomedical science into one of three types: that which is **categorical**, that which is **continuous**, and data described as a **titre**.

Categorical data

This type of data fits into one of any number of discrete boxes or categories—there are no 'in-betweens'. An example of this is the number of men and the number of women in a particular group. It is important that data lies only in one of a discrete number of categories—and sex is a good example of this. Leaving aside rare conditions such as Turner's and Klinefelter's syndromes, the presence or absence of a single Y chromosome can be used to define an individual's sex; there are generally no in-betweens.

Another example of categorical data is the consequence of sex—children. In real life you can't have a fraction or proportion of a child, only whole numbers. However, this is often overlooked, it being suggested that the average family has 2.4 children!

This concept can be extended to more than two discrete groups. Just about everyone belongs to one of the four ABO blood groups, A, B, AB, and O. There are a few exceptions (one of which is the Bombay phenotype) which, if they do occur, are so rare as to be ignored in the analysis of the frequency of blood groups in different populations (unless you are an expert in blood transfusion!).

Microbiologists can say with a good degree of confidence that an organism (such as MRSA) is present or absent; there should be no in-between, as this information may drive treatment (with an antibiotic). Immunologists can say 'yes' or 'no' to the presence of antibodies, to organisms such as a bacterium or a virus, or if the patient has high levels of certain antibodies to a particular infective agent (which implies recent exposure to that agent). Histologists will tell us that a tissue either is or is not invaded by a cancer, or perhaps that a portion of tissue from a patient with colorectal cancer is best placed into one of four precise Dukes' stage categories which refer to the severity of the tumour. Similarly, cytologists report the presence or absence of certain abnormal cells, such as in cervical cancer. All of these results will be considered when the question of treatment is broached.

Continuous data

Information of this type includes factors such as height, weight, blood pressure, and almost all haematology and biochemistry results (such as the number of white blood cells and levels of triglycerides and sodium in the serum). The data, consisting of individual numbers, can be described by almost any figure in a given scale or range (such as age, which can be anywhere between 0 and 120 years or so).

Semi-quantitative data

A third type of data can be viewed as the mid-point between categorical and continuous data. An example is to quantify how much of an antibody is present in a sample of serum. The serum may first be diluted one part in ten, to give 1/10; a series of increasingly weak dilutions—such as 1/20, 1/40, 1/80, 1/160, 1/320, and 1/640—can then be prepared. This series of dilutions is called a titration (from which **titre** is derived). The scientist will determine which sample still gives a positive result (that is, still shows evidence of the antibody being present when a suitable probe is used). If the sample is still active at a dilution of 1/160, but there is no activity at a dilution of 1/320, then the titre is 1/160. We call this system semi-quantitative, because there is variety from dilution to dilution, but there are no intermediate titres such as 1/30, 1/50, or 1/100.

This system can be used to define the strength of an antiserum but can also help to define disease—for example, if it is likely that a particular disease is present only when the titre exceeds

1/80. It follows that a patient with a result of 1/40 may not (under these conditions) be likely to have the disease, but a patient with a titre of 1/320 is very likely to have disease.

The presentation and analysis of continuous data is more complex than if the data is categorical. When the data is continuous we need to consider three aspects: the central point or tendency, the variation (sometimes called the measure of dispersion), and the distribution. Let us now consider each of these in turn.

SELF-CHECK 1.2

We can classify almost all quantitative data into one of two major classes. What are these groupings and what are their key characteristics?

The central point

The most important statistical aspect of continuous data is the central point of that particular data set, which may be seen as the 'middle', or some loosely related number that can be grasped quite easily. Some statisticians prefer to use the word 'tendency' (which may seem rather inexact) instead of 'point'.

The central point can be defined in three ways—as the mean, as the median, or as the mode.

The **mean point** of a set of data is obtained by simply adding up all the data, and then dividing by the number of individual data points. As such, the mean is equivalent to the 'average' number.

Mean
the 'average' of a set of data points

Consider a group of eleven data points with individual values of 25, 30, 21, 27, 35, 28, 22, 25, 29, 27, and 24. These data (which may be the ages of members of a football team) add up to 293. Thus the mean value is the sum (293) divided by the number of data points (11), which is 26.6.

By placing the eleven data points in order from lower to highest, i.e.,

21, 22, 24, 25, 25, 27, 27, 28, 29, 30, 35,

the mean value (26.6) is indeed very close to the middle value of the whole set (the sixth, with a value of 27). The importance of 'closeness' of these two figures will be discussed further in this section.

This way of arranging a set of numbers by placing them in order, i.e. from the lowest to the highest, is called ranking. The data point that is in the middle of the entire rank is the **median**, and this value is arrived at using a completely different set of rules than for the mean value.

Median
the single data point that occupies the middle position of a set when all of the data points are ranked in order

Consider the data set, 21, 60, 15, 32, 19, 25, 113, 24, 29, 35, and 22. If these are arranged, or ranked, in order from the smallest (on the left) to the largest (on the right), the following series is obtained, i.e.,

15, 19, 21, 22, 24, **25**, 29, 32, 35, 60, 113.

In this ranked series of 11 data points, the one in the middle is the sixth one—i.e. with five data points larger (greater than 25) and five data points smaller (less than 25). This central point is the value 25—also known as the median (highlighted in bold). Note that it is not the same as the mean. In this data set the mean (average) would be the sum of all of the individual points (395) divided by the number of individual data points (11) to give 35.9—a very different figure from 25. This difference is important and will be discussed later in this section, and is a recurrent theme in the analysis of data.

If the data set is of an even number of data points, when there is no exact middle point, this is taken as the average of the two middle points. For example, in the series

10, 11, 12, 14, 18, 26, 39, 60

there is no one single middle point, so we must create one, which in this series is the mid-point between 14 and 18, i.e. 16.

Mode
the single data point that occurs most frequently

The **mode** is that number which appears most frequently in a data set. It can be any number in the data set, and often bears little relationship to the mean and the median.

Consider the data set, 24, 29, 35, 28, 14, 24, 36, 22, 24, 40, 32, 28, 30. The number 24 appears three times, the number 28 twice. Therefore, the mode is 24 to one decimal place. As it happens, the mean of this data set is 28.1, and the median is 28. So it is entirely possible for a data set to have a different mean, median, and mode.

To conclude this section, we see that each set of data has a mean, a median, and a mode. Sometimes they are the same number (or are very close together), but in other data sets they may be very different. The mean is certainly the index used most frequently, whilst the mode is very rarely discussed and so is unlikely to trouble us in the remainder of this volume.

SELF-CHECK 1.3

What are the mean, median, and mode of the following data set?

156, 217, 199, 267, 306, 224, 256, 278, 172, 267, 317.

Variation

Variation
the degree of variation or dispersion in a data set

A second important concept in coming to grips with a set of data is its **variation** (derived from its root word 'variety')—that is, how much variety there is within a given data set. Others may use the word 'dispersion' for the same factor. The variation of a data set tells us about the highest and lowest points in a particular group of data, and the extent to which the data clusters tightly around the central point of the mean or the median. If it is tightly clustered, the data is said to have low variation, and the data has a low degree of dispersion. Conversely, if the data set is more diverse, or is more dispersed (spread out), it has a high degree of variation.

Standard deviation
the degree of variation in a data set when the central point is the mean

Inter-quartile range
the degree of variation in a data set when the central point is the median

We use one of two particular measures of variation depending on the nature of the central point (the mean or the median). When the central point of a data set is the mean, we use **standard deviation** (SD) to describe its variation. However, if the data is such that the central point is the median, then the variation is described in terms of the **inter-quartile range** (IQR).

The standard deviation (SD)

The standard deviation (SD) helps us to evaluate the degree to which the data in a particular data set is clustered closely about the mean value, or is more spread out. In other words, it gives a measure of the variation present in that data set. For example, suppose a data set has a mean of 100 and a SD of 20. The SD is telling us that most of the data points (in fact, about two-thirds of them) can be found within one SD either side of the mean—that is, between 80 and 120. However, we can go further, and we often find that nearly all of the data points (actually, about 95% of them) are between 60 (that is, the mean minus two SDs) and 140 (that is, the mean plus two SDs). This figure of 95% is important, and we will return to it later, and in more detail, in Chapters 6 and 7. Going one step further, the mean plus and minus three standard deviations includes over 99% of the data points (Figure 1.1).

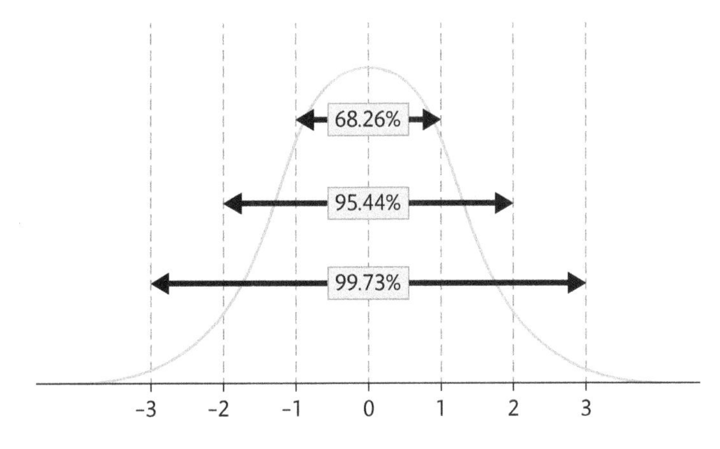

FIGURE 1.1

The mean and standard deviation. Slightly over two-thirds of data points are within one standard deviation either side of the mean, slightly over 95% are within two standard deviations, and almost all are to be found within three standard deviations.

Now consider another data set, which has a similar mean of 100 but a much smaller SD, such as 8. This time, the SD tells us that slightly over 95% of the data points should be in the range between 84 (that is, 100 minus 16) and 116 (that is, 100 plus 16). So a small SD tells us that the set of data is tightly clustered near to the mean, and when the SD is large, this means that the data is more spread out. This is illustrated in Box 1.1.

BOX 1.1 The standard deviation

Take the data set we have already seen:

21, 22, 24, 25, 25, 27, 27, 28, 29, 30, 35.

The mean value is 26.6, but some numbers (such as 35) are greater than this and some (such as 21) smaller. The 'range' of the points of data is therefore 14 (that is, 35-21). Now consider a second set of data:

15, 20, 24, 25, 25, 27, 27, 28, 29, 32, 41.

The mean value of this set is also 26.6, but the two lower numbers (15 and 20) and higher numbers (32 and 41) are different from the upper set and give it a greater 'range', which is 26 (that is, 41 minus 15). The variation of a data set is simply the degree to which individual data points are near or far from the mean and is called the deviation, and can be formally defined mathematically as the standard deviation. This can be calculated using a hand-held calculator, or using some statistical software on a personal computer. Accordingly, the standard deviations of these two data sets are 3.9 and 6.6 respectively.

When describing the first set of data we would generally say that the mean and SD are 100 (20), although some use the notation 100 ± 20. Similarly, the second data set can be written as 100 (8), or perhaps 100 ± 8. Technically, it is inappropriate to describe a data set as 100 + 20: the appropriate descriptor is 100 (20). The SD is quite hard to calculate on a simple hand calculator, but is easier to obtain using a programmable calculator and, of course, on a computer loaded with a statistical software package. Figure 1.2 illustrates the degree of variation between two different patterns.

Cross reference

Chapter 7 has the method for determining the standard deviation with a simple hand calculator

The word 'distribution' is used to describe the 'shape' of a data set, as is illustrated in Figures 1.1 and 1.2. Most data sets (such as height and weight) naturally have this pattern—where the

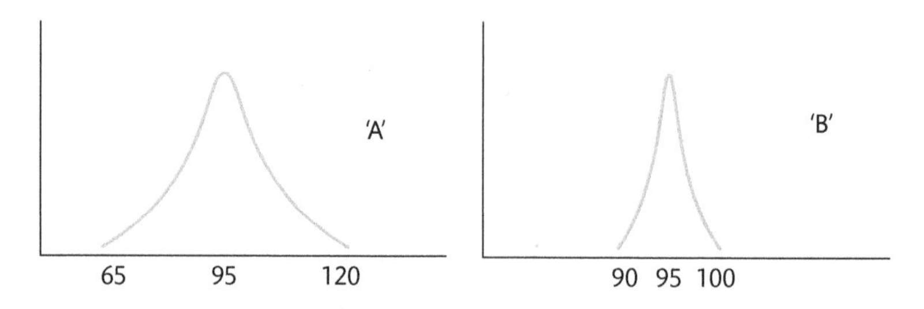

FIGURE 1.2

Data sets with different SDs. This figure shows two sets of data—pattern A and pattern B—which both have the same central point value of about 95. However, the two patterns have very different variations—one of them (pattern A) covers a much greater range (from about 65 to 120) compared with the other (pattern B), which runs from 90 to about 100. So in both cases the mean is 95, but the SD of pattern A is 9, whilst the SD from pattern B is only 2.

peak is roughly in the middle with the same slow and steady curves to the left and right. Data that conform to this pattern are said to have a **normal distribution**, often also described as a Gaussian distribution, or most simply, as a bell-shaped curve. This concept is very important in the statistical analysis of data, and forms a major part of Chapter 7.

Normal distribution
a pattern of data that is best described in terms of mean and standard deviation

SELF-CHECK 1.4

Which of these data sets of mean and (standard deviation) have the largest and the smallest variation: 75 (30), 100 (20), and 150 (25)?

The inter-quartile range (IQR)

The second method for describing variation within a data set, the inter-quartile range (IQR), is derived from the data points that are one quarter and three quarters of the way into the complete data set when it is ranked from lowest to highest. So starting from the lowest point, and working our way up, after we have reached 25% of the data points, we have the 25th percentile. Continuing to work up the data set, the halfway point (i.e. after 50% of the data points) is the 50th percentile, i.e. the median.

Continuing up the rank, after 75% of the data points have been passed, the 75th percentile is achieved. Finally, the highest value is the 100th percentile, as 100% of the data points will have been assessed. So unlike the SD, the IQR is relatively easy to obtain from the raw data. The IQR, like the SD, also gives us an idea of the spread of the data. We would write a summary of the data in Box 1.2 of the median and IQR as 77 (49–148).

BOX 1.2 The inter-quartile range

Consider the data set of fifteen points:

65, 43, 79, 250, 82, 49, 148, 55, 77, 40, 516, 101, 59, 47, 156.

In order to obtain the median and IQR, we must unravel, or sort out, the data and place it in a rank order, from lowest on the left to highest on the right, as follows:

40, 43, 47, **49**, 55, 59, 65, 77, 79, 82, 101, **148**, 156, 250, 516.

The middle value, the eighth one (i.e. one with seven points above and seven points below) is 77—the median—and is underlined. Similarly, the middle value of the lower group of 7 (from 40 to 65) is 49 (highlighted in bold), and the middle value of the higher group (from 79 to 516) is 148 (also highlighted in bold). So the IQR is 49 to 148.

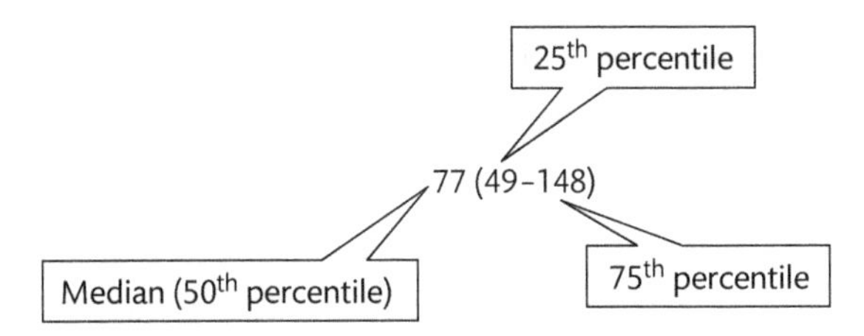

Note that the median value here (77) is certainly not in the middle of the IQR of 49 to 148. Clearly, 77 is much closer to 49 than it is to 148, and this fact is an important feature of this type of data. This mathematical feature is illustrated in Box 1.2.

Just as two sets of data can have the same mean but different SDs, so two sets of data can have the same median and different IQRs. In both cases, the data set with the greater SD or IQR has the greater variation. Two such patterns are illustrated in Figure 1.3, with patterns C and D.

This type of data, with a median and IQR, is therefore different from that expressed as mean and SD (as in Figures 1.1 and 1.2), and is encountered much less frequently in biomedical science. So because the data set is different from the 'normal' shape of Figures 1.1 and 1.2, we describe data in this form as having a **non-normal distribution**, as is explained in Figure 1.3. Once more, the importance of this is described in Chapter 6.

Non-normal distribution a pattern of data that is best described in terms of median and inter-quartile range

SELF-CHECK 1.5

What are the two ways in which we can describe the variation of different types of data?

The reference range

A great deal of the work of a scientist is to provide an accurate and reliable result on an index (such as haemoglobin or serum potassium) that will help clinical practitioners to diagnose and

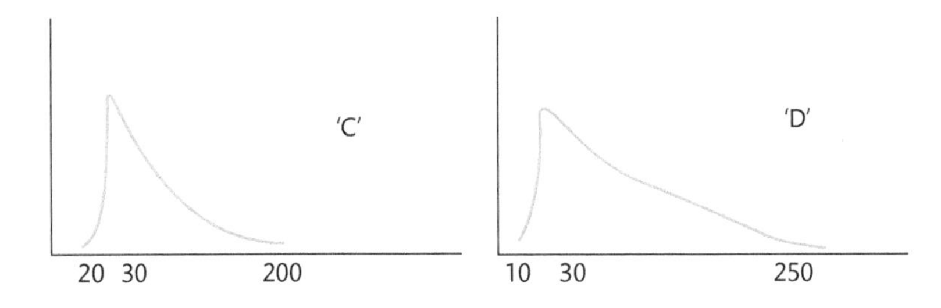

FIGURE 1.3

Data sets with different IQRs. This figure shows two sets of data—pattern C and pattern D—which both have the same central point value, a median, of about 30 units. However, the two patterns have very different degrees of variation—one of them (pattern C) covers a much smaller range (from about 20 to 200) compared with the other (pattern D), which has a larger range, running from 10 to about 250. So in both cases the median is about 30, but the IQR of pattern C is perhaps 25 to 90, whilst the IQR in pattern D is approximately 20 to 150.

manage disease. However, the practitioner needs to know whether the result from a particular patient has pathological significance (that is, has something to do with a certain disease or condition). This is generally done by comparing the result from the patient with that from a group of people who are presumed to be healthy. The results from these apparently normal people make up the reference range (sometimes called a normal range, or in some cases, a target range), which we have already briefly mentioned.

The reference range is made up of data points from hundreds or perhaps thousands of people. As described in Figure 1.1, if the distribution of the index is normal, then results from 95% of these people will lie within almost two standard deviations either side of the mean (hence mean + two SDs), whilst nearly all of them (over 99%) will be found within three standard deviations (i.e. mean + three SDs). So a result found to be four or five standard deviations away from the mean is a long way outside this range, and so is increasingly likely to be abnormal, and therefore reflective of a particular pathology. In the same way, an index with a non-normal distribution (such as serum triglycerides) will also have its own reference range.

The understanding of reference range is not merely an exercise in academic mathematics; it is fundamental to many aspects of biomedical science. Suppose a family member comes to you with the news that the result of a particular blood test of theirs shows the amount of Substance X to be 15 units/mL. Should they be concerned?

If the reference range is 10–20 units/mL, the result is right in the middle, and so is effectively normal, and the risk of a disease or other pathological condition minimal. However, if the reference range is 10.5–14.9 units/mL, then the result is only marginally outside the reference range, and so the probability that it reflects a pathological condition is not that likely. But if the reference range is 2–8 units/mL, or 30–45 units/mL, then the result of 15 units/mL is far from either reference range, and it is difficult to avoid the likelihood that it does indeed reflect an abnormality (Figure 1.4).

The importance of the reference range is demonstrated by its frequent occurrence throughout this book and in daily laboratory practice.

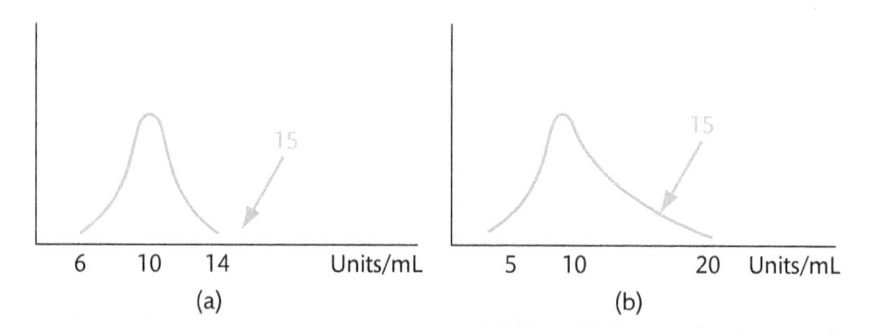

FIGURE 1.4

Likelihood that a particular result is within the reference range. The two panels illustrate the likelihood that a result of 15 units/mL brings concern. In (a), the data has a distribution that is non-normal, with a median of 10 and a reference range of 6 to 14. In this example, the result of 15 is outside this range, and therefore worthy of attention. In (b), the distribution is wider: the median is again 10, but the reference range is 5 to 20, so that the value of 15 is well within this range. These panels demonstrate the importance of correctly determining a reference range.

1.4 Qualitative information

Qualitative information is gathered and analysed primarily as words (single or in phrases), possibly in narrative form, or with descriptive quotations. The proponents of qualitative research quite reasonably argue that not all information can or should be reduced to a number. For example, how can you reasonably quantify (place a numerical value on) someone's feelings, attitudes, or opinions, such as belief, fear, and love? Qualitative information is often obtained from people in surveys, interviews, or observations, ideally from a validated or structured questionnaire. An alternative would be to observe and record verbal or written comments collected freely (that is, without a structure) from a defined collection of people.

A further distinction between qualitative and quantitative approaches to data collection is less to do with the collected information itself, but more the purpose to which that information is directed. The tradition from which quantitative research has evolved is of prediction and the laws of cause and effect, both of which are central to asking and answering research questions. For example, we can suppose that, in a given group of patients, there will be a change in a defined index (blood pressure, cholesterol) following treatment with a particular drug.

In contrast, qualitative research generally does not generate numerical information that can easily answer a precise research question. Instead, the approach it often takes aims to develop a theoretical explanation for the subject of study, the strength of which is therefore not necessarily in predicting an outcome, but in providing an increased understanding of the topic being investigated. Thus the qualitative researcher may attempt to gain insights into an individual's feelings and thoughts, often within their own natural setting, such as at home or at work.

Qualitative information is to the fore in sociology, psychology, and other studies, but does not have as great a role in biomedical science as does quantitative data. However, there are instances where the patient experience is important, and so, in some circumstances, a qualitative approach may be the best way to collect information, and in these cases a questionnaire is most commonly used.

The questionnaire

The questionnaire is often at the centre of many qualitative research projects. Considerable effort must be directed to ensure, as far as is possible, that questions are appropriate, unbiased, and cannot be misunderstood. For example, whilst the word 'permit' may be seen as the opposite of 'forbid', the latter word is viewed by many as pejorative, whilst 'not permit' elicits more responses.

Alternatively, the questioner may offer a dozen or more alternative words, and let the subject choose those he or she is happiest to use. In closed questions, the respondent is offered only a small number of answers, such as yes, no, and don't know, or may be asked to provide a response from a scale such as agree strongly, agree mildly, neither agree nor disagree, disagree mildly, and disagree strongly. A weakness of this latter scale of five is the temptation to sit on the fence and give the middle response, and be neutral, a position easily changed by simply removing the middle expression and reducing the number of options to four, thus forcing a choice. An additional problem with questionnaires is that respondents sometimes give a different response if re-tested, giving rise to fears of poor reproducibility and so questionable reliability.

Open-ended questions

One way to determine the response of a group of individuals is to ask them questions to which there is no exact answer—the individual is free to make any response they feel moved to make. An advantage of this is that the information is genuine and unadulterated. However, a disadvantage is that there may be 20 different responses from 20 different people, and this therefore gives us no clear common theme.

Consider a simple question such as 'How was your day at work?' In asking for a one-word answer, responses from 20 of your colleagues may be: great, boring, exciting, stressful, frustrating, fine, no problem, tiring, standard, terrible, horrible, enjoyable, pressured, stimulating, nice, irritating, lousy, and satisfying; and some may prefer not to give an answer (Table 1.1). Of course, it may be that several of your colleagues use the same words. These 20 colleagues of yours constitute a **focus group**.

Focus group
a group of individuals called upon to give an opinion on a particular topic or topics

In many cases the composition of a focus group is crucial, and the responses are applicable only to the make-up of the group: its conclusions cannot be generalized. A further problem is the extent to which the word is truly representative—it is very likely that a different day will elicit a different word, so that the information is far from reproducible. On the other hand, the chosen word or words are very genuine. From the human resources viewpoint, the words summarized in Table 1.1 could be classified as 'good' or 'bad', and as it happens your colleagues have responded with an equal number of each.

Leaving aside the implications of the choice of these words, they can be used to formulate a new set of questions, perhaps with some editing. For example, do the words 'exciting' and 'stimulating', and the words 'terrible' and 'horrible', carry the same weight? One way forward is to offer a second group of individuals a limited number of words to choose from in answering the same question.

SELF-CHECK 1.6

Name an advantage and a disadvantage of the value of information obtained from a focus group.

Semi-structured questions

In these settings the individual is faced with a finite number of choices, and they pick the one they feel most at ease with. For example, you may ask, "Please choose from the following list up to three 'bad' words and up to three 'good' words that best fit your thoughts on your day

TABLE 1.1 **Responses to the question 'How was your day at work?'**

'Bad' responses		'Good' responses	
Tiring	Boring	Great	Exciting
Stressful	Frustrating	No problem	Fine
Terrible	Horrible	Standard	Enjoyable
Pressured	Irritating	Stimulating	Nice
Lousy	Pointless	Satisfying	Fabulous

TABLE 1.2 **Potential response to the question "Please choose from the following list up to three 'bad' words and up to three 'good' words that best fit your thoughts on your day at work".**

'Bad'		'Good'	
Terrible	Horrible	Standard	Enjoyable
Pressured	Irritating	Stimulating	Nice
Lousy	Pointless	Satisfying	Fabulous

at work"; such a list is shown in Table 1.2. Note the request to choose up to three, as some may prefer to choose none, or only one, or two words from each group. This is important as it allows some of your colleagues to give more or less emphasis on their side—they may have neither anything good nor bad to say!

These questions may also be described as 'closed-ended' (that is, the opposite of open-ended) as the respondent is directed to give one or more of a series of answers that have been provided by the questioner. So although the respondent has less freedom, the information is easier to collate.

The ultimate closed question is where the respondent can answer only yes or no. The 'loaded' question is one where the respondent is almost forced into a particular answer, such as asking 'Have you ever driven a car whilst drunk?'. This is a difficult question to answer, not merely because of the implications, but because of the differing perceptions of being 'drunk'.

The results of this question can then be collated, as in Table 1.3. With a large enough number of colleagues responding (the sample), the most representative words will inevitably come to the fore, and so reflect the views of the respondents. Information of this nature is clearly important and informative as it provides a snapshot of the view of the workforce in how they perceive their daily work.

The 'good' word most frequently chosen is 'enjoyable', with five choices, followed by 'satisfying' and 'standard' with four each, all of which may indicate a happy workforce. However, as many chose the latter two 'good' words as also chose a 'bad' word in that they considered themselves to be 'pressured'. Consequently, interpretation can be difficult, especially if the sample size (the number of respondents) is small.

If the number of respondents is large enough (with perhaps dozens of completed questionnaires), then the information can also be sub-classified according to factors such as sex and the grade of the worker. This can also provide interesting and possibly useful information that one would expect would be used to improve the working environment.

Cross reference
We discuss the concept of appropriate sample sizes in Chapters 6 and 7

TABLE 1.3 **Responses to the question "Please choose from the following list up to three 'bad' words and up to three 'good' words that best fit your thoughts on your day at work".**

	Five responses	Four responses	Three responses	Two responses	One response
'Bad'	(no responses)	Pressured	Irritating	Pointless	Horrible
'Good'	Enjoyable	Satisfying, Standard	Stimulating, Nice	(no responses)	(no responses)

Scaling methods

Scaling methods ask the respondent to choose only one of a small but well-defined scale of words that are linked in an order—that is, are ordinal. An example of this may be to seek the response of the patient to the amount of pain they are feeling, asking them to choose one from:

- None
- Mild
- Moderate
- Severe
- Excruciating

In this ordinal scale, the pain gets worse step by step. Another scale might ask the opinion of the respondent on giving their agreement or disagreement to a particular question. For example, in answering the statement, "You are enjoying reading this chapter", do you:

- Strongly disagree
- Disagree
- Neither agree nor disagree (or perhaps are 'undecided')
- Agree
- Strongly agree

Likert scale
a series of discrete or categorical choices that are linked in an ordinal pattern

Naturally we hope the latter are chosen more frequently! This scaling system is named after its inventor, Rensis Likert. As briefly mentioned, one of the problems with this five-point **Likert scale** is a tendency for many respondents to adopt the mid-point of neither agree nor disagree. A solution to this is to reduce the scale point to four by removing the mid-point so that the only options are:

- Strongly disagree
- Disagree
- Agree
- Strongly agree

which forces the respondent to choose a positive or negative response. Other choices for a Likert scale include: very frequently, frequently, occasionally, rarely, and never, although there are of course many different variations.

Converting qualitative information to quantitative data

Visual analogue scale
a method for converting the subjective opinion of a respondent into a number within a defined scale that permits a quantitative analysis

It is entirely possible to assign a numerical value to many types of qualitative information, and so perform a formal quantitative analysis. An example of this is the **visual analogue scale**, where the respondent subjectively selects any point on a scale that has no units, such as a ruler with no markings, but which clearly has two defined extremes. Taking the 'pain' model used above, and defining the left-hand extreme as no pain, and the right-hand extreme as the worst pain that can be imagined, the respondent may mark with an 'X' the level of pain they experience, giving:

No pain --------------X------------------------ Worst pain

It is then merely a matter of the researcher attaching an arbitrary scale to give a value to the point 'X' chosen by the respondent:

No pain ---------------X-------------------------Worst pain
0–1–2–3–4–5–6–7–8–9–10–11–12–13–14–15–16–17–18–19–20–21

So that in this case, the level of pain the respondent is experiencing is perhaps 10 from a scale of 0 to 21. Repeat this process with nine other respondents and clearly there will be a consensus, which may be the mean or median level of pain. This method is reasonably reproducible (that is, the respondent tends to score a similar result on different occasions), and can also be used to test the effect of a drug designed to relieve pain, where the score may fall from 16 to 8.

In the same way, the five-point fixed scale can be given a numerical value, the most obvious being none and strongly disagree = 1; mild and disagree = 2; moderate and neither agree nor disagree = 3; severe and agree = 4; and excruciating and strongly agree = 5. With a large enough sample size, a mean and standard deviation can also be obtained, and by sub-classifying, the views of different groups (perhaps men and women) can be collected. The precise method by which we determine whether differences are truly significant is discussed in Chapter 7.

Chapter summary

- Information can be as letters and words (i.e. qualitative) or as numbers (i.e. quantitative, and is described as data).
- Quantitative data can be categorical or continuously variable.
- The central point of a continuously variable data set is the mean (the average) or the median (the middle).
- Variation, or dispersion, is used to describe the extent to which the individual points of a data set are tightly bunched together around the central point, or are widely spread out.
- The standard deviation is used to quantify the variation when the central point is the mean.
- The inter-quartile range is used to quantify the variation when the central point is the median.
- When data is normally distributed, the central point is the mean.
- When data is non-normally distributed, the central point is the median.
- The reference range is a group of numbers used to provide an idea of expected or desirable results.
- The focus group and questionnaire are key components of the collection of qualitative information.
- Methods such as the Likert scale and the visual analogue scale provide the opportunity to convert qualitative information into quantitative data for formal analysis.

Suggested reading

- Altman, D.G. *Practical Statistics for Medical Research*. Chapman & Hall, London, 1991.
- Daly, F., Hand, D.J., Jones, M.C., Lunn, A.D., and McConway, K.J. *Elements of Statistics*. The Open University/Addison Wesley, 1995.
- Holmes, D., Moody, P., and Dine, D. *Research Methods for the Biosciences*. Oxford University Press, Oxford, 2006.
- Petrie, A. and Sabin, C. *Medical Statistics at a Glance*, 2nd Edition. Blackwell, Oxford, 2004.
- Swinscrow, T.D.V. *Statistics at Square One*, 9th Edition (Revised by M.Y. Campbell). BMJ Books, London, 1996.

Questions

1.1 In the data set 25, 37, 29, 30, 43, 32, and 22, what (to the nearest whole number) is the mean and what is the median?

1.2 Consider the two sets of data:

Set A: 124, 99, 346, 194, 115, 140, 77

Set B: 120, 148, 81, 96, 143, 118, 205

Which of the following statements is true?

a. Both sets of data are normally distributed.

b. The median value of Set A is 124.

c. The inter-quartile range of Set B is 96 to 148.

1.3 What methods can be used for converting qualitative information to quantitative data?

1.4 In the following paragraph, what information is qualitative and what is quantitative?

> *"I've really got a terrible bladder problem. In all my 46 years I've never known anything like it, even my wife of 21 years says she hardly gets a wink of sleep with all my tossing and turning most of the night, I must be up to the toilet 5 or 6 times before morning."*

2

Handling quantities: mass, volume, and concentration

Learning objectives

After studying this chapter, you should be able to ...

- Explain the importance of the Système International convention
- Outline the structure of the atom
- Describe key features of the periodic table
- Understand the concepts of atomic number and atomic mass
- Recall the key features of the determination of the mole and molarity
- Grasp key concepts of acids, bases, and buffers

The human body is a complex collection of organ systems, composed of different types of cells and tissues. In order to grasp the failure of these organs, and how this leads to disease (pathology), we must first study how they work correctly in the first place (physiology). However, to achieve even this we must grasp key concepts of cell biology, and how the cell itself operates.

In moving to ever smaller levels of complexity, cells are composed of subcellular components such as the nucleus and Golgi apparatus, themselves constructed of proteins, carbohydrates, lipids, and nucleic acids. These four fundamental forms of biological materials are formed from combinations of the 92 naturally occurring chemical elements, the study of which is chemistry.

Laboratory scientists are interested in chemistry because the chemical elements come together to form molecules, cells, tissues, organs, and so ultimately an entire organism. We also need to understand chemistry because differences in the levels of chemicals and molecules in the blood, other fluids, and tissues can give us clues to the presence of particular diseases, and often how they can be managed. An example of this is the bicarbonate

ion, consisting of atoms of carbon, oxygen, and hydrogen. Levels of the bicarbonate ion are important in helping to regulate the pH of the blood. A knowledge of bicarbonate levels is important in investigating and treating the consequences of abnormal pH, as in the clinical conditions acidosis and alkalosis. We will consider these issues later in the chapter.

The field of chemistry has also given us tools for measuring and characterizing different substances—also of central importance to laboratory-based diagnosis. Examples of these tools include electrophoresis, ion-specific electrodes, colorimetry, and high-performance liquid chromatography.

The objective of this chapter is to introduce a number of important aspects of chemistry—and particularly the measurement of chemical compounds—that are central to biomedical science practice.

2.1 Units and quantities

We know from everyday life that units are essential to understanding quantities. We might buy one *pint* of milk, or a laptop with a 15-*inch* screen. The unit—in italics here—is a vital part of communicating a particular quantity or measurement. If we saw an advert for a laptop that said it had a screen with a width of 15, it would be meaningless to us; it could be *centimetres*. Units make all the difference to our understanding.

The history of science has borne witness to slow and steady agreement about the best units to use when assessing key attributes such as weight and distance. These units need to function in all areas of biology, chemistry, and physics across national and international borders. The Imperial system of measurement, which was widely used in the UK (and still remains in use today), features units of length including the inch, foot, yard, and mile; whilst units of weight include the ounce, pound, stone, and ton. For decades, the standard unit to quantify the amount of mechanical work being performed was that which could be done by a horse, hence 'horse power'. But from the scientific perspective, these systems of measurement are inefficient, and a better, more coherent system is called for. The major structure which brings these together is that of the **Système International** (SI).

Système International
a convention for assessing key physical features

SI units are the ampere (the unit of electric current), the candela (the unit of luminosity or brightness), the kilogram (mass or weight), the metre (distance or length), the second (time), the kelvin (temperature), and the mole (the amount of a substance). Details of those that are most relevant to biomedical science are shown in Table 2.1.

The metre is defined as the distance travelled by light in a vacuum in 1/299792458 of a second, the kilogram as the mass of the international prototype kilogram. Note therefore that the kilogram, not the gram, is the base unit of mass.

When considering chemical quantities, we use units of the mole (mol). Moles are often the source of much confusion—but they are simply a specific quantity: 6×10^{23}. This quantity is defined as the amount of a substance that contains as many elementary 'entities' as there are atoms in 0.012 kilograms of the isotope 12 of carbon (carbon-12 or ^{12}C). As such, 'one mole' is a shorthand way of saying '6×10^{23}', just as 'one dozen' is a shorthand way of saying '12'.

When we use units of mol we must specify the species that we are quantifying—i.e. moles of atoms, molecules, ions, electrons, other particles, or specified groups of particles. (In principle, we can have a mole of anything—apples, eggs, cats—but the quantities involved would be so

TABLE 2.1 **Key features of the SI convention in biomedical science**

	Unit (abbreviation)	Example
Mass (weight)	kilogram (kg)	The weight of an adult male is approximately 80 kg
Distance (length)	metre (m)	The height of an adult male is approximately 1.7 m
Time	second (s)	The speed of a certain nervous impulse is 2 metres per second
Thermodynamic temperature	kelvin (K)	The boiling point of water is 373 K
Amount of a substance	mole (mol)	An average human body contains 25 moles of calcium

huge that it only makes sense, in reality, to quantify species in moles when we are considering things at the scale of atoms, molecules, etc.)

SELF-CHECK 2.1

Give examples of the use of the five major SI units in biomedical science other than those in Table 2.1.

Derived units

By combining any of the primary SI indices shown in Table 2.1, we can derive several other units. For example, a newton is the force required to give a mass of one kilogram an acceleration of one metre per second per second. Major derived units in biomedical science are shown in Table 2.2.

Other derived SI indices of occasional use in biomedical science include the gray (the absorbed dose of ionizing radiation, units of which are J/kg), the sievert (the equivalent dose of ionizing

TABLE 2.2 **Key features of derived SI units in biomedical science**

	Unit (abbreviation)	Example
Frequency	Hertz (Hz): units are s^{-1}	The frequency of the heart beat is approximately 60–70 beats per minute (that is, about 1 beat s^{-1})
Pressure	Pascal (Pa): units are $N.m^{-2}$	The partial pressure of oxygen in the blood is approximately 12–14 Pa
Temperature	Celsius (°C): temperature relative to 273.15 K	The temperature of the body is approximately 37 °C
Radioactivity	Becquerel (Bq): decays per second	The rate of decay of a radioisotope is 17 events s^{-1}

radiation, also in units of J/kg), and the katal (catalytic activity, kat, generally in units of moles of substrate consumed or product generated per second, i.e. mol s^{-1}).

SELF-CHECK 2.2

Give an example of the practical use of derived SI units in a biomedical science laboratory.

Non-SI units

In spite of the efficient and comprehensive nature of the SI convention, it is imperfect, and there are several instances where old non-SI units have been retained, almost always for convenience. In other cases it is simply not worth redefining a unit that is widely recognized worldwide. For example:

- The litre (l or L) as a unit of volume, generally of a fluid, is often preferred over the SI unit of dm^3 (the volume occupied by a cuboid of sides 0.1 (a tenth) of a metre (hence decimetre, dm)). However, another commonly used unit of volume is the cm^3. This is a cuboid of sides 0.01 of a metre (hence centimetre, cm). A litre is a volume that can be accommodated in a regular cube of dimension $10 \times 10 \times 10$ cm.
- The correct SI unit of the mass of an atom or molecule is the kilogram, but for convenience we use a dalton, shortened to Da, to be detailed more comprehensively shortly. One dalton is equal to 1.6605×10^{-27} kg.
- Time, where increasingly large units of the base SI unit (the second) are expressed to the base of 60, 24, 7, 12, 10, 100, and 1000. That is, from the second to the minute, hour, day, week, month, year, decade, century, and millennium. However, increasingly small units of time of less than a second adopt the base ten: a millisecond is one thousandth of a second.
- Blood pressure is measured in units of millimetres of mercury. Although there is no mercury in today's devices for measuring blood pressure, the unit is retained.

Prefixes

The SI system works very well, but does not in itself cover all contingencies. For example, the length of an adult thumb is approximately 0.04 m, whilst the diameter of a red blood cell is approximately 0.000 007 5 m. These numbers are clearly often inconvenient, so a second, complementary system has been developed to enable large and small numbers to be described with ease. This system uses multiples of one hundred, one thousand, and one million, etc. Thus, 0.04 m can be converted to 40 mm, where m = milli, and represents one thousandth (or 1/1000, or 10^{-3}), and 0.000 000 000 75 becomes 0.75 nm, where n = nano and is equivalent to one thousand millionth (or 1/1 000 000 000, or 10^{-9}).

A parallel system is used to make large and small numbers such as 10 000 and 0.0001 more accessible. The system uses powers of ten, so that 100 and 0.01 become 10^2 and 10^{-2} respectively, whilst 1000 and 0.001 become 10^3 and 10^{-3} respectively. The system can be extended by bringing in other numbers, so that 2500 becomes 2.5×10^3 and 0.0004 becomes 4×10^{-4}. This system is summarized in Table 2.3.

SELF-CHECK 2.3

What is the best prefix for 0.000 005 units?

TABLE 2.3 **Conventions for large and small values of the SI units**

	Factor	Prefix	Example
Increasingly large			
Kilo-	1000; 10^3	k	The molecular weight of albumin is 67 kDa
Mega-	1 000 000; 10^6	M	The molecular weight of von Willebrand factor can be as great as 20 MDa
Increasingly small			
Centi-	0.01; 10^{-2}	c	The length of the adult thumb is perhaps 4 cm
Milli-	0.001; 10^{-3}	m	A normal healthy diet provides about 25 mmol of calcium per day
Micro-	0.000 001; 10^{-6}	µ	A typical serum concentration of creatinine is 80 µmol/L
Nano-	0.000 000 001; 10^{-9}	n	The wavelength of the visible spectrum ranges from around 385 nm to 725 nm
Pico-	0.000 000 000 001; 10^{-12}	p	The concentration of serum parathyroid hormone is about 3.5 pmol/L
Femto-	0.000 000 000 000 001; 10^{-15}	f	The volume of a blood platelet is approximately 7 fL

SELF-CHECK 2.4

Convert 450 000 to a number between 1 and 10, and a factor of 10.

Now we have introduced some concepts of mass (the gram and kilogram) and concentration (the mole), we can begin to explore basic concepts in chemistry.

Proton
a charged particle and component of the nucleus of the atom. It has an atomic weight of one unit

Electron
a negatively charged particle and component of an atom. It may be seen as being in orbit around the nucleus

Neutron
an uncharged particle and component of the nucleus of the atom. It has an atomic weight of one unit

2.2 Quantities at the chemical scale

Individual atoms of the chemical elements comprise three subatomic particles—the **proton**, **electron**, and **neutron**. The protons and neutrons associate to form the nucleus at the centre of an atom; the electrons occupy a 'cloud' around the central nucleus. A proton carries a single positive charge, while an electron carries a single negative charge; a neutron is electrically neutral—it carries no charge. An atom contains an equal number of protons and electrons; consequently, atoms carry no overall electrical charge—the positive and negative charges of the protons and electrons cancel each other out. Figure 2.1 illustrates how the three fundamental particles come together.

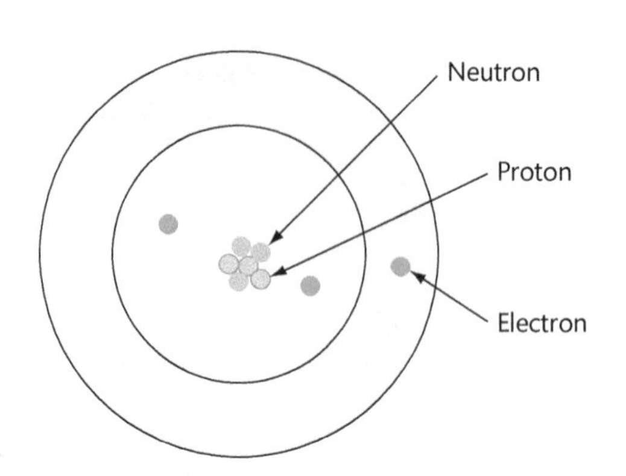

FIGURE 2.1

The structure of the atom. At the centre of the atom is the nucleus, which in this example consists of three neutrons (black circles) and three protons (white circles). The positive charge of the nucleus is balanced by the negative charges of three electrons (blue circles), which in this case occupy two separate orbitals.

Atomic number
the number that defines the place of an element in the periodic table. It equates to the number of protons in the nucleus of a given element

Periodic table
a system in which all the elements are arranged in a formal pattern according to their physical and chemical characteristics

Mass number
the sum of the number of protons and neutrons in the nucleus of a given atom

The identity of each chemical element is defined by the number of protons it contains. An atom containing six protons is an atom of carbon; an atom containing 12 protons is an atom of magnesium. The number of protons present in an atom is represented by the **atomic number**. We have just said that an atom has an equal number of protons and electrons. Consequently, the atomic number is also equal to the number of electrons present in an atom of a given element. Hence the element represented in Figure 2.1 has an atomic number of three: it is lithium.

The chemical elements were first organized into a coherent scheme, the **periodic table**, by Dmitri Mendeleev in 1869. The position of an element in the periodic table is determined by its atomic number (the number of protons it contains); as we move across the periodic table, the atomic number increases in increments of one. Look at Figure 2.2, which shows an extract of the periodic table. Notice how the atomic number increases in steps of one as we move from carbon to nitrogen, and from nitrogen to oxygen.

The sum of protons and neutrons in an atom's nucleus is called its **mass number**. So, an atom with six protons and six neutrons has a mass number of 12. The mass number can be used to predict either the number of protons or the number of neutrons present in an atom. For example, the mass number of sodium is 23, and the atomic number is 11. Consequently, there must be 11 protons and 12 neutrons in the nucleus (23 protons and neutrons i.e., 11 protons + 12 neutrons). In Figure 2.1, lithium has three protons and three neutrons, and so a mass number of six.

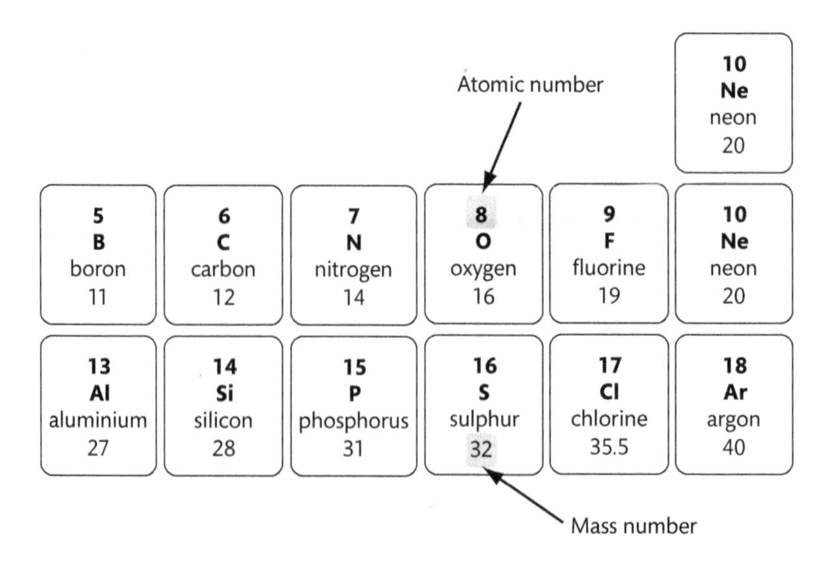

FIGURE 2.2

An extract of the periodic table. The arrangement of elements in part of the periodic table. Notice how elements are organized according to their atomic number, which increases in increments of one as we move from left to right along a row (which, in this context, is called a period).

Atomic mass and relative atomic mass

Each element in the periodic table has its own particular atomic mass, which is equal to the sum of the number of protons and neutrons present in the nucleus. For example, carbon-12—with six protons and six neutrons in its nucleus—has an atomic mass of 12 atomic mass units (or amu). In fact, carbon-12 is used as the reference element when it comes to measuring atomic mass: 1 amu is defined as one-twelfth of the mass of one atom of ^{12}C. By contrast, hydrogen, with a nucleus containing just one proton, has an atomic mass of 1 amu.

1 amu is equivalent to 1 dalton (Da), the unit most widely used in biology. However, it is important to note that the dalton is a non-SI unit.

It is important to note that the atomic mass of an element relates to a single atom of a specific element. So, an atom of ^{12}C has an atomic mass of 12 amu, while ^{13}C has an atomic mass of 13 amu. In biological systems, however, elements such as carbon exist as a mixture of different isotopes, so it isn't really meaningful to talk of atomic masses in that context. We consider **relative atomic masses** instead.

Relative atomic mass
the 'average' mass number of an atom when the relative contributions of its different isotopes are taken into account

Details of the most common elements in biomedical science are presented in Table 2.4.

Isotopes

Close attention to Table 2.4 shows that chlorine has a mass of 35.5. Since protons and neutrons have a mass of 1, how can this be? The answer is that, in contrast with the number of protons, the number of neutrons isn't always the same in all atoms of a given element. Atoms of the same element containing a different number of neutrons (and so different relative atomic masses) are called **isotopes**. Different isotopes are distinguished by writing the mass number as a superscript before the chemical symbol. For example, carbon exists as several isotopes, including ^{12}C and ^{13}C. ^{12}C contains six protons and six neutrons (6 + 6 gives a mass number of 12); ^{13}C contains six protons and seven neutrons (6 + 7 gives a mass number of 13).

Isotope
alternative forms of an element that possess different numbers of neutrons in the nucleus

The relative atomic mass of an element is the weighted average mass of all isotopes of that element. For example, chlorine exists as two isotopes, ^{35}Cl and ^{37}Cl. 76% of naturally occurring chlorine is ^{35}Cl, with an atomic mass of 35 amu, while 24% is ^{37}Cl, with an atomic mass of 37 amu. The weighted average of these two atomic masses—the relative atomic mass (or atomic weight)—is 35.5, as shown in Table 2.4. Note that the relative atomic mass is dimensionless: it doesn't carry units. (So we don't quote the relative atomic mass of chlorine as 35.5 amu; we simply write 35.5.)

TABLE 2.4 **Details of common elements**

Element	Symbol	Relative atomic mass	Element	Symbol	Relative atomic mass
Hydrogen	H	1	Carbon	C	12
Nitrogen	N	14	Oxygen	O	16
Sodium	Na	23	Magnesium	Mg	24
Phosphorus	P	31	Sulphur	S	32
Chlorine	Cl	35.5	Potassium	K	39
Calcium	Ca	40	Iron	Fe	56

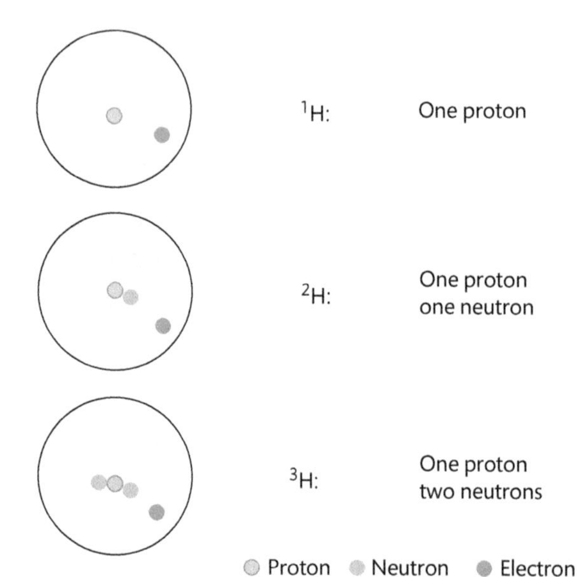

FIGURE 2.3

Isotopes of hydrogen. The structure of the three isotopes of hydrogen. Each isotope carries a single electron. With a single proton, each isotope has an atomic number of 1, but each isotope has a different mass number of 1, 2, or 3 depending on the number of particles in the nucleus.

Hydrogen has three isotopes: ^{1}H consists of a proton and an electron, ^{2}H (also known as deuterium) is composed of a proton, an electron, and a neutron, ^{3}H (also known as tritium) is formed from a proton, an electron, and two neutrons. The structures of each of these isotopes are shown in Figure 2.3.

The stability of isotopes

Combinations of only a certain number of protons and neutrons in the nucleus are stable. The two isotopes of chorine (with 17 protons plus 18 neutrons [=35] and 17 protons plus 20 neutrons [=37]) are stable. However, the combination of 17 protons and 19 neutrons [=36] is unstable, and undergoes radioactive decay to a form of sulphur or argon.

Perhaps the best example of this radioactive decay is uranium, defined by 92 protons in the nucleus, and so possessing the atomic number 92. There are several isotopes of uranium, all of which are unstable, the most common being uranium-238 (^{238}U, making up approximately 99.28% of all uranium). Uranium-238 has 146 neutrons, and uranium-235 (0.71%) has 143 neutrons. These isotopes emit alpha particles (two protons and two neutrons) and eventually degrade to lead. The radioactive isotopes of other elements emit beta particles (an electron, as seen in the decay of sodium-22 to neon-22) and/or gamma radiation (as seen in the decay of cobalt-60 to nickel-60).

Radioactive isotopes have a wide range of uses in modern medicine. However, their use comes at a price: they may cause mutations in DNA that could lead to cancer. In the laboratory, this has driven the replacement of radioactive tracers (such as those used in radioimmunoassays) with other markers, such as enzymes (as in enzyme-linked immunosorbent assays, ELISAs).

SELF-CHECK 2.5

Phosphorus has an atomic number of 15. Use the information in Table 2.4 to determine the number of neutrons in its nucleus.

Relative molecular mass

Two or more atoms linked together form a molecule; a molecule can comprise the same or different elements (such as O_2 and H_2O respectively). By adding up the relative atomic masses (as in Table 2.4) of those individual elements that comprise a molecule, we obtain the **relative molecular mass** (or molecular weight). Thus for methane (CH_4) this is $12 + (4 \times 1) = 16$. Hence the relative molecular mass of methane is 16. By contrast, the relative molecular mass of water is 18.

Relative molecular mass
the sum of the relative atomic masses of the component atoms of a molecule

We often see relative molecular masses quoted in daltons (such that, for example, water has a mass of 18 Da). To equate molecular masses (in daltons) with relative molecular masses like this isn't strictly correct (not least because relative molecular masses—like relative atomic masses—have no units), but is an approximation you are likely to see in widespread use.

The same mathematics can be used to determine the molecular weight of increasingly large molecules, such as the amino acid glycine, which has a molecular weight of 75 daltons. Amino acids combine to form small proteins, such as the hormone insulin, with its molecular weight of 5808 daltons. The most abundant protein in the blood, albumin, has a mass of 67 000 daltons, which is more usually described as 67 kilodaltons, or 67 kDa.

SELF-CHECK 2.6

Use the information in Table 2.4 to determine the relative molecular mass of potassium carbonate (K_2CO_3).

2.3 Quantities in aqueous solution

We see in Section 2.1 how the mole is a particular amount: 6×10^{23}. This is a huge number—but what does it relate to, in reality? If we imagine that we have an element in solid form, a mole of that element is equal to the relative atomic mass of that element, in grams. For example, carbon has a relative atomic mass of 12, so 12 g of carbon contains 1 mole of carbon atoms. We say that carbon has a **molar mass** of 12 g mol^{-1}.

Molar mass
the amount of a substance (an element or compound) expressed as its relative atomic mass or relative molecular mass

The same is true for molecules. Glucose ($C_6H_{12}O_6$) has a relative molecular mass of 180 (the sum of the relative atomic masses of six C, twelve H, and six O); there is one mole of $C_6H_{12}O_6$ in 180 g. We say that glucose has a molar mass of 180 g mol^{-1}.

Now let us imagine that we weigh out 36 grams of $C_6H_{12}O_6$ on an appropriate balance. How many moles would we have? Since one mole of $C_6H_{12}O_6$ weighs 180 grams, then we can deduce that 36 grams contains 0.2 moles. More formally, the number of moles of a substance is related to its molar mass by the following relationship:

$$\text{Amount of substance (mol)} = \frac{\text{Mass (g)}}{\text{Molar mass (g mol}^{-1}\text{)}}$$

So, using our example of $C_6H_{12}O_6$:

$$\text{Amount of substance (mol)} = \frac{36\text{ g}}{180\text{ g mol}^{-1}} = 0.2\text{ mol}$$

SELF-CHECK 2.7

Use the information in Table 2.4 to determine the amount of substance (mol) in 40 grams of calcium carbonate ($CaCO_3$).

Very often, we have to work with substances that are in solution—that is, dissolved in water or some other solvent. When working with substances in solution we quantify the amount of substance present by considering its concentration—the amount of substance present per unit volume. We quote concentrations in units of moles per litre (mol L^{-1}).

For example, sodium hydroxide, NaOH, dissolves easily in water. NaOH has a molar mass of 40 g mol^{-1}. If we weigh out 8 g we have 0.2 mol. So if we dissolve 8 g NaOH in water and make up the volume to 1 L, we have a solution of NaOH with a concentration of 0.2 mol L^{-1}. You may sometimes see the concentration of a solution referred to as its **molarity**; a 0.2 molar solution (often written 0.2 M) is simply another way of saying a 0.2 mol L^{-1} solution.

Molarity
the concentration of a substance when dissolved in a fluid such as water

It is also possible to dissolve different molecules in the same volume of water. One of the most common fluids in biomedical laboratories is **phosphate buffered saline** (PBS). The underlying solution is of sodium chloride (NaCl), often to a concentration of 135 mmol L^{-1}. To this is added a second compound, such as sodium hydrogen phosphate (Na_2HPO_4), although there are variants that include other compounds, such as potassium phosphate and/or potassium chloride.

Phosphate buffered saline
the most common buffer solution used in biomedical science, composed of a mixture of sodium chloride (saline) and a second molecule that includes a phosphate group

SELF-CHECK 2.8

Using the relative atomic masses in Table 2.4, what is the molarity of 9.5 g of magnesium chloride ($MgCl_2$) when dissolved in 250 mL (0.25 L) of water?

Ions, acids, and pH

Some substances, when dissolved in water, undergo dissociation into a negatively charged particle and a positively charged particle. A charged particle is called an **ion**, and we can use the theoretical molecule XY composed of an atom of 'X' and an atom of 'Y'. When this molecule dissolves in water it dissociates into X^+ (a cation, which is positively charged) and Y^- (an anion, which is negatively charged).

Ion
a charged particle

For example, the compound sodium acetate (CH_3COONa) dissolves easily in water to form the cation Na^+ and the anion CH_3COO^-. Another example is the theoretical molecule HA, and in water this too forms two ions: an anion (A^-) and a positively charged hydrogen ion (H^+, which is a cation):

$$HA \rightleftharpoons A^- + H^+$$

Compounds that liberate hydrogen ions in this way are called **acids**. Some compounds dissociate very readily to yield a relatively large number of hydrogen ions; these are called strong acids. By contrast, other compounds dissociate to a very small degree and so release relatively few hydrogen ions. Such compounds are called weak acids. For example, hydrogen chloride is an acid; when dissolved in water it dissociates to liberate a hydrogen ion and a chloride ion as follows:

Acid
a compound that increases the concentration of hydrogen ions in a solution

$$HCl \rightleftharpoons H^+ + Cl^-$$

In fact, HCl is a strong acid, which means that almost all of it dissociates into its component ions of H^+ and Cl^-. The concentration of hydrogen ions present in solution is termed the **pH**. The relationship between the pH and the concentration of hydrogen ions (written as $[H^+]$) is given by the following expression:

pH
a convenient way of evaluating the concentration of hydrogen ions in solution, and so its acidity

$$pH = -\log[H^+]$$

For example, a solution containing hydrogen ions at a concentration of 1×10^{-3} mol L^{-1} has a pH of -log $(10^{-3}) = 3.0$. By contrast, a solution containing hydrogen ions at a concentration of 0.1 mol L^{-1} has a pH of -log 0.1 = 1.0. Notice how the value of the pH gets smaller as the hydrogen ion concentration increases. So, a solution with a low pH contains a higher concentration of hydrogen ions—we say it is more acidic—than a solution with a higher pH (which contains a lower concentration of hydrogen ions).

SELF-CHECK 2.9

The concentration of hydrogen ions in a solution is 1×10^{-8} mol L^{-1}. What is its pH?

Measuring the degree of dissociation: the dissociation constant

We have just seen how different compounds undergo dissociation in water to different extents: some dissociate readily, others less so. If a compound is a strong acid and dissociates readily when dissolved in water, the equilibrium $HA \rightleftharpoons A^- + H^+$ lies heavily to the right; much of the HA dissociates to form A^- and H^+, and little HA remains.

By contrast, the equilibrium for a weak acid, which dissociates less readily, lies more to the left; little of the HA dissociates to form A^- and H^+, so a relatively large amount of HA remains. The degree of dissociation exhibited by a particular compound at equilibrium is given by its **dissociation constant**, *K*. We calculate the dissociation constant using the following equation:

Dissociation constant
a number that reflects the degree to which a compound dissociates in solution

$$K = \frac{[A^-][H^+]}{[HA]}$$

For example, if HA is a strong acid, then $[H^+]$ and $[A^-]$ will be large compared with [HA]; perhaps a thousand-fold. So *K* will be large, typically 1.5×10^3 or thereabouts. Examples of strong acids include hydrochloric acid, nitric acid, and sulphuric acid. Other acids do not dissociate as readily, and so much more of the acid remains in the HA state, and the dissociation constant is small. Accordingly, they are called weak acids, examples of which include hydrocyanic acid (HCN) and acetic acid (CH_3COOH). For instance, in the case of the latter:

$$K = \frac{[CH_3COO^-][H^+]}{[CH_3COOH]}$$

Here, $K = 1.76 \times 10^{-5}$, meaning that there are perhaps 100 000 (or so) molecules of CH_3COOH for each H^+. Therefore, in general, strong acids have large values of *K*, and weak acids have small values of *K*.

Pure water dissociates to a very small extent such that the equilibrium $H_2O \rightleftharpoons H^+ + OH^-$ lies heavily to the left. In other words, pure water mostly comprises molecules of H_2O with very few hydrogen ions (H^+) and hydroxide ions (OH^-). Consequently, the dissociation constant for water is very small—approximately 1×10^{-14}:

$$K = \frac{[H^+][OH^-]}{[H_2O]} = 10^{-14}$$

Given the tiny degree of dissociation, the concentration of hydrogen ions in pure water is just 10^{-7} mol L^{-1}. Consequently, we calculate the pH of pure water to be -log $(10^{-7}) = 7.0$. A pH of 7.0 tells us that pure water is neutral—that is, neither acidic nor basic.

Conjugate acids and bases

In reality, we don't actually see 'free' hydrogen ions in solution. Instead, we see the hydronium ion, H_3O^+, formed from the combination of a water molecule with a hydrogen ion. So, when hydrogen chloride dissociates in water, we're actually seeing the following:

$$HCl + H_2O \rightleftharpoons Cl^- + H_3O^+$$

So, the HCl has dissociated to liberate a hydrogen ion (which has combined with a water molecule to form H_3O^+) and Cl^-. In this instance, the water is acting as a base: a substance that can *accept* a hydrogen ion (in contrast with an acid, which *liberates* a hydrogen ion).

Specifically, acids and bases operate as **conjugate acid-base pairs**. So what does this mean? In our example of the dissociation of hydrogen chloride in water:

- Hydrogen chloride acts as an acid and liberates a hydrogen ion to become its conjugate base, Cl^-.
- Water acts as a base and accepts a hydrogen ion to become its conjugate acid H_3O^+.

This is an important point: acids can't act in isolation; they must have a base to which to donate the hydrogen ion they are liberating. And so we see them operating as a pair: the acid donates its hydrogen ion to the base, with the acid becoming its conjugate base, and the base becoming its conjugate acid.

2.4 Buffers

Hydrogen ions are constantly being generated as a side product of normal cell biology. Consequently, increased acidity can be a problem. If excess hydrogen ions accumulate we see the condition acidosis. To avoid this, excess hydrogen ions need to be removed, which is one of the functions of the kidney. Therefore, renal dysfunction naturally leads to problems of pH imbalance.

Even the slightest deviation of the blood pH from its reference range (in the region of pH 7.35–7.45) can have adverse biochemical consequences. However, we have another system designed to keep the pH of the blood within acceptable limits—a system of **buffers**.

Buffer
a combination of ions and molecules that resists a change in the pH of a fluid

Buffers are ions and molecules that resist changes in pH outside this range, often by absorbing any excess hydrogen ions. Most buffers are a combination of a weak acid in solution with its conjugate base, the major plasma buffer being bicarbonate; others include phosphate, ammonia, and proteins.

Unlike plasma, urine pH may vary considerably between pH 4.8 and 7.8, but is more likely to be acidic. Urinary buffers are required in acid–base homeostasis as they provide the major mechanism by which hydrogen ions are excreted from the body; they are also essential in the generation of bicarbonate. Ammonia, phosphate, and bicarbonate are all involved in urine buffering.

Bicarbonate

A product of aerobic respiration is carbon dioxide, which is generally excreted at the lungs. However, it cannot move from the tissues to the lungs as a gas. Instead, it does so in combination with water, in the form of carbonic acid, which is generated by the enzyme carbonic anhydrase (also referred to as carbonate dehydrase):

$$CO_2 + H_2O \rightleftharpoons H_2CO_3$$

However, carbonic acid undergoes dissociation in water according to the following equilibrium:

$$H_2CO_3 \rightleftharpoons HCO_3^- + H^+$$

Therefore, the dissociation of carbonic acid generates its conjugate base, the bicarbonate ion, HCO_3^- (often referred to simply as bicarb), and a hydrogen ion.

The bicarb system is a very important buffering system, providing almost three-quarters of the blood's buffering capacity. Indeed, we have a relatively large amount of bicarbonate in the plasma (around 24–25 mmol/L) which is available to soak up hydrogen ions. However, if there are too many hydrogen ions present, the buffering capacity of the bicarb system will eventually be overwhelmed (all of the free plasma bicarbonate will be consumed by its reaction with hydrogen ions). Consequently, a failure to maintain an adequate level of bicarbonate ions in the blood can lead to a falling pH, and so acidosis, which can have undesirable clinical consequences.

Phosphate

Phosphoric acid, H_3PO_4, like carbonic acid, is a weak acid, and dissociates to give the phosphate ion and up to two hydrogen ions:

$$H_3PO_4 \rightleftharpoons H_2PO_4^- + H^+ \rightleftharpoons HPO_4^{2-} + 2H^+$$

At physiological pH (7.4) most phosphate within plasma exists as monohydrogen phosphate (HPO_4^{2-}). This form of phosphate can accept H^+ to form dihydrogen phosphate ($H_2PO_4^-$). Because of its relatively low concentration, phosphate tends to be a minor component of buffering plasma, but it contributes significantly to the buffering of urine.

Ammonia

The metabolism of the amino acid glutamine to α-ketoglutarate by the cells of the kidney results in the generation of ammonia (NH_3). NH_3 can diffuse out of the cells into the developing urine, where it acts as a base, absorbing a hydrogen ion to become the ammonium ion:

$$NH_3 + H^+ \rightleftharpoons NH_4^+$$

In so doing, this small pathway eliminates two toxins at the same time. The remaining α-ketoglutarate within the cell can also act as a buffer by absorbing H^+, eventually resulting in the generation of glucose, which can act as an additional source of energy. This mechanism is remarkably efficient, as ammonium ion formation accounts for over half of the hydrogen ion excretion derived from metabolic acids.

Protein buffers

Proteins, especially albumin, account for 95% of the non-bicarbonate buffering capacity of plasma. They act as buffers both inside cells and in plasma. Albumin behaves as a weak acid, due to its high concentration of negatively charged amino acids (mainly carboxyl groups in glutamic and aspartic acids).

The Henderson–Hasselbalch equation

Each buffer has its own particular pH that it will try to maintain in the face of extremes of acidity or alkalinity. The exact pH of a buffer can be determined from the concentrations of the undissociated acid [HA] and the concentration of the conjugate base [A^-]. We also need a third component that is obtained from the dissociation constant K for the acid. However, the numerical value of K is likely to be very small—that for acetic acid is 1.76×10^{-5}, and is effectively unmanageable.

Fortunately, we can convert this number to something we can get to grips with by simply converting to pK in exactly the same way that we convert $[H^+]$ to pH. Therefore, for example, the negative $\log_{10}$ of 1.76×10^{-5} becomes 4.75. This gives us the components of the Henderson-Hasselbalch equation:

$$pH = pK + \log_{10}\left(\frac{[A^-]}{[HA]}\right)$$

A worked example: an acetate buffer

What this means in practice is that we can construct a buffer that will try to maintain a certain pH. Suppose we are measuring the activity of an enzyme that is most efficient at pH 6.45, and that a buffer based on acetic acid is appropriate. The dissociation of acetic acid (CH_3COOH) gives the anion CH_3COO^- and H^+, and we already have the dissociation constant from the previous paragraph, where $pK = 4.75$. So feeding these values into the Henderson-Hasselbalch equation we have:

$$6.45 = 4.75 + \log_{10}\left(\frac{[CH_3COO^-]}{[CH_3COOH]}\right)$$

which we can simplify to:

$$\log_{10}\left(\frac{[CH_3COO^-]}{[CH_3COOH]}\right) = 1.7$$

And since the antilog of 1.7 is approximately 50, this tells us we need the concentration of acetate ions $[CH_3COO^-]$ to be 50 times that of the concentration of acetic acid $[CH_3COOH]$. A convenient source of acetate ions is sodium acetate (CH_3COONa), so the buffer may be formed from a concentration of sodium acetate of (say) 0.05 molar and a concentration of acetic acid of 0.001 molar.

Biochemistry and the Henderson-Hasselbalch equation

This exploration of 'pure' chemistry has a direct application in the clinical biochemistry laboratory. As previously mentioned, the pH of the blood is important. Deviations from the reference range of pH 7.35-7.45 lead to the clinical conditions of acidosis and alkalosis. This also involves analysis of gases in the blood—the levels of oxygen and carbon dioxide—which are measured as their partial pressures (pO_2 and pCO_2 respectively).

The major buffering system of the blood involves the bicarbonate ion (HCO_3^-) and carbonic acid (H_2CO_3). As the dissociation constant of carbonic acid is known ($pK = 6.1$) we can therefore estimate the pH of the blood given the concentrations of carbonic acid and the bicarbonate ion from the Henderson-Hasselbalch equation:

$$pH = 6.1 + \log_{10}\left(\frac{[HCO_3^-]}{[H_2CO_3]}\right)$$

The concentration of the bicarbonate ion can be obtained from standard biochemical methods. However, the concentration of carbonic acid (in mmol/L) in arterial blood is strongly related to the amount of carbon dioxide (as pCO_2), according to the equation $[H_2CO_3] = 0.03 \times pCO_2$. So the equation given above can be amended to:

$$pH = 6.1 + \log_{10}\left(\frac{[HCO_3^-]}{[0.03 \times pCO_2]}\right)$$

A practical example of the importance of this follows in the coming section.

2.5 Chemistry and the practice of laboratory science

In Section 2.2 we were introduced to the concepts of atomic and molecular mass. The latter is an important component of many aspects of clinical science for laboratory scientists and many of the practitioners that work face to face with patients. This is particularly pertinent in biochemistry, where experience gained from laboratory practicals and experiments in undergraduate courses in biomedical science come to real-life use in clinical practice.

Urea and electrolytes

The most-requested blood test in biochemistry is called 'urea and electrolytes', which tests for sodium, potassium, urea, and creatinine. The former two elements are present as ions in the blood, that is, Na^+ and K^+, but in clinical practice the charge is often ignored, so they are written as Na and K. Similarly, when chloride levels are requested, the anion Cl^- is measured. With atomic masses of 23 and 39 respectively, the reference range for serum sodium is often around 135–145 mmol/L, whilst that for potassium is generally 3.8–5.0 mmol/L. An important function of the kidney is to regulate levels of these ions. Urea and creatinine have relative molecular masses of 60 and 113 respectively, their serum reference ranges being 3.0–8.3 mmol/L and 44–133 μmol/L respectively. Levels of all four indices are influenced by renal function, whilst serum urea is also often low in chronic liver disease (as this is where the molecule is synthesized).

SELF-CHECK 2.10

Comment on a serum sodium result of 148 mmol/L and a serum potassium result of 3.5 mmol/L.

Acidosis and alkalosis

The kidney also regulates the hydrogen ion concentration of the blood, and so the degree of acidity and alkalinity, and ideally the blood pH will be in the range 7.35–7.45. Therefore, renal dysfunction may well lead to pH problems. However, we have another system designed to keep the pH of the blood within its limits—the system of buffers we have already discussed—and in many cases, these provide essential support in maintaining acid/alkali homeostasis. We have already met one of these buffers—albumin—in preceding sections. This blood protein, with a relative molecular mass of around 67 000, is synthesized by the liver, and has a reference range in serum of 35–50 g/L. Note here that some laboratory indices are quantified in moles (sodium, potassium), others as grams (albumin, and most other proteins).

Acidity is a problem because hydrogen ions are constantly being generated as a side product of normal cell biology. In biological terms, a falling blood pH leads to the pathological condition of **acidosis**. So if acidity of the cell, organ, or the blood is to be avoided, the hydrogen ions need to be removed, which is done through the kidney, and this is why urine is acidic. Furthermore, increasing levels of hydrogen ions in the blood are bad because rising acidity has many adverse effects on the body, such as interfering with enzyme function in a number of metabolic processes.

Acidosis
a clinical term describing the acid nature of the blood; often present when the pH is less than 7.0

While it is true that the reverse of acidity, that is, alkalinity, is also to be avoided, normal body metabolism does not directly generate negatively charged ions as its does the positively charged hydrogen ion (H^+). It does, however, generate a great deal of carbon dioxide (CO_2), and this molecule, although it does not ionize directly, does have profound effects on pH and pathology, as we shall see in the following worked example. Carbon dioxide is important in the setting because it contributes to the bicarbonate buffer system. Nevertheless, a high pH, and so **alkalosis**, is an important clinical problem. Both acidosis and alkalosis can be caused by problems with metabolism, but are also caused by disease of the lungs.

Alkalosis
a clinical term describing the alkaline nature of the blood; often present when the pH is greater than 7.6

A worked example

A 75-year-old woman presents to the Emergency Department in a confused state, and is unable to give any helpful details about her condition. Arterial blood gas analysis shows her pCO_2 to be 35 mmHg (reference range 35–45 mmHg), and serum bicarbonate is 17 mmol/L (reference range 24–29 mmol/L). Putting these data into the Henderson-Hasselbalch equation gives ...

$$\begin{aligned} pH &= 6.1 + \log([17]/[0.03 \times 35]) \\ &= 6.1 + \log(17/1.05) \\ &= 6.1 + \log 16.2 \\ &= 6.1 + 1.21 = 7.31 \end{aligned}$$

The result of pH 7.31 is outside the reference range of 7.35–7.45. This supports the diagnosis of acidosis, and is likely to trigger additional investigations and direct treatment.

The anion gap

Several of the concepts in the final part of this chapter come together in the anion gap. The pH of the blood (generally 7.36–7.44) is slightly alkaline because, once the buffering system has been addressed, the sum of the major cations (sodium and potassium) is less than the sum of the anions (chloride and bicarbonate). In practice it is calculated from the molar concentrations of each component, that being ...

$$\text{Anion gap} = ([Na^+] + [K^+]) - ([Cl^-] + [HCO_3^-])$$

This gap can both be calculated and also measured directly in the laboratory. However, in practice, because the potassium concentration is very low compared to that of sodium, it is often omitted, so that a reference range for the anion gap is often cited as 6–12 mEq/L, which is effectively all the 'missing' ions (mostly anions such as phosphates and sulphates) not accounted for by the equation. The importance of the anion gap is that it helps diagnose a number of conditions. Should the measured anion gap be markedly different from the calculated anion gap, there must be a high concentration of unaccounted-for ions. For example, a high anion gap indicates acidosis, which if present in someone with diabetes, could indicate diabetic ketoacidosis.

Chapter summary

The Système International (SI) method defines the units for the vast majority of biological analytes.

- The major SI units are mass/weight, distance/length, time, thermodynamic temperature (kelvin), and the amount of a substance.
- Other units that are derived from these are frequency, pressure, temperature (Celsius), and radioactive decay.

- Commonly used non-SI units are the litre (measuring volume), the dalton (atomic and molecular weights), and millimetres of mercury (blood pressure).
- Fundamental particles are the proton, neutron, and electron. The atomic number tells us the number of protons in an atom of an element; the mass number tells us the sum of the number of protons and neutrons.
- The sum of the relative atomic masses of all atoms in a molecule gives us its relative molecular mass. If we measure out the mass, in grams, of a substance to a value equal to its relative atomic or molecular mass we will have one mole of that substance.
- The concentration of a substance in a solution is often expressed in terms of moles per litre (mol L^{-1}).
- The acidity or alkalinity of a solution can be determined from the concentration of hydrogen ions, and expressed as pH. Acids have a pH less than 7, alkalis have pH greater than 7.
- Buffers, such as those based on carbonates and phosphates, resist changes in pH. The Henderson–Hasselbalch equation can be used to determine the pH of a buffer given the concentration of the acid or alkali, and its respective conjugate base or acid.

Questions

2.1 In Table 2.1 it is stated that the average body contains 25 moles of calcium. What does this weigh?

2.2 What mass of sodium hydroxide (NaOH) is required to make 500 mL of a solution of concentration 0.25 molar?

2.3 An element has 26 protons and 30 neutrons. What are its atomic number and mass number?

2.4 Use the information in Table 2.4 to determine the mass of 0.1 moles of potassium chloride (KCl).

2.5 A solution of hydrochloric acid has a pH of 5. What is the concentration of hydrogen ions?

2.6 A young woman with diabetes is brought to the Emergency Department late in the evening by friends. She does not appear to be drunk, but is confused and difficult to engage in conversation. She has a high pulse rate (100 beats per minute), a high temperature, and her skin feels clammy. Blood results show pH 7.65 (reference range 7.35–7.45) and serum bicarbonate 31 mmol/L (reference range 24–29). Comment on these results.

3

Obtaining and verifying data

Learning objectives

After studying this chapter, you should be able to ...

- Appreciate pre-analytical error
- Explain the importance of accuracy and precision
- Describe the differences between intra- and inter-assay coefficients of variation
- Understand how operator performance can be assessed
- Recall the principles of sensitivity and specificity
- Outline the methods for comparing a new method with an existing method

All of our work starts with the individual. They may come to our profession because they complain of symptoms (such as chest pain or aching joints) that imply a particular condition or disease (perhaps angina or rheumatoid arthritis, respectively). Alternatively, they may be free of specific symptoms (that is, they are asymptomatic) and have provided a sample from a screening programme. These people are not always patients, but may become so. Indeed, pregnant women should not immediately be seen as, or described as, patients but, unfortunately, some may become so.

In this chapter we explore how we obtain data from machines and from people themselves, how we ensure the result is correct, and how that result should be interpreted. We will first address, in Section 3.1, sources of potential error. We then look, in Section 3.2, at crucial factors in laboratory practice–optimizing assay performance, ensuring reliability, and monitoring reliability. In Section 3.3 we will consider operator performance, and the clinical value of data generated in the laboratory will be discussed in Section 3.4. A key part of this final section will consider the formal comparison of different techniques.

3.1 Error

Biomedical analysis typically begins with some kind of sample from an individual. This sample may be body fluid or a piece of tissue, and may be obtained in a variety of ways, including venepuncture, swabbing, or surgery. Once obtained, the sample is transported to the laboratory where analysis can begin.

Errors can creep into the analytical process at various points, even before the sample arrives at the laboratory. **Pre-analytical errors** arise before analysis is carried out, **analytical errors** happen during the process of analysis, and **post-analytical errors** arise after the analysis is completed. We consider each of these types of error in this section.

Having arrived at the laboratory, the sample is generally analysed by a machine (common in biochemistry and haematology), although some analyses demand specialist attention from a skilled scientist (such as in histology). However, each assay or process will have a **standard operating procedure** (SOP), which is essentially a series of steps to be followed in order to produce a good result. The purpose of an SOP is to minimize the likelihood of the introduction of error by the scientist. Nevertheless, in almost all cases, techniques are imperfect and so call for checking for potential errors by a number of processes which include quality assurance and quality control.

Pre-analytical error
mistakes made in obtaining and/or processing a sample before it arrives in the laboratory

Analytical error
mistakes made during the process of analysis

Post-analytical error
mistakes made once the analysis is complete

Standard operating procedure
a series of steps in a method that must be followed if a good result is to be obtained

Pre-analytical error

There are several types of error that can happen during the journey of the sample from the patient to the laboratory where it is to be analysed. The primary source of error is simply taking the sample from the wrong person (which the laboratory cannot detect)—another is a discrepancy between the information written on the container carrying the sample and that on the accompanying form (which the laboratory *can* detect). A third is failure to correctly collect the sample, such as blood failing to be anticoagulated, or use of an inappropriate anticoagulant, or perhaps an incorrect or insufficient fixative or preservative.

A common problem is the length of time between the sample being obtained and its arrival at the laboratory, and again this cannot be addressed by scientific staff. Delays can lead to errors such as:

- increased size of the red blood cell
- increased serum potassium
- excessive bacterial growth
- necrotic changes to a sample of tissue.

These errors may call for the rejection of the sample, and a phone call to the practitioner for another sample, which may not always be possible.

Fortunately, errors made in the pathology reception are rare, and if present, may be correctable. The most common include instructing the particular laboratory to perform an analysis other than that which is requested, or the omission of an analysis. Another is a failure to correctly prepare serum or plasma by centrifugation, although this may be done in the analytic laboratory itself. Samples may even be passed to an inappropriate laboratory.

Analytical error

The most common sources of error in the laboratory are, unfortunately, human. These include errors in sample preparation and in presenting the sample for analysis. Highly technical analysers demand great care, and incorrect results will result if they are mistreated—for example, if the user fails to provide the correct reagents, or fails to perform essential maintenance. However, even the most careful laboratory may from time to time experience problems with their procedures. We explore some of these problems in Section 3.2.

TABLE 3.1 **Sources of error**

Pre-analytical (50-65%)	Insufficient sample, poor sample condition, incorrect sample, incorrect/inadequate patient details, delay in transport.
Analytical (8-16%)	Equipment malfunction, programming error, sample tracking error, reagents problem.
Post-analytical (15-45%)	Reporting of analysis, improper data entry, delay in verification/release.

Cross reference

We discuss calibration further in Section 3.2.

Less frequent are errors due to faulty equipment. All analysers need to be regularly checked and serviced if they are to provide accurate and reproducible results. A key aspect of ongoing maintenance is the external calibration of equipment to ensure good performance. Issues of calibration and servicing will be the responsibility of senior staff.

Post-analytical error

Once the sample has been analysed, and the accuracy of the result confirmed, it is passed to the requestor, generally a clinical practitioner. Electronic transfer is least likely to produce error; by contrast, the verbal transmission of results, often over a telephone, is a common source of error. The source may lie on both sides of the transfer, as the scientist in the laboratory may 'read' an incorrect result, whilst the practitioner taking the call may transcribe incorrectly. For this reason a common policy is to have both parties repeat back to each other the identity of the patient and the result, and to record that transfer in a log book.

The culture of checking is highly developed in blood transfusion, as an incompatible transfusion may be fatal. Sources of error are summarized in Table 3.1.

3.2 Assay performance

Although biomedical science encompasses many disciplines and techniques, several rely heavily on the personal technical skill of the scientist, whereas others rely on autoanalysers. However, central to any method or technique—whether carried out by an autoanalyser or by a laboratory scientist—is the **accuracy** and **precision** of that method.

Accuracy
the degree to which the result comes close to that from the reference sample

Precision
the degree to which a method can deliver the same result several times from the same sample

Accuracy

This property describes the ability of a method to give a result that is as close as is feasibly possible to reality—for example, when the concentration of a particular substance in a sample determined by a particular method mirrors the actual concentration that is present. The definition of the 'actual' result is generally derived elsewhere by a reference sample, often provided by an agency independent of the laboratory. In many cases, this outside agency may be a body such as the National External Quality Assurance Scheme (NEQAS), or it may be provided by the manufacturer of an autoanalyser, in which case it is described as a **calibrator**. A result very close (for example, 101 arbitrary units/litre) to that expected of this calibrator (such as 100 arbitrary units/litre) suggests the method is very accurate, whereas a result of 120 arbitrary units/litre suggests an inaccurate method.

Calibrator
a sample that contains a known amount of a certain analyte, and so is used to ensure that an analyser is producing a correct result

Although the accuracy of a method can be determined with a single sample, in practice, several samples are often analysed, and the average (mean) value taken.

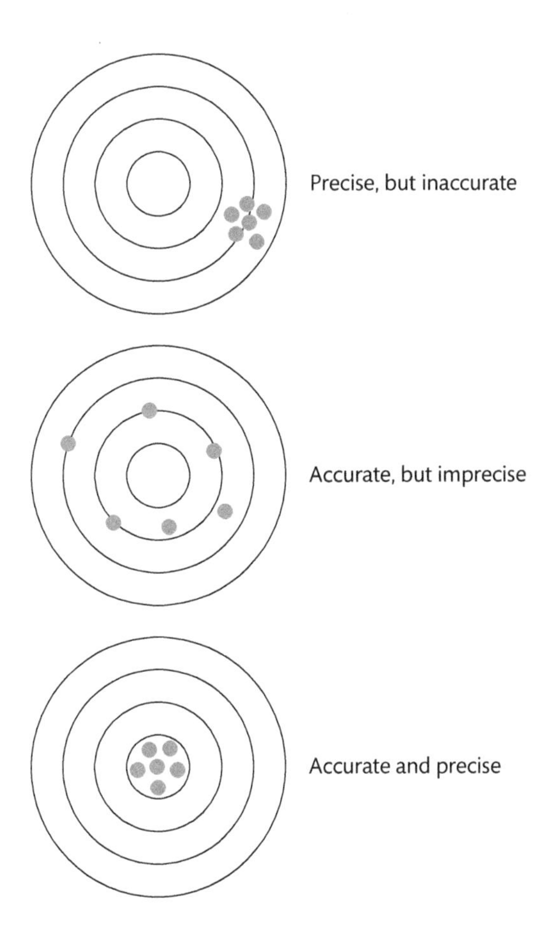

FIGURE 3.1

Accuracy and precision. We can view accuracy and precision in terms of trying to hit the centre of an archery target.

- In (a), the results of the analysis are tightly clustered together, and so reflect good reproducibility. However, they are far from the centre of the target (the gold standard) and so are inaccurate.
- In (b), the results are widely dispersed, and so have poor reproducibility, but their average centres on the middle of the target, and so they are mathematically accurate.
- In (c), the results are tightly clustered together around the centre of the target. Consequently, the results are both precise and accurate.

Precision

This property refers to the closeness of a number of repeated analyses of the same sample, and is independent of the calibrator. Let us say that a calibrator has a value of 40 units/litre. If a method gives results of 40, 38, 45, 35, and 42 (which average to 40) we would probably say it is imprecise, as the higher and lowest results vary by over a quarter. Results of a second method, which are 39, 41, 39, 40, and 41 (which also average to 40), but which vary by 5%, imply a more precise method. We can go a step further in saying that the **reproducibility** of the first method is poor (it gives a result anywhere between 35 and 45), whereas the reproducibility of the second set is much better, as the results it gives vary between 39 and 41. Figure 3.1 summarizes these points.

Reproducibility the mathematical expression of precision

SELF-CHECK 3.1

Consider substance X, a molecule found in the blood and related to a certain pathology. The established method for measuring this molecule gives a result of 125 units/mL.

Three new methods have been developed. The same sample measured ten times by each of these methods gives the following results:

Method A: 130, 122, 125, 129, 133, 126, 126, 129, 122, 130

Method B: 129, 132, 121, 129, 134, 125, 119, 126, 127, 125

Method C: 128, 127, 129, 130, 129, 128, 127, 129, 130, 128

For each method, calculate the mean value, and look at the range of values. Use these to give an account of their accuracy and precision, and then make a preliminary recommendation about which is best.

Coefficients of variation

As part of establishing a new method, its reproducibility must be determined. This is the extent to which a method gives consistent, reliable results, and there are a number of ways of doing this. In an ideal world the repeat analysis of a sample will give exactly the same result each time it is analysed. However, in the real world we recognize that this is almost impossible to achieve, and we are prepared to accept a small degree of variation. This variation can be quantified and expressed as the **coefficient of variation** (CV), and it gives us a feel for the precision of a method. The CV is obtained by dividing the standard deviation of a set of data by its mean, and then expressing this as a percentage.

Coefficient of variation
the mathematical expression of the degree of variation of a method

Consider the data set:

29, 32, 29, 31, 28, 30, 32, 29, 31, 29, 30, 34

The mean and standard deviation of this data set are 30.3 (to one decimal point) and 1.7 respectively. This data set consists of 12 repeat measurements of the sample, and is called the **sample size**. The CV is 1.7 divided by 30.3, ×100, which is 5.6%. Now consider another data set:

Sample size
the number of individual data points in a particular analysis

27, 30, 35, 27, 30, 25, 34, 32, 28, 33, 31, 29

The mean and standard deviation of this data set are 30.1 and 3.1 respectively. Hence the CV is 10.3%. The larger the CV, the more variation there is in the results, and so the less reproducible (and therefore reliable) is the method. Conversely, the smaller the CV, the less the variation, which brings a higher degree of reliability and, therefore, confidence in the procedure.

Cross reference
We have discussed the standard deviation, a measure of the degree of variation of a data set, in Chapter 1

We would hope that the CV of an assay would be as small as possible. However, the CV of a method depends a great deal on its nature. Methods that are highly automatable and which rely on simple chemistry are expected to have a smaller CV than those that rely on antibody binding (that is, immunoassay). Assays involving live cells (such as leukocytes and platelets) generally have larger CVs, implying a lower degree of reliability. This reduced reliability usually reflects the natural variation in the cells themselves.

We can look at CV in two ways: the intra-assay CV and the inter-assay CV.

Intra-assay CV

This type of CV assesses the reproducibility of an assay in the short term, generally in a single day or in a single batch. Accordingly, it is also described as within-assay, or run–run variability. It is conducted by the repeated assay of the same sample within a single process, which may take minutes or hours, depending on the method. For an immunoassay such as enzyme-linked immunosorbent assay (ELISA), we expect the intra-assay CV to be <5%, whereas many biochemistry methods have a CV <3%, or even lower. It follows that what is considered acceptable in one setting may be unacceptable in another.

Intra-assay CV
the degree of within-batch (or within-run) variation of a method. Assessment is generally in a single 24-hour period

Inter-assay CV

This method assesses the reproducibility of an assay over a longer term, such as from day to day, or week to week. It may also be called between-assay variability. The inter-assay CV is obtained by repeating the analysis of a sample in consecutive batch runs of the process.

Inter-assay CV
the degree of batch to batch (or run to run) variation of a method. It may be performed at intervals of days or weeks

It is expected that the inter-assay CV of an assay will be larger than the intra-assay CV. Accordingly, the inter-assay CV of an ELISA is generally acceptable if <10%; whereas in biochemistry, <5% is expected.

Self-check question 3.2 provides the opportunity to put the principles of CVs into practice.

SELF-CHECK 3.2

In self-check 3.1, we looked at the accuracy and precision of three methods:

Method A: 130, 122, 125, 129, 133, 126, 126, 129, 122, 130

Method B: 129, 132, 121, 129, 134, 125, 119, 126, 127, 125

Method C: 128, 127, 129, 130, 129, 128, 127, 129, 130, 128

Now work out the CV of each method. For this you need to derive the SD of each set, which can be obtained from a programmable calculator, or perhaps mathematical software on a computer, such as Excel or Minitab.

Importance of the sample size

Although in theory the standard deviation of a method does not rely on the size of the sample, in practice a small sample size may well include one or more unusual results (often called outliers) that will disproportionately influence the standard deviation and so lead to a large CV. As a broad guideline it is likely that a sample size of at least 12 is called for in clinical laboratory science in order to generate a representative standard deviation and so the likelihood that the true CV of the method has been obtained, as in the examples given above. The sample size is a feature of several other aspects of laboratory science.

- Those researching a new analyte will inevitably need to determine if it has relevance in a clinical setting, and if so may enter routine laboratory practice. Step one is to establish a reference range of the analyte in healthy people, step two is to compare it with levels in people with a relevant disease. The problem arises in finding a sufficiently large group of people known to be healthy and so free of any disease. Such a group would be likely to include equal numbers of the sexes, and of the full adult lifespan (18–80 years of age). In order to accommodate these criteria, a sample size of over 200 would be expected. Of course, finding enough healthy persons in their seventh decade will be a challenge.
- A sample size determination is needed to establish the exact number of people to be recruited to a research project so that a particular hypothesis can be correctly tested. The process of determining the minimum sample size is complex and is best done with the help of some statistical software such as Minitab. This process is called a **power calculation**.
- When validating a new method for a continuously variable index (such as haemoglobin, serum albumin, or the number of microbes in a sample of urine), the new method must be compared to the existing, standard method. The most reliable comparison would be to determine the analyte in at least 100 samples. This is explained in Section 3.5 that follows.

Power calculation
the process of determining the correct sample size needed to test a particular hypothesis

Cross reference
The concept of the sample size and power calculation are developed in Chapter 6, Section 6.2

Monitoring long-term reliability

A further aspect of developing confidence in a method (that is, in its CV) lies in assessing the reliability of an established process over the long term, which may be weeks or months. Here, we refer to the quality of the data produced by a technique, of which there are two forms: quality control and quality assurance. These two methods are an important part of ensuring the quality of the results of the laboratory, and are explained in Chapter 5.

3.3 Operator performance

Theoretically, if two scientists are analysing the same samples using the same SOP, there should be no meaningful difference between the results they obtain. However, if there is such a difference, then the two operators are likely to be doing something different. The way to find out if this is indeed the case depends on the nature of the result. If the method returns a result that can take any value, such as somebody's weight, the data is said to have a **continuous variation**. In the haematology and biochemistry laboratories, examples of this type of data include haemoglobin and serum potassium respectively.

Cross reference

Continuous and categorical variation: these two different types of data are discussed in Chapter 1

An alternative type of data is one where the result can have only one of a small number of exclusive options (or categories), such as present/absent. If so, the data is said to have **categorical variation**. Examples of this in biomedical science include the presence or absence of an infection, or whether or not some normal tissue has been invaded by a cancer.

Continuously variable data

A typical procedure may see two operators testing the same number of samples, and then comparing results. Consider the data in Table 3.2, gathered by two scientists following the same SOP on 14 different samples of plasma. They may be measuring levels of a particular molecule or ion.

TABLE 3.2 **Differences in analysis reported by two scientists**

Sample	Scientist 1	Scientist 2
1	106	110
2	95	97
3	124	122
4	89	92
5	56	59
6	100	102
7	115	119
8	76	80
9	89	84
10	121	125
11	69	72
12	101	104
13	78	80
14	121	122
Mean result	95.7	97.7

The mean results from the two scientists seem to be very close, and indeed differ by only 2.1%. However, close scrutiny of the data shows that Scientist 2 has found higher levels of 12 of the 14 samples. If there was complete agreement between the results one would suspect a degree of collusion, but in a random setting, we would expect only 7 to be higher (and 7 lower).

The difference between the two sets of data can be tested formally with an analysis called a paired t test, which indeed found the difference to be statistically significant. The mathematics of the test are reasonably complex, and so are best done on a computer loaded with statistical software. Complete details of the use of this test are presented in Chapter 7.

Cross reference

t test: Complete details of the use of this and allied tests are presented in Chapter 7

Categorical data

Data of this sort may be deciding upon one of a small number of choices, such as mild, moderate, or severe. This may take the form of the degree of invasion or infiltration of some normal tissue by a tumour, the density of a cytochemical stain, or perhaps the likelihood of an infection.

Let us suppose that two scientists are both reporting on the same group of 71 slides of some stained tissue. They need not have actually stained the slides themselves, but they have to decide whether or not the intensity of the colour of the stain on each slide is minor, moderate, or marked. Let us pass over the precise definition of these three possible outcomes, and compare results in Table 3.3.

Focusing first on where there is agreement, both scientists found 10 slides to have minor staining, 25 to have moderate staining, and 15 to have marked staining, giving a level of agreement in 50 of the 71 slides (that is, 70.4%). However, there were disagreements in all other categories, comprising a level of disagreement in 21 (29.6%) of the slides. For example, Scientist A found one slide to have marked staining, but his or her colleague found that slide to exhibit only minor staining—a clear disagreement. Conversely, Scientist A reported three slides as minor, but Scientist B found the staining of these slides to be marked.

A statistical test can be applied to the proportion of agreements and disagreements, which provides the **kappa (k) statistic**. Depending on the degree of agreement (or not), the kappa statistic will return a result anywhere between 0 and 1. The mathematics of this test are far easier than those of the paired t test, and can be condensed to a small number of steps as follows:

Kappa statistic
a method for determining the agreement between two operators or techniques

1. Add together all the instances where there is complete agreement: in this example, as discussed above, this is $10 + 25 + 15 = 50$, giving 70.4%, or 0.704 as a decimal.
2. Form a product of the totals from the two minor staining totals (that is, $14 \times 18 = 252$), and divide this by the square of the complete total (that is, $71 \times 71 = 5041$), hence $252/5041 = 0.05$
3. Similarly, form a product of the totals from the two moderate staining totals (that is, 34×33, $= 1122$), and again divide this by the square of the complete total (that is, 71×71, $= 5041$), hence $1122/5041 = 0.223$ (to three significant figures).
4. Finally, form a product of the totals from the two marked staining totals (that is, 23×20, $= 460$), and divide this once more by the square of the complete total (that is, 71×71, $= 5041$), hence $460/5041 = 0.091$.
5. Now add these three products: $0.05 + 0.223 + 0.091 = 0.364$
6. $\text{Kappa} = \dfrac{0.704 - 0.364}{1 - 0.364} = \dfrac{0.340}{0.636} = 0.535$

TABLE 3.3 **Comparing results from two scientists**

	Scientist A			
Scientist B	Minor staining	Moderate staining	Marked staining	Total
Minor staining	10	3	1	14
Moderate staining	5	25	4	34
Marked staining	3	5	15	23
Total	18	33	20	71

By consensus, we consider the level of agreement to be moderate (Table 3.4). It is for senior staff to determine whether the level of agreement is acceptable, and if not, what steps need to be taken in order for the degree of agreement to improve.

Interpretation

These two tests, of a data set with a continuous variation and a data set with a categorical variation, compare the performance of two operators, making no judgement about which is the better scientist, or whether they are equally proficient, especially if they are of the same professional grade and experience. However, if one is markedly more experienced than the other, and there is a difference in outcomes, it may reflect their relative technical expertise. This process of comparisons can be developed into a test of competency, which may be formal and have professional implications.

SELF-CHECK 3.3

A scientist ('A') reports that 43 of 105 samples are positive, and that 62 are negative. These samples are analysed by a second scientist ('B'), who finds different results. Both scientists find the same 32 samples to be positive, and the same 54 samples to be negative. However, of the 43 samples found to be positive by scientist A, 11 are found to be negative by scientist B. Furthermore, of the 62 samples found to be negative by scientist A, scientist B found 8 to be positive.

Derive the kappa statistic for this data and so determine their level of agreement.

TABLE 3.4 **The kappa statistic**

Value of kappa	Strength of agreement
0.2	Poor
0.21–0.4	Fair
0.41–0.6	Moderate
0.61–0.8	Good
0.81–1.0	Very good

3.4 Clinical value of data

Our professional colleagues rely on us to give them accurate and relevant information to help make clinical decisions. Indeed, it has been estimated that 75% of such information comes from the laboratory. Put another way, the laboratory provides three times as much relevant information as do all other investigations (imaging, signs, symptoms, history etc.) combined.

Implicit in this reliance is the assumption that the result itself is precise and accurate, as we have already discussed. However, excellent precision and accuracy is irrelevant if the result is unrelated to aspects of clinical medicine and a particular disease or condition. Key features of the relationship between the result of a test and its relevance include **sensitivity**, **specificity**, **positive predictive value**, and **negative predictive value**, which we will now examine.

Sensitivity
the proportion of people with the disease who are correctly identified by the test

Specificity
the proportion of people without the disease who are correctly identified by the test

Positive predictive value
the proportion of patients with a positive test result who are correctly diagnosed

Negative predictive value
the proportion of patients with a negative test result who are correctly diagnosed

Let us suppose that researchers have discovered a new molecule in the blood that may be useful in determining the presence or absence of a certain cancer. The value of the blood test for this new molecule, shall we say Substance X, must be tested in two groups:

- A group of people in whom there is absolute confidence that the disease is present; these people are called 'cases'.
- A group of people in whom there is absolute confidence that the disease is absent; these people are referred to as 'controls'.

The value of the test must be ascertained in a representative group of cases, and compared with a group of controls. Let us suppose that the mean level of Substance X in a group of 52 cases with cancer is 100 units/mL, with a standard deviation of 15 units/mL. Similarly, in a group of 73 controls free of cancer the results are 44 units/mL and 18 units/mL respectively. This difference is statistically significant, and seems a very good finding, as the levels in cancer patients are well over twice that in the healthy controls. It therefore supports the view of the researchers that their test may be useful. These results can be plotted graphically, which allows us a rapid assessment of the data (Figure 3.2).

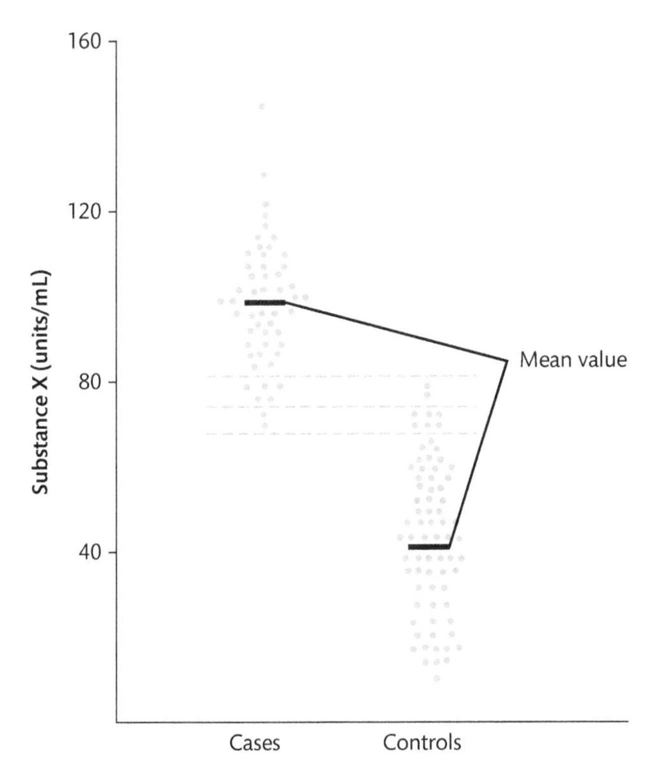

FIGURE 3.2

Levels of Substance X. Levels of Substance X in the blood of cases and controls. The black bar is the mean value for each data set. The upper dashed line exceeds the maximum result in the controls. The lower dashed line is less than the minimum result in the cases. The middle dashed line is the mean of the upper and lower dashed lines.

Although levels of Substance X are, on the whole, significantly higher in the cases than in the controls, there is some overlap, with several controls having higher levels than the cases, and vice versa. We can use this data in two ways:

- We can say that the risk of actually having the disease increases with the level of Substance X.
- We can determine a cut-off point at which we can make predictions about whether or not an individual has or has not got the cancer.

The difficult choice is now to determine a concentration of Substance X that gives useful information. If we choose a high level, such as 82 units/mL (the upper dashed line in Figure 3.2), we will exclude all the controls (as the highest result is 79 units/mL), but include only 45 of the 52 cancer cases, as seven of these have levels below 82 units/mL.

Conversely, if we choose 68 units/mL (the lower dashed line), we will include all of the cases we know to have cancer, but this number will also include eight of the 73 cases who do not have cancer. The compromise is the mid-point of these two points, 75 units/mL (the dashed line shown in blue). This will include 50 of the 52 (96.1%) cases with cancer and 71 of the 73 (97.3%) controls free of cancer, but it will also incorrectly assign four people (two with cancer and two without) into the wrong group.

We can derive a classification system for this as follows:

- those cases that the test has correctly identified as having cancer are true positives (TPs);
- those controls that the test has incorrectly identified as having cancer are false positives (FPs);
- those controls that the test has correctly identified as being free of cancer are true negatives (TNs);
- those cases that the test has incorrectly identified as being free of cancer are false negatives (FNs).

Cross reference

We will return to the concept of false positive and false negative from the research perspective in Chapter 6, Section 6.2

By comparing these four groups we can derive four descriptors of the value of the test: the sensitivity, specificity, positive predictive value, and negative predictive value of the test.

Sensitivity

The sensitivity of a method is the proportion of people known to have a disease or condition that are correctly identified by that method. This is the number of true positives (TPs) compared with the sum of the true positives and false negatives (that is, all the cases combined, hence TPs + FNs). Ideally, we would find that the test correctly identifies (by inclusion) everybody with the disease, and correctly identifies (by exclusion) everybody without the disease.

$$\text{Sensitivity} = \frac{\text{true positives}}{\text{true positives} + \text{false negatives}}.$$

Specificity

The specificity of a method is the proportion of people who are known not to have the disease or condition that are correctly identified by that method. This is the number of true negatives (TNs) compared with the sum of the true negatives and false positives (that is, all the controls combined, hence TNs + FPs). Ideally we would hope that the test correctly identifies all of the controls who are free of the disease, and none of the cases who have the disease.

$$\text{Specificity} = \frac{\text{true negatives}}{\text{true negatives} + \text{false positives}}.$$

Positive predictive value

An extension of sensitivity and specificity gives the ability of the test to predict the presence or absence of the disease. The positive predictive value (PPV) is, essentially, how good a test is in determining whether or not the subject has the particular disease. It is defined as the number of TPs divided by the total number of positive results, i.e. TPs plus FPs.

$$\text{Positive predictive value} = \frac{\text{true positives}}{\text{true positives} + \text{false positives}}.$$

Negative predictive value

This estimate is essentially the reverse of the PPV, and refers to the ability of the test to determine whether or not the subject does not have the particular disease. The negative predictive value (NPV) is obtained from TNs divided by the total number of negatives, i.e. TNs plus FNs.

$$\text{Negative predictive value} = \frac{\text{true negatives}}{\text{true negatives} + \text{false negatives}}.$$

A worked example

Let us now apply the data from our hypothetical test above to determine these indices, based on the cut-off point of 75 units/mL, which we hope will give the best overall outcome. This is best done by constructing a table, as shown in Table 3.5. Despite the fact that there are four columns and four rows, it is called a two by two contingency table, as the analysed data takes up the middle two rows and columns.

Feeding these data into the equations gives:

- Sensitivity = TP/(TP + FN) = 50/(50 + 2) = 50/52 = 96% or 0.96. Another way of looking at this is that the true positive rate is 96%, and therefore the false positive rate is 4%. An ideal true positive rate is 100%, so many would consider a rate of 96% to be acceptable.
- Specificity = TN/(TN + FP) = 71/(71 + 2) = 71/73 = 97% or 0.97. Hence the true negative rate is 97%, and so the false negative rate is 3%. Again, in an ideal world the true negative rate would be 100% (that is, excludes everybody who has not got the disease).
- PPV = TP/(TP + FP) = 50/(50 + 2) = 50/52 = 96% or 0.96. This number tells us how many of the people that the test considers to be positive for the disease do actually have it.
- NPV = TN/(FN + TN) = 71/(71 + 2) = 71/73 = 97% or 0.97. This refers to the proportion of people that the test considers to be free of the disease who are actually free of the disease.

TABLE 3.5 **A two by two contingency table**

	52 patients with cancer	**73 controls free of cancer**	
New test: positive result	50 true positives (TP)	2 false positives (FP)	$PPV = \frac{TP}{TP+FP}$
New test: negative result	2 false negatives (FN)	71 true negatives (TN)	$NPV = \frac{TN}{FN+TN}$
	$\text{Sensitivity} = \frac{TP}{TP+FN}$	$\text{Specificity} = \frac{TN}{FP+TN}$	

For an ideal method, all of these would be 100%. The actual value of the test therefore depends on the degree to which all these indices are close to (or distant from) 100%. However, in practice this is almost impossible to achieve, so there must be a degree of acceptance of whether a test really can be useful in clinical practice. For this reason, clinical judgements are often made on empirical grounds, and will generally be made alongside other information, such as the result of an X-ray or an ultrasound investigation, or signs and symptoms of the patient.

Note that Table 3.5 has striking similarities with the answer to self-check question 3.3. However, the latter considered the derivation of the kappa statistic for comparing two different methods, whereas the former looks at the ability of a single test to differentiate two groups of individuals.

SELF-CHECK 3.4

In assessing the ability of a certain cut-off point of a test to differentiate between 106 patients and 133 controls, there were 95 true positives, 13 false positives, 11 false negatives, and 120 true negatives. Use these data to calculate the specificity, sensitivity, positive predictive value, and negative predictive value.

In many cases, these four measures can be difficult to apply, and can be condensed down to (yet) another index. Many would like the clarity of having access to an index for comparing the probability of the test getting the diagnosis right (that is, correctly diagnosing an illness or condition), compared with the probability of the test giving a positive result when the individual is, in fact, healthy. One way of achieving this is with the **likelihood ratio**.

Likelihood ratio
a comparison of a test getting the correct result with it producing an incorrect result

Likelihood ratio (LR)

The likelihood ratio is often described as the gold standard for the value of a diagnostic test. There are two forms, mathematically derived from the sensitivity and specificity. The positive likelihood ratio is given by the formula, positive LR = sensitivity/(1-specificity), whilst the negative likelihood ratio is similarly given by the equation, negative LR = (1-sensitivity)/specificity.

Following through our example, the positive LR is therefore 0.96/(1-0.97), which is 32.0. A result greater than 1 implies that the test is associated with the disease, so our example test scores very well. The larger the score, the more valuable the result. Similarly, the negative LR is (1-0.96)/0.97, which is 0.04. A result less than 1 indicates that the test is associated with the absence of the disease, so the example again scores well. The smaller the score, the more valuable the result. These figures are scrutinized closely by authorities that decide whether a new test is adopted into clinical practice.

Choice of cut-off point

The example we have been working on assumes that the best cut-off point is 75 units/mL. However, other alternatives of 78 units/mL and 71 units/mL may be better. Using these latter two indices, we can derive different values for the sensitivity, specificity, PPV, and NPV of the test, which condense down to the positive and negative LRs (Table 3.6).

The trend is clear: there is an inverse relationship between the positive and negative likelihood ratios, and these depend strongly on the exact cut-off point. The formal method for determining the highest positive LR and concurrent lowest negative LR is to plot the sensitivity value with the 1-specificity value for each proposed cut-off point.

TABLE 3.6 **Likelihood ratios for different cut-off points**

Cut-off point	Positive LR	Negative LR
78 units/mL	65.9	0.078
75 units/mL	35.6	0.04
71 units/mL	14.2	0.02

Relatively poor positive and negative LRs do not necessarily imply that the test has no value. The highest standards are demanded by tests purporting to define the high likelihood of a disease being present and so of value in diagnosis. However, some tests may be useful in population or community screening, where a quick and simple first step may be valuable, but which might incorrectly identify some healthy people as having the particular disease (that is, they are false positives). This then requires a more demanding test that has higher standards and which can therefore define the presence or absence of the disease with much more confidence.

SELF-CHECK 3.5

Determine the positive and negative likelihood ratios for the data in self-check question 3.4.

Receiver-operating characteristic analysis

Useful as these measures of clinical value can be, they are to a degree clumsy and can be difficult to determine and interpret. As indicated in the preceding section, what most practitioners want to know is what particular laboratory value has the best sensitivity and specificity for diagnosing a particular disease: in Table 3.6, is it 71 units/mL, 75 units/mL or 78 units/mL? Fortunately, many statistical software packages now provide a measure called **receiver-operating characteristic** (ROC) analysis, which effectively brings together the two key indices of sensitivity and specifity.

Receiver-operating characteristic analysis
a visual method for rapidly assessing the clinical validity of a particular technique

An ROC analysis is a graph that plots the true positive rate (sensitivity) on the vertical axis against the false positive rate (1-specificity) on the horizontal axis, and provides a statistic called 'area under the curve', or AUC (Figure 3.3). If the laboratory value being tested identifies, for example, as many true positives as having the disease as true negatives having the disease, the ROC plot will be a diagonal straight line from bottom left to top right, and the AUC will be 0.5 (that is, 50% of the area of the square). In this case this particular test is without value. Alternatively, if the laboratory value correctly identifies many more true positives and true negatives, the ROC line will bend towards the top left-hand corner, as in the figure.

The strength of the software is that it can calculate the AUC of this new line, and in addition will generally provide a 95% confidence interval, and a linked probability that the outcome is genuine. Applying this to the data in Figure 3.2 and Table 3.6, a ROC analysis of the choice of 71 units/mL may give an AUC and its associated confidence interval of 0.85 (0.81–0.93). This result tells us that this particular laboratory index gives an AUC of 85%, far greater than the 'failure' of 50% (that is, an AUC of 0.5). The number of data points that made up this analysis provides us with a confidence interval (that is, that we are 95% confident that the real result lies between an AUC of 81% and 93%).

Cross reference

Probability, p values, and confidence intervals are explained in Chapter 6, Section 6.3

The software will also give us an estimate of the probability that the result is, statistically speaking, significant. This probability, abbreviated by the letter p, is deemed to be genuine if it has a value of <0.05. The p value associated with the AUC result of 0.85 (0.81–0.93) is 0.0046, considerably less than the required cut-off value of 0.05. The smaller the p value, the more likely it is that the data is genuine. Thus data where p = 0.0002 is more reliable than data where p = 0.0035, which in turn is more reliable than data where p = 0.017.

However, although the result of 71 units/mL gives a perfectly acceptable probability value (p = 0.0046), perhaps the other possible cut-off points are better at predicting the presence of the disease, with an even smaller p value. It may be that 75 units/mL gives an AUC of 0.89 (95% CI 0.84–0.96, p = 0.0021) and that 78 units/mL gives 0.86 (0.79–0.91, p = 0.0055). Thus all three could be acceptable, but 75 units/mL gives the best discrimination (that being p = 0.0021, the smallest). A further advantage of this type of analysis is that it can provide an AUC for different laboratory tests altogether, such as comparing the ability of amylase and lipase to diagnose pancreatitis (see the research paper by Zweig and Campbell listed in the suggested reading at the end of this chapter).

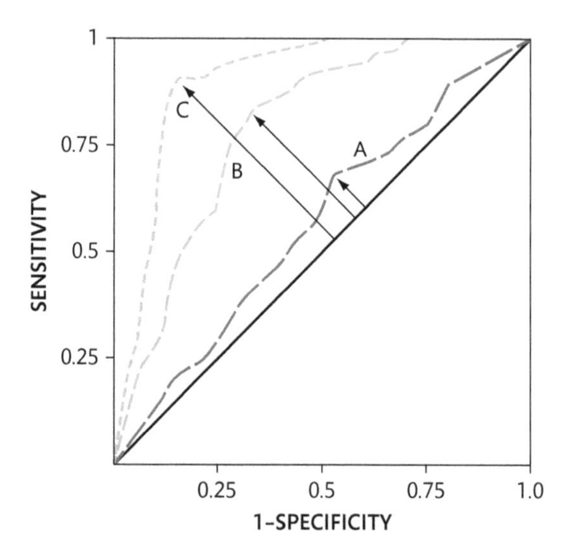

FIGURE 3.3 **Receiver-operating characteristic analysis.**
The graph plots sensitivity against 1-specificity for different values of a laboratory index (or different laboratory indices). The solid diagonal line indicates a result that would have no overall benefit. The line of long dashes (A) represents a value where the optimum results are a sensitivity of 0.67 and a 1-specificity of 0.52, as indicated by the small arrow. This line deviates a little from the diagonal line, giving an AUC of perhaps 0.57. However, we require a measure of the accuracy/precision of this index, best provided as a 95% confidence interval (CI), which in this case is 0.48–0.66. As this interval includes 0.5, it is not significant at the p < 0.05 level. Line B represent a second set of data much further away from the diagonal, and where the maximum distance from that diagonal (the middle arrow) gives a sensitivity of 0.83 and a 1-specificity of 0.32, indicating an AUC of 0.78 (95% CI 0.68–0.88). As this interval does not include 0.5, it is significant at p < 0.05. Line C represents a third data set, the point most distant from the diagonal (the longest arrow) giving a sensitivity of 0.90, a 1-specificity of 0.15, and so an AUC (95% CI) of 0.94 (0.89–0.99), which is significant at p < 0.01.

3.5 Evaluating a new method

An extension of the principles of sensitivity and specificity can be applied to the assessment of a new method, or in comparing two different methods. Technical developments, often driven by industry, inevitably feed through to the routine laboratory, such that laboratory managers may be faced with the decision of whether to adopt a new method, or to introduce the use of a new autoanalyser. In spite of rigorous checks, any new method must be assessed in-house against the existing method to determine whether or not it represents a better (more effective) choice. The overall principle is the same whether the technique is reporting a categorical index such as the presence or absence of cancer cells in a tissue biopsy, or a continuously variable index such as the number of bacteria or concentration of a metabolite.

Evaluating a method with a categorical outcome

An example of a method that delivers a categorical outcome (such as yes/no) is the presence or absence of the invasion of a cancer into a normal tissue such as the skin or a lymph node. Perhaps there is a new antibody or chemical stain that, it is claimed, gives better definition of tissues that have been invaded (or not) by the cancer. The method here is the same as in the comparison of two operators, as described above with the kappa statistic (Table 3.4). Depending on the technique, the sample size required for this comparison may be at least 100 different specimens of tissue, and these must be processed by the same scientist to eliminate operator bias.

Consider an example where the standard method assesses 123 samples, finding 48 to be positive and 75 to be negative. A second, alternative method found 60 to be positive and 63 to be negative. However, there are 26 samples where the methods are not in agreement. Because, by definition, the standard method must be correct, we can classify all results as being true or false positives, or true or false negatives (Table 3.7).

The kappa is derived from the following mathematics:

1. There is total agreement in 41 (true positives) and 56 (true negatives), giving full agreement in 97 of the 123 samples, a proportion of 0.79
2. The total proportion of positive column results is $60 \times 48/(123)^2 = 0.19$
3. The total proportion of negative column results is $63 \times 75/(123)^2 = 0.31$
4. Add these two proportions: $0.19 + 0.31 = 0.50$

$$\text{Kappa} = \frac{0.79 - 0.5}{1 - 0.5} = \frac{0.29}{0.5} = 0.58$$

According to Table 3.4, a kappa value of 0.58 suggests that the level of agreement is 'moderate', being in the range for kappa values 0.41-0.60. Kappa values 0.61-0.80 are considered good,

TABLE 3.7 **Comparison of two methods**

	Standard method positive	Standard method negative	Total
Alternative method positive	41 (true positives)	19 (false positives)	60
Alternative method negative	7 (false negatives)	56 (true negatives)	63
Total	48	75	123

whilst 0.81–1.00 denote very good agreement. It is for senior staff to determine whether the moderate level of agreement with the existing method is good enough, or perhaps the agreement should be better. It may be that such staff might consider that the level of agreement must be 'very good' in order for the new test to supersede the established test.

Evaluating a method with a continuously variable outcome

An example of this type of technique may be the concentration of a molecule in the blood. The assessment of the new method (let us say, Method B) is compared with an existing method (Method A) on a sample size of perhaps hundreds of separate samples. If the sample size is too small, the likelihood of a false positive or a false negative would rise, and so the comparison would be invalid.

Limits of agreement
a method for comparing results from two methods for measuring the same index

The gold standard analytical method is called '**limits of agreement**'. The key point is that the scientist must decide what is the minimum degree of disagreement that there must be between the two tests in order for the new technique to be thought of as sufficiently comparable with the existing technique. These limits of acceptability must be set before the analysis proceeds. One often-used criterion is that 95% of the data from the new technique (Method B) must lie within 5% of the existing method (Method A). So in this case, the limits of agreement are 105% and 95% (where complete agreement is 100%). Other criteria use the actual difference in the concentration of the molecules (perhaps in mmol/L), whilst some use a difference based on the standard deviation of the data of the existing method.

Once the technical analysis of the samples is complete, two new sets of data are required. The first is the relative difference between individual pairs of results (often expressed as a percentage of the result by the new method divided by that of the existing technique, that being A/B), so that a result of 50 by the existing method A and 52 by the new method B give a relative difference of 96% (50/52 = 0.96). Another pair of results may be 78 by the existing method A and 75 by the new method B, giving a relative difference of 104% (78/75 = 1.04). The second new data set is the mean of the particular pair of results (that is, [50 + 51]/2 and [78 + 75]/2), which are 50.5 and 76.5 respectively. The two new sets of data are then plotted with the relative difference on the y-axis and the mean on the x-axis (Figure 3.4).

The plot of the analysis of a sample size of 100 pairs of data sets shows that, overall, there are eight cases where relative difference between the two methods (that is, A/B × 100) exceeds the limits of agreement: one above and seven below. Although there are ten instances where the A/B differences are exactly on the 5% limit of agreement, these are permissible under the predefined limits of agreement, as the criterion is for the difference to exceed 5%.

However, these instances of a large difference between A and B are mostly clustered towards the right-hand side of the plot where the mean of A and B is high. Indeed, below a level of 100 units, there is only one instance where the A/B difference exceeds the upper 5% limit (= 105%) of agreement (where the mean difference is around 50 units/mL). Conversely, above 125 units/mL, there are seven A/B differences that are lower than the 95% limit, and at above 140 units/mL there are no cases where the result by the new method is less than by the existing method. This suggests that the two methods are comparable up to around 125 units/mL, but beyond this point the new method tends to produce higher levels that the existing method.

Correlation

Correlation
a method for assessing the strength of the association between two linked sets of data

It is not uncommon to see comparisons between two methods plotted together in a **correlation** graph. Correlation is a process where two indices are compared. However, these two indices have to be linked, and a good example of this is the relationship between height and

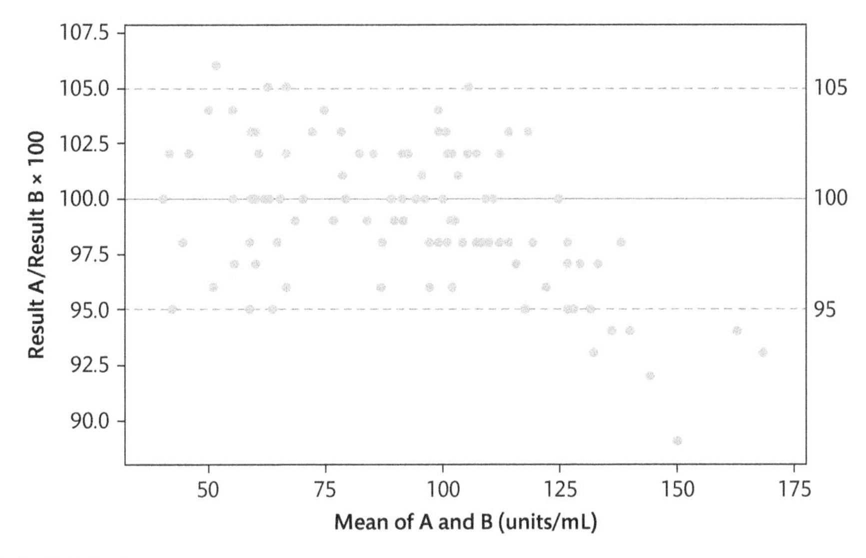

FIGURE 3.4

Limits of agreement. A plot of the difference between the results from the two methods, given in the y-axis, against the average of the results from the two methods on the x-axis ([A + B]/2). The units of the difference on the y-axis are the value of new test (B) divided by the value of the old test (A) multiplied by 100. The middle line (100%) is where both data sets give the same result (that is, where A = B). The upper dotted line is where the new method B result exceeds that of the existing method A by 5%. The lower dotted line is where the new method B is 5% less than that of the existing method A. Graphic from Minitab software.

weight. Generally, the range of weights of a group of people is linked to their heights—the taller someone is, then the heavier they are likely to be.

The degree to which these two indices are linked can be expressed by the correlation coefficient, which can vary between 0 and 1. A perfect relationship scores 1, whereas if the score is 0, then there is no meaningful relationship. It is likely that the correlation coefficient of the relationship between height and weight in a large group of people may be perhaps 0.7, indicating a strong relationship.

Correlation can also be used to examine the relationship between two sets of data that are linked in that the data is derived from measurement of the same substance by two different techniques. If we plot the values in the data given in Figure 3.4, we can obtain a graph (Figure 3.5). At first sight this looks good—many of the points lie on a line that originates from the bottom left and rises in a straight line to the top right. The correlation coefficient of this straight line is 0.99, which is very close to the perfect score of 1.0.

However, the straight line of this scatterplot enables us to look at a fundamental property of graphs of this nature, the equation $y = mx + c$. This equation allows us to calculate any one of the component parts if given the other three parts. The letter m is the gradient or slope of the line, the letter c is the point where the line crosses the y-axis. The equation of the straight line for the data set illustrated in Figure 3.5 is $y = 0.92x + 6.34$, where y is the Method A result and x the Method B result. The key aspect of this equation is that the gradient of the line is 0.92, and if there is a good linear relationship between A and B we would expect the gradient to be 1.0.

But close attention to Figure 3.5 shows that at levels of Method B above around 140 units/mL, seven data points do not lie on the straight line, but are below it, indicating a lack of agreement. Is this a coincidence, or does it reflect a methodological glitch? We can investigate this with a further analysis looking at the data set above and below 100 units/mL. Below this point, the correlation coefficient is 1.0, indicating a perfect relationship between A and B.

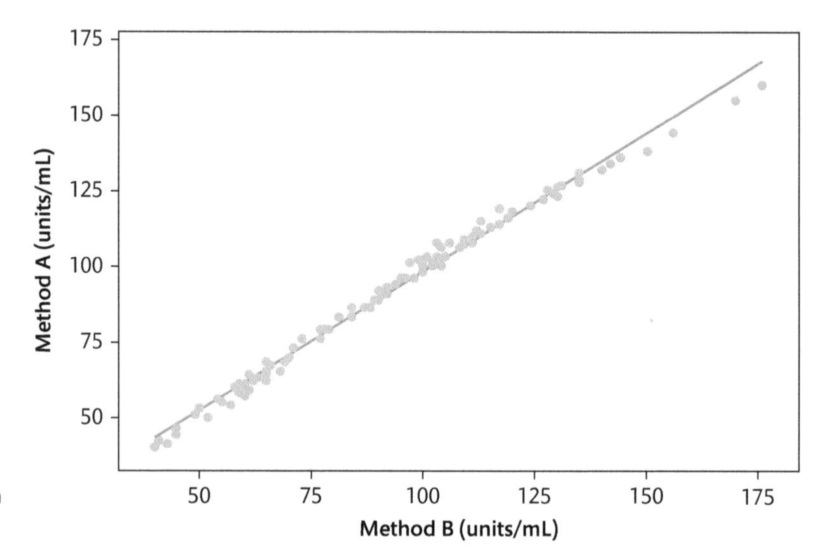

FIGURE 3.5

Correlation. **The relationship between results for Methods A and B. The line is mathematically constructed to give the best fit between the two sets of data (graphic from Minitab software).**

The equation of the straight line for data sets below 100 units/mL is $y = 0.999x + 0.18$, so that the gradient is much improved from that of the entire line, which has a gradient of 0.92. However, the equation of the straight line for data sets above 100 units/mL is $y = 0.79x + 21.4$; the gradient of 0.79 represents a less equal relationship between A and B. These analyses above and below 100 units/mL point to a difference in the method. We may conclude that the methods are comparable up to perhaps 130 units/mL, but above this point Method B records a consistently higher result than does Method A.

Limits of agreement or correlation?

The problem with using the correlation technique to compare results from two different methods is that it almost invariably gives a correlation coefficient that is very significant, indicating a good level of agreement. A weakness of this method (as can be determined by simply looking at the data, and as is evident in Figure 3.4) is that several poor results will not necessarily influence this correlation, leading to a false sense of security. The four data points on the far right (Method B results 150–175) are all relatively far from the line of best fit; the previous three data points are slightly below the line of best fit. For these reasons, the limits of agreement method is preferred as it readily shows the degree of disagreement of each pair of data points, and that this can be quantified with precision.

Chapter summary

- Error in laboratory testing can be pre-analytical (where the error is introduced before the sample arrives at the laboratory), analytical (the error is made during analysis), and post-analytical (which often involves incorrect information leaving the laboratory).
- The concepts of accuracy and precision tell us about the ability of a process to produce good and reliable information.
- Coefficients of variation tell us about the reproducibility of a method: intra-assay within a batch or a run of samples, and inter-assay, between different batches, which are often analysed on consecutive days.

- The performance of two different operators can be compared using the paired t test or the kappa statistic.
- Sensitivity is the proportion of people with the disease who are correctly identified by the test. Specificity is the proportion of people without the disease who are correctly identified by the test.
- The positive predictive value is the proportion of patients with a positive test result who are correctly diagnosed. The negative predictive value is the proportion of patients with a negative test result who are correctly diagnosed.
- Evaluation of a new method calls for techniques such as the kappa statistic and limits of agreement. Correlation alone should not be used.
- A receiver-operating characteristic plot gives a useful visual of the extent to which a method can determine the ability to differentiate two groups.

Suggested reading

- **Bland, J.M., and Altman, D.G. (1986). Statistical methods for assessing agreement between two methods of clinical measurement.** *Lancet* **327(8476):307-10.**
- **Zweig, M.H., and Campbell, G. (1993). Receiver-operating characteristic (ROC) plots: A fundamental tool in clinical medicine.** *Clin Chem* **39:561-77.**
- **Christenson, R.H., and Committee on Evidence Based Laboratory Medicine of the International Federation for Clinical Chemistry Laboratory Medicine (2007). Evidence-based laboratory medicine—a guide for critical evaluation of in vitro laboratory testing.** *Ann Clin Biochem* **44:111-30.**

Questions

3.1 Consider the data set 45, 49, 47, 42, 46, 48, and 47. What is the coefficient of variation?

3.2 Study Figure 3.4. It shows data from a comparison of two methods plotted according to the 'limits of agreement' analysis. Comment on the relationship between the methods.

3.3 Describe the strengths of receiver-operating characteristic analysis.

3.4 Kappa value for a new test is 0.58. Discuss whether it should be accepted into the laboratory in place of the existing method.

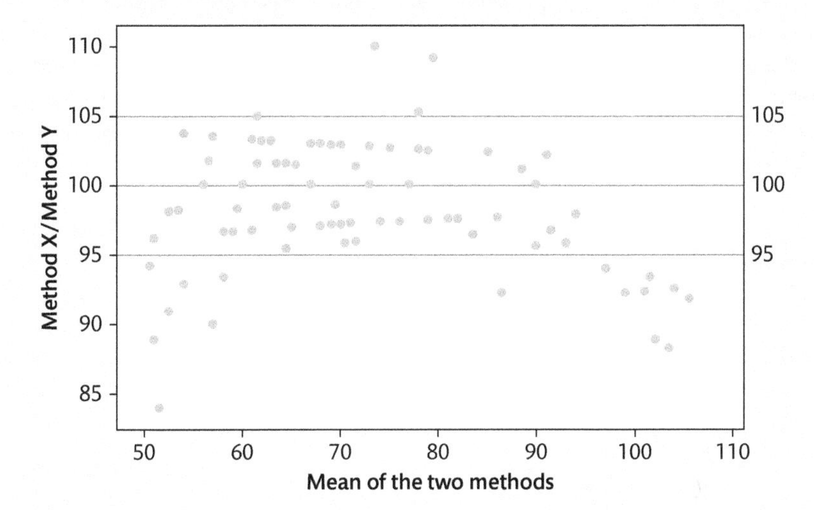

4

Presenting data in graphical form

Learning objectives

After studying this chapter, you should be able to describe and interpret numerical data when they are presented graphically in the following forms...

- Normal and non-normal distribution
- Scatterplot
- Histogram
- Pie chart
- Dot plot and box and whisker plot
- Line plot
- As a graph

This chapter consists of seven sections that will provide a firm background into the most common formats (pie charts, graphs etc.) in which data can be presented visually, and how such data can be interpreted. In so doing, we will look at methods of presenting both categorical and continuously variable data. This will be reinforced by self-check questions where you will have the opportunity to plot some data on graph paper. The chapter will conclude with Section 4.8, which gives advice on which formats to choose. Fortunately, many software packages (such as Excel, Graph-Pad, and SPSS) can produce excellent figures suitable for presentation and publication, so that hand-drawn artistic skills are rarely needed. Merely because of the author's experience, figures in this (and other) chapters are modified from those provided by Minitab.

Categorical data
data that can only exist in a small number of discrete categories, or 'boxes'

Continuously variable data
data that can (within reason) take any value along a scale

4.1 The distribution of data

In Chapter 1 we looked at two ways in which we can get a feel for the way in which the individual numbers in a set of data are distributed—hence the 'distribution'. There are several types of distribution, by far the most common being a normal distribution and a non-normal distribution.

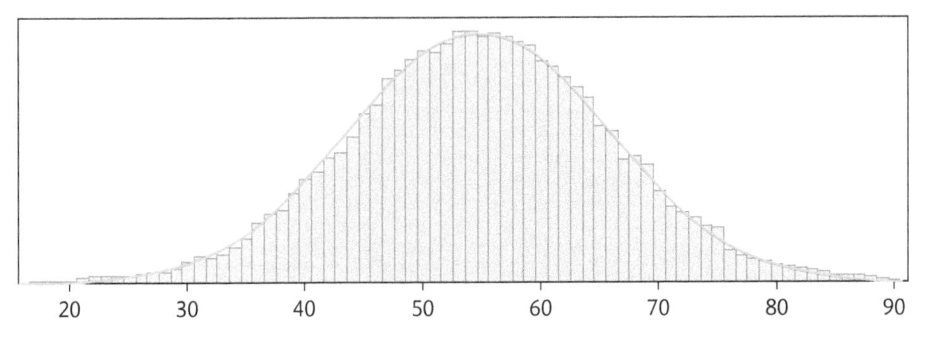

FIGURE 4.1

A normal distribution. A normal distribution of the type shown in this figure is a summary of the individual points of Substance A, levels of which vary from 20 units to 90 units. The vertical axis is the number of data points. The greatest number of data points are in the middle (between 50 and 60), with smaller numbers of data points to the left and right. The blue line shows the best fit of the data for a normal distribution.

The normal distribution

The normal distribution is the most important and most frequently encountered distribution (hence 'normal'), although you may see it referred to as a Gaussian, or parametric, distribution. In a normal distribution (as is illustrated in Figure 4.1), most of the individual data points are clustered around the middle of the entire data set, and this is generally where the **mean** value is to be found. Indeed, the value with the highest frequency is inevitably the mean value; in the case of the data in Figure 4.1, this value is 55.

Mean
the 'average' of a set of data points

A second aspect of a normal distribution is that there are an equal number of data points above and below the mean value, that is, to the left and right of the highest value in Figure 4.1. Furthermore, the data can be fitted to a curved line, and so may be called a 'bell-shaped curve'.

Cross reference
Chapters 1 and 7 have additional details of the mean value of a data set, and introduce the concept of distribution

This type of distribution is by far the most common in biomedical science, and indeed in almost all areas of biology. Key indices that follow this distribution in a normal healthy population include the number of red blood cells in a certain volume of blood, the concentration of sodium in the serum, the height of a group of people, and their weight.

The non-normal distribution

The second most frequent type of distribution is called non-normal, and is present when the greatest number of data points are found to either the right or left of the centre of the data set; such a data set is said to be skewed. The main summary index of this type of data is not the mean, but the **median**. Figure 4.2 shows such a distribution, where most of the data is skewed to the left. A key aspect of data with a non-normal distribution is that the mean and median are different values, in this case they are 10 and 8.5 respectively. A further index, the **mode**, has a value of 7. This is simply the single value that occurs most frequently.

Median
that single data point that occupies the middle position when all of the data points are ranked in order

Mode
that single data point from a set of data that is present most frequently

Examples of indices from biomedical science that adopt this distribution in a healthy population include serum triglycerides, the erythrocyte sedimentation rate, and the marker of inflammation, CRP. The data point that is in the middle of the entire set is called the median.

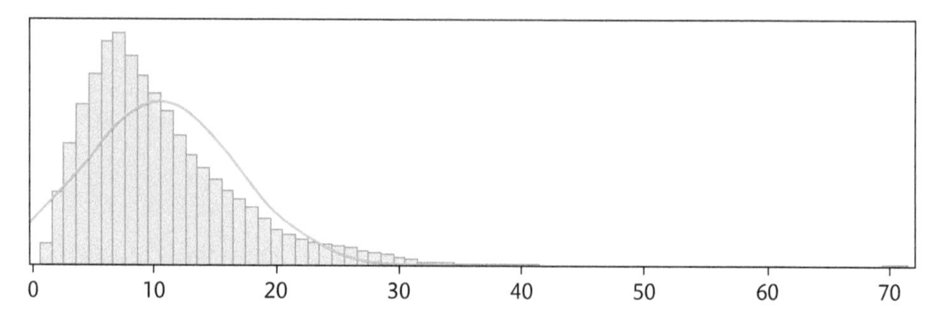

FIGURE 4.2

A non-normal distribution skewed to the left. In this summary plot of a non-normal distribution of individual points of Substance B, which ranges from 1 unit to 75 units, the vertical axis is the number of data points. The greatest numbers of data points are to the left or right of the mid-point of the entire data set. In this case, the data set is skewed over to the left with no 'tall' columns on the right. Note that the blue line (supposed to be tracking a bell-shaped curve) fails to overlap with the grey columns of data, which is further evidence that the data has a non-normal distribution.

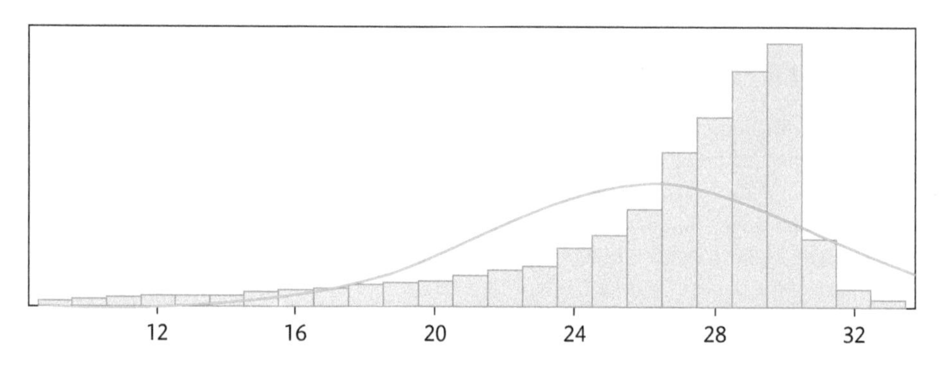

FIGURE 4.3

A non-normal distribution skewed to the right. In this non-normal distribution of individual points of Substance C, the greatest numbers of data points are skewed with many 'tall' columns on the right; there are no 'tall' columns on the left. As with the left-skewed data in Figure 4.2, in this data set the blue line of best fit for a normal distribution fails to follow the grey-coloured distribution of the blocks of data.

Cross reference

Chapters 1 and 7 have further details of the median and mode values of a data set

Very rarely, a data set with a non-normal distribution has its median value skewed over to the right-hand side of the entire data set. Figure 4.3 shows the distribution of levels of Substance C in the blood, which has a marked right shift. As is required of data with a non-normal distribution, the mean and median values are different (27 and 28.5 respectively). Furthermore, these numbers also differ from the mode, which in this case is 30 (the single value that is present most often, and represented by the 'tallest' individual column).

A bimodal distribution

Occasionally, a set of data may contain two different populations, as illustrated in Figure 4.4. The two populations are reasonably easy to identify, one on the left which peaks at 19, and a second on the right with a peak value of 29. There seems to be a cut-off point between the two distributions.

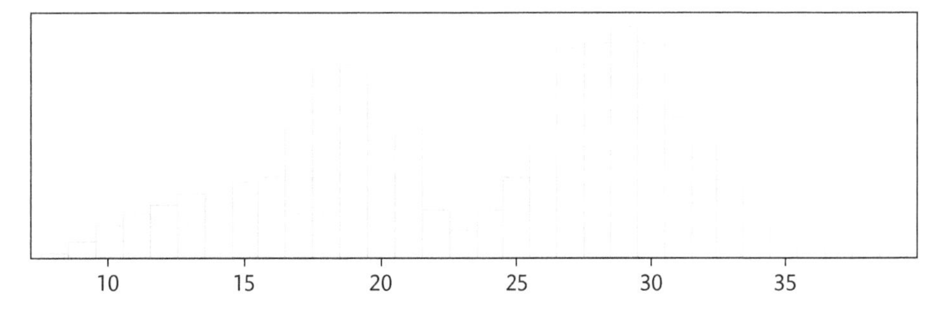

FIGURE 4.4

A bimodal distribution. In this non-normal distribution, there are two separate sets of data which both seem to have a normal distribution, but which have been pushed together into one plot.

A bimodal distribution is often encountered when comparing data from the two sexes, so that Figure 4.4 may be illustrating differing levels of sex hormones in males and females, or perhaps (with a change of unit) the heights or weights of groups of men and women.

4.2 Scatterplots

Let us now go on to consider another way of representing data, the scatterplot. The most common use of a scatterplot is to determine whether there is a relationship between two indices such as the height and the weight of each individual in a group of people. In so doing we can determine whether or not such a relationship exists by looking for a **correlation** between them. There are two opposing instances where scatterplots provide such an opportunity: where there is a positive correlation (where both indices increase in value), or where there is a negative correlation (where as one index increases, the other decreases).

Correlation

a method for determining the presence or absence of an association between two linked sets of data

The strength of the relationship between two indices can be defined statistically as the **correlation coefficient**, for which the appropriate symbol is the letter *r*. When there is a very strong relationship between two indices, the correlation coefficient approaches 1; whereas if the relationship is weak, the coefficient is closer to zero. In the case of an inverse relationship, *r* takes a negative value, so that an *r* value of −0.85 tells us of a strong inverse relationship, but if $r = -0.15$, the inverse relationship is weak.

Correlation coefficient

an index that provides a numerical estimate of the strength of the association between two linked sets of data

The mathematics of the generation of the correlation coefficient seems daunting, as the appropriate equation is:

$$r = \frac{\sum xy - \left[\left(\sum x\right)\left(\sum y\right)/n\right]}{\left[\left(\sum x^2 - \left(\sum x\right)^2/n\right)\left(\sum y^2 - \left(\sum y\right)^2/n\right)\right]^{1/2}}$$

where Σ means 'sum of', and x and y is the pair of indices we are focusing upon. However, the equation can be broken down into small parcels that are eminently manageable, given patience and a hand-held calculator (Box 4.1).

Cross reference

The concept of correlation is introduced in Chapter 3 and will be expanded upon in Chapter 7

A positive correlation

When a positive correlation exists between two indices, as one of the indices increases, so does the other index. Look at Figure 4.5, which depicts a data set exhibiting a positive correlation between two indices, the height of an individual, and their weight. Notice how, as height increases (y-axis), the weight also increases (x-axis): there is a positive correlation between the two. Examples of a positive correlation include the relationship between height and weight, and the relationship between systolic blood pressure and diastolic blood pressure.

BOX 4.1 Calculating a correlation coefficient

In order to calculate the correlation coefficient for a set of paired data, we need to generate five separate indices. Consider the data set:

X value	Y value
8	6
12	14
10	12
12	11
15	15

- **We first need to multiply together each X and Y value (i.e. $8\times6+12\times14+10\times12+12\times11+15\times15$), then add them together, giving Σxy, which is 693.**
- **The second index is the sum of all the X values ($\Sigma x=8+12+10+12+15=57$).**
- **The third is the sum of all the Y values ($\Sigma y=6+14+12+11+15=58$).**
- **We then need the sum of all the X values when they are squared, hence $\Sigma x^2=8^2+12^2+10^2+12^2+15^2=677$.**
- **Finally, we need the sum of all the Y values when they are squared, hence $\Sigma y^2=6^2+14^2+12^2+11^2+15^2=722$.**

Returning to the equation ...

$$r=\frac{\sum xy-\left[\left(\sum x\right)\left(\sum y\right)/n\right]}{\left[\left(\sum x^2-\left(\sum x\right)^2/n\right)\left(\sum y^2-\left(\sum y\right)^2/n\right)\right]^{1/2}}$$

...we simply substitute in the five indices we have just calculated:

$$r=\frac{693-\left[(57)(58)/5\right]}{\left[\left(677-(57^2/5)\right)\left(722-(58^2/5)\right)\right]^{1/2}}$$

which becomes...

$$r=\frac{693-661.2}{\left[27.2\times49.2\right]^{1/2}}=\frac{31.8}{36.6}=0.87$$

Hence the correlation coefficient r between X and Y is 0.87, which is strong. However, it is based on only five pairs of data, so we must be cautious in interpretation.

The correlation coefficient

It is well known that the relationship between height and weight is very strong, but note that in the example in Figure 4.5, the subjects are weighed fully clothed, and their height is measured with footwear. Since the individual has considerable choice of what clothes and shoes to wear

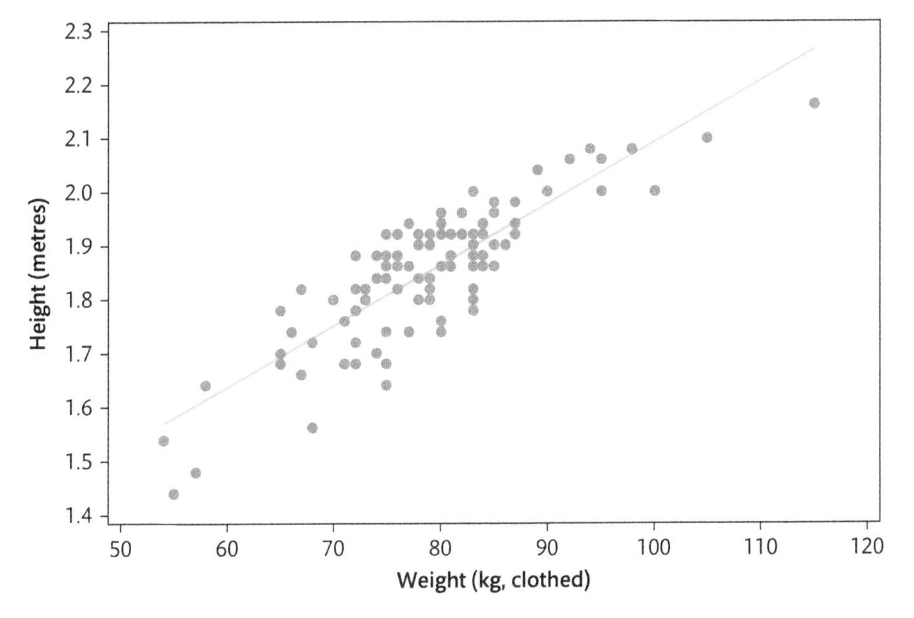

FIGURE 4.5

Correlation between height and weight when clothed. There is a clear positive correlation between the height and weight of a group of people when they are fully clothed. The correlation coefficient is 0.85.

(and how heavy these are), and this is very variable, then the weight of an individual free of clothes and shoes (that is, naked) is likely to provide a more accurate relationship between true height and true weight; this relationship is shown in Figure 4.6.

Unsurprisingly, Figures 4.5 and 4.6 appear similar. But the average weight when naked is 3 kg less than when clothed, and this reduction means that the correlation coefficient is now 0.89, which is an improvement. The explanation is likely to be that, once naked, the weights of the subjects have lost the variability attributable to the clothes they were wearing, and their heights have lost the variability attributable to their choice of footwear.

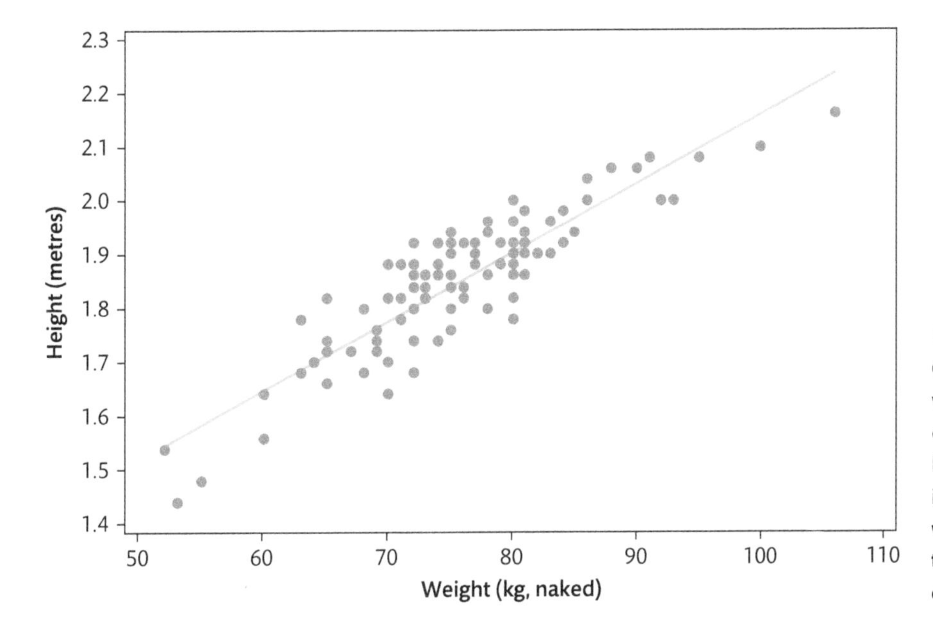

FIGURE 4.6

Correlation between height and weight when naked. There is a clear positive correlation between height and weight when naked. The individual data points are, on the whole, closer to the line of best fit than in Figure 4.5. The correlation coefficient is 0.89.

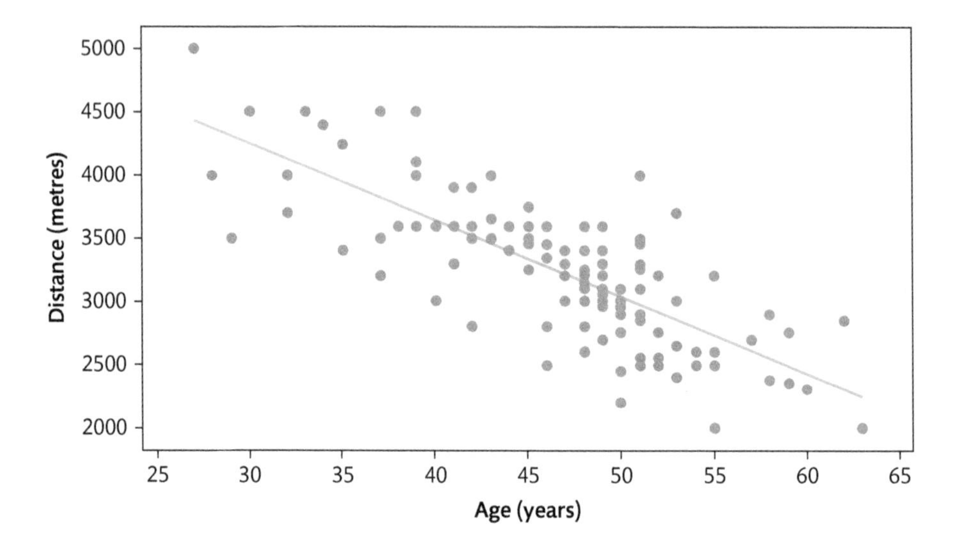

FIGURE 4.7

A negative correlation. There is a clear negative correlation between the age of an individual and the distance he or she can run in a given time period.

A negative correlation

A negative correlation is present when one index falls whilst the other index rises. It may also be described as an inverse relationship. We see an example of a negative correlation when we consider the relationship between the distance an individual can run in a set period of time, and their age, as illustrated in Figure 4.7. Understandably, as the physical strength of the human body deteriorates with age, so running ability generally falls. However, this relationship is not perfect, as close attention to the figure will identify. Notice how individual data points to the right of the graph show how some older individuals can run further than those several years younger than themselves.

As with other scatterplots, the software has drawn a line of best fit, allowing the calculation of a correlation coefficient (r); in this case the value is –0.77. As this correlation coefficient is less than that of previous examples (where $r = 0.85$ and 0.89), we can say that the relationship between age and running is weaker than those between height and weight.

No clear correlation

There will of course be instances where two indices show no clear relationship. An example of this is shown in Figure 4.8, which plots age versus the HbA1c (an index of the effect of glucose on haemoglobin in red blood cells) in a group of individuals with diabetes. The line of best fit shows a very slight downwards slope, so the correlation coefficient, although extremely small, is negative, at –0.006, indicating absolutely no significant relationship between the two.

Cross reference

The importance of correlation as a statistical tool is developed in Chapter 7

SELF-CHECK 4.1

Plot the X and Y data in Box 4.1 as a scatterplot, and then draw what you feel to be the line of best fit.

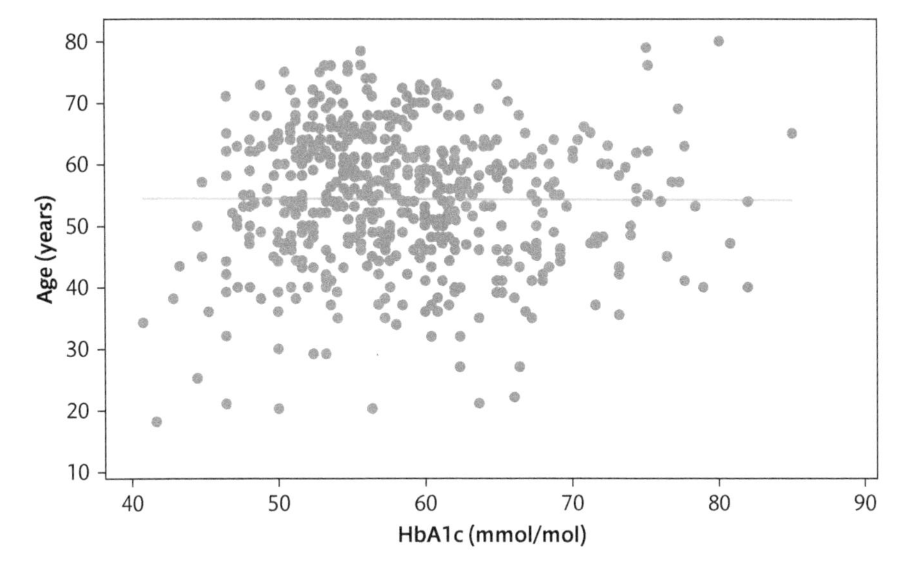

FIGURE 4.8

No clear correlation. In this plot there is no clear correlation between the age of a group of diabetics and their HbA1c. The data points seem randomly scattered, and the line of best fit is horizontal.

4.3 Histograms

Histograms are often used to illustrate as a series of columns the number (or frequency) of discrete events or units in a given data set. Figure 4.9 presents the number of children in 60 families, with increasing numbers of children from left to right. As you can't have a fraction of a child the data take the form of whole numbers. Five families have no children, eight families have one or four children, but the most frequent number of children is two, which is the case in 20 families.

Note that the data set as a whole is skewed to the left, reminiscent of the data in Figure 4.2. It could therefore be argued that this tells that the data carries a non-normal distribution, and that the median value is 2 (children). However, we cannot make this kind of statement: we can only apply the concepts of normal and non-normal distributions to data that has a continuous variability. By contrast, the data in Figure 4.9 is categorical.

One way we can summarize the data set is to note which number of children per family is the most frequent. Twenty families have two children, but 12 have three children. Thus we can rapidly and easily state that the most common number of children per family is two; this number is the mode.

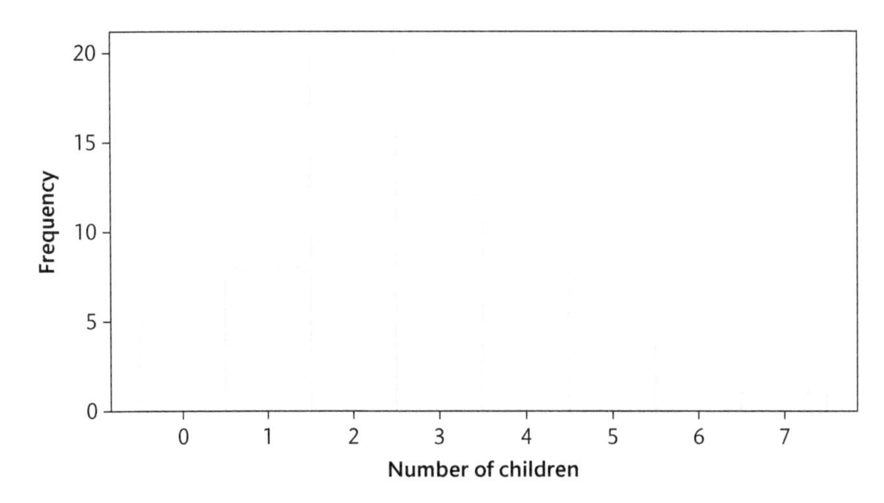

FIGURE 4.9

A histogram. A plot of the number of children present in a group of 60 families. The highest frequency (the tallest column) is two children, present in 20 families, whereas only one family has seven children.

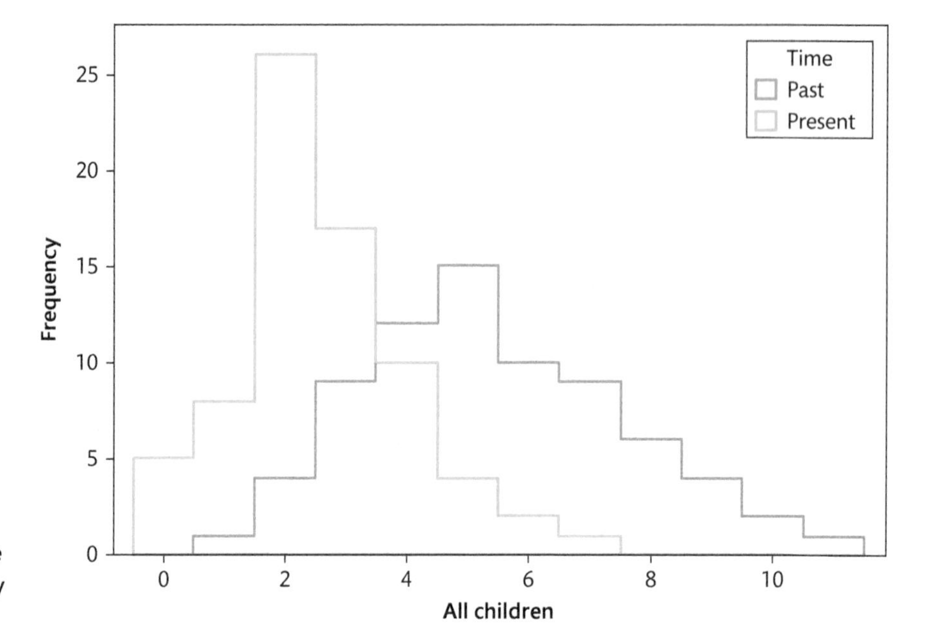

FIGURE 4.10

A dual histogram. A plot of the number of children present in two groups. On the left, a contemporary data set where the mode number of children is two per family. On the right, data from the 1950s shows the mode number of children per family was five.

A variation of this form of data presentation places histograms of two different sets of data alongside each other. In Figure 4.10 there are two histograms, both of which also represent the distribution of the number of children in a family. The data on the left, representing a contemporary data set of 73 families, again shows the mode number of children per family to be two. On the right is a second set of data from another 73 families, taken from a sample from the middle of the twentieth century, when family planning advice was less freely available than at present. Accordingly, the number of children per family is larger, with the mode number being five children and the maximum being 11 children.

4.4 Pie charts

We use pie charts to represent the number or proportion of a particular index as sectors of a circle. Consider a group of 50 people, of whom 25 are men, 15 are women, and 10 are children. The total number of people (50) can be represented by a circle of 360°; and the subgroups (men, women, and children) are represented by individual sectors (or 'slices of the pie') of different colour (Figure 4.11).

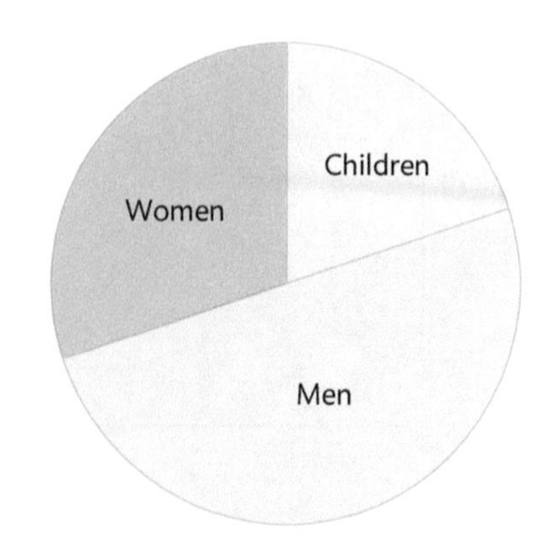

FIGURE 4.11

A pie chart. This figure plots the proportions of men, women, and children in a group of 50 people. The major group is easily identifiable on the bottom right, occupying a half of the entire circle.

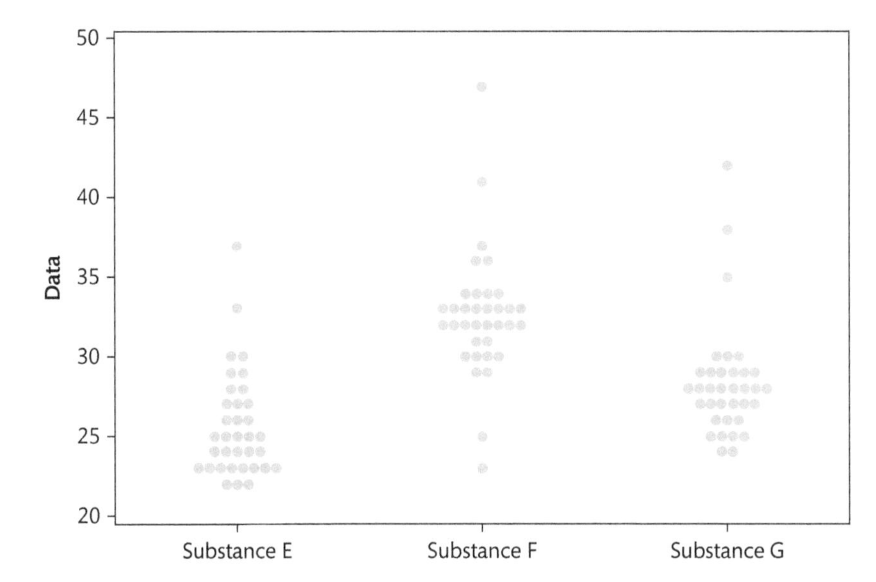

FIGURE 4.12
The dot plot. This figure shows individual data points (vertical axis) of the levels of Substances E, F, and G.

Starting at the top of the circle (the 0°/360° point), and moving clockwise, the first segment is the 10 children, followed by the 25 men. This is the largest segment, which makes up a half of the entire group. The remaining segment (top left) represents the 15 women.

4.5 Dot plots and box and whisker plots

The dot plot is probably the best-known form of data presentation, an example of which is shown in Figure 4.12. This shows 35 dots, indicating levels of Substances E, F, and G in the blood, according to the scale on the left-hand (y) axis, which runs from 20 to 50. For each of the three substances, each single data point can be seen, and so the general distribution of each substance can be visualized and compared. Levels of Substance E are lowest, and are also the most tightly clustered together. Levels of Substance F are highest; these, alongside those of Substance G, are more spread out than those of Substance E.

A modification of the dot plot is a box and whisker plot, as shown in Figure 4.13, which summarizes key aspects of the data. In this form, the box represents the upper (from the median point to the 75th rank point) and lower (from the median to the 25th rank point) quartiles, and the line across the box is the median: this being 25, 32, and 28 for Substances E, F, and G respectively. Further details of the median and inter-quartile range are presented in Section 1.3 of Chapter 1. The whiskers (the straight lines running up and down from the box) generally represent the remaining lower and higher quartiles (from the lowest to the 25th rank point, and from the 75th rank point to the highest level respectively).

Figure 4.13 has been generated by some statistical software. It has identified that in some cases there are unusually high or low values—these are called outliers, shown as an asterisk (*). In Substance E there is a single outlier above the upper whisker, whereas Substance F has four outliers—two high and two low. Substance G has three high outliers.

Both dot plots and box and whisker plots allow us to physically view the distribution of the data, and so draw conclusions as to whether or not it is likely to have a normal or a non-normal

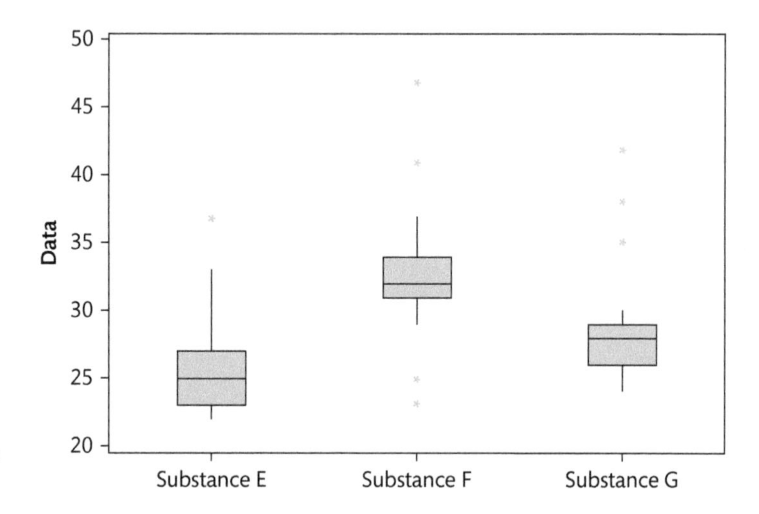

FIGURE 4.13
A box and whisker plot. This figure also shows a summary of the levels of Substances E to G, each with a box, two whiskers, and some outliers (*).

distribution. It could be argued that Substance F has a normal distribution, as individual points seem to lie equally either side of the central box. By contrast, both substances E and G may have a non-normal distribution, as the bulk of the data points are clustered around the lower end of each set, with a smaller number of data points at the high end of each set. However, formal definition of the type of distribution is made by a statistical test (analysis of variance), details of which are presented in Chapter 7.

SELF-CHECK 4.2

Using some standard graph paper, plot the data on levels of substances H and I, and make a comment on the likely distribution of each.

Substance H: 64, 76, 58, 79, 60, 45, 75, 70, 54, 64, 53, 56, 64, 72, 70, 58, 60, 67, 68, 61.

Substance I: 67, 120, 58, 150, 58, 43, 100, 85, 53, 62, 50, 55, 64, 92, 77, 55, 58, 70, 73, 62.

4.6 Line plot

There are many instances where it is important to be able to determine whether there is a change in a particular measurement. For example, we might want to explore how the levels of a particular molecule in the blood of a series of people differ over time. Or we may want to consider the effect of systolic blood pressure (SBP) in a group of people before and after they start to take a drug designed to treat high blood pressure (hypertension), as illustrated in Figure 4.14.

In this example, the SBP before the intervention (data on the left vertical axis) from each of the fifteen people are presented in a vertical column. In each case there is a line joining this point to that person's SBP after the intervention (vertical column of data on the right), hence the use of the term 'line plot'. Close attention to the plot shows that in 13 of the 15 patients, SBP has fallen; but in two cases, the blood pressure increased. The line plot therefore gives us an immediate view as to whether the intervention is effective. However, this needs to be confirmed or denied with a precise statistical test (such as a paired t test), details of which are in Chapter 7.

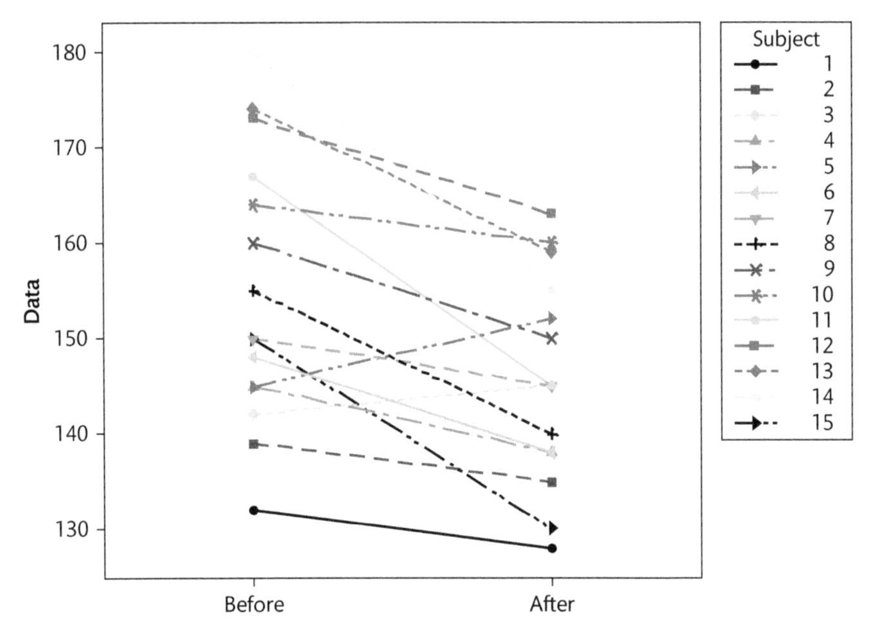

FIGURE 4.14

A line plot. A line plot of systolic blood pressure (SBP) in 15 people before and after use of an antihypertension drug.

4.7 Presenting data as a graph

Sets of data measured under several different conditions are often represented as a graph. Two such examples are changes in a certain variable (such as levels of an ion, molecule, or cell in the blood) under different biochemical conditions, or perhaps the changes in a particular variable over time. In the case of the former, the different conditions may be increasing concentrations of a drug, in which case the experiment would be carried out to determine a 'dose-response' curve. The latter may be the response of a physiological measurement, such as heart rate, at set periods of time, perhaps over five minutes. Data of this type may be from an individual source, or from several sources combined.

Data from a single source

An example of data from a single source is the change in concentration of a component of the serum over time, as illustrated by the graph in Figure 4.15. The figure shows a reduction in serum creatinine from an initially high level in a patient with acute renal injury. This data can be interpreted to indicate that the patient experienced a steady fall in creatinine levels over a three-week period.

As the upper limit of the reference range is in the region of 133 μmol/L, then it is safe to say that the acute renal injury steadily resolved, and by day 14, serum creatinine was within the reference range. After three weeks, the creatinine level was stable at approximately 85 μmol/L, by which time it seems likely that the renal crisis was over.

SELF-CHECK 4.3

Using some standard graph paper, plot the following data of temperature (degrees Celsius) recorded on the first day of each month: January, 2; March, 4; May, 7; July, 15; September, 10; November, 5.

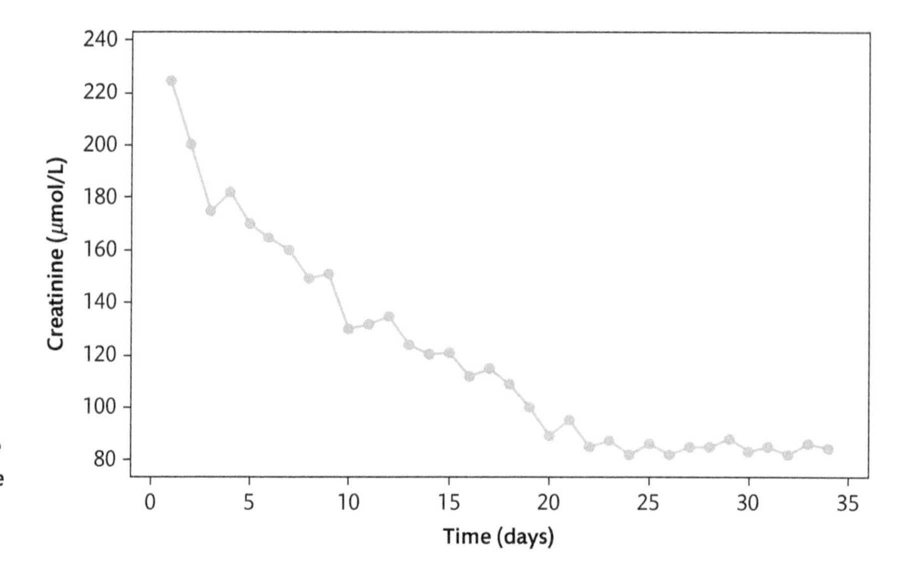

FIGURE 4.15

Graph of change in serum creatinine over time. The graph plots the change in levels of serum creatinine over a 34-day time period.

Data from several sources

A similar graph to that shown in Figure 4.15 can be used to plot changes in an index in a group of subjects over a period of time, or under a series of changing conditions (such as increasing concentration of a drug, for example). However, in situations where there are several measurements, there will be a degree of variability in each data set. This variability can be represented as the mean and standard deviation (SD), discussed in Chapter 1.

Standard error
an index for assessing the reliability of the mean value

An additional index often used to show variability in data is the **standard error** (SE). This index may be used to provide an estimate of the degree of certainty of the accuracy of the mean, and is derived from the SD and the number of data points in that particular set (the 'n' value). Unlike the SD, the SE tells us nothing about the distribution of a data set. Mathematically:

$$SE = SD/\sqrt{n}$$

If a data set has a SD of 2.5, and the sample size (n) is 64, then the SE is 2.5/8 = 0.312. However, if the sample size is larger, perhaps consisting of 121 individual points, and has the same SD, then the SE is 2.5/11 = 0.227. Therefore, as the SE gives us an idea of the accuracy of the mean, we are more confident in the reliability of a data set that is derived from a larger sample size. In the case of the data given here, notice how the standard error reduces from 0.312 to 0.227 when the sample size is increased from 64 to 121. Intuitively, the larger the sample size (and so the smaller the SE), the more we are assured that the mean is valid.

When presenting data in a graph, the reliability of a mean value can be represented with small bars above and below the mean value. These bars often reflect the SE, and so are described as 'error bars'. However, these bars may also represent the SD, so care must be taken.

A worked example

Consider an experiment to test the ability of a drug to influence levels of haemoglobin in an animal model. Four groups of animals, each containing five subjects, are given an increasing dose of the drug. A fifth group receives no drug. At the conclusion of the experiment, haemoglobin is measured in all animals to give the results summarized in Table 4.1, which shows the mean, SD, and SE of the haemoglobin data in the five groups of animals.

TABLE 4.1 **Haemoglobin levels in five groups of animals**

Group	Drug dose (mg/kg)	Mean haemoglobin (g/L)	SD	SE	**Mean ± SE***
1	0	123	7.5	3.3	120–126
2	2	127	4.3	1.9	125–129
3	4	131	5.2	2.3	129–133
4	7	134	5.1	2.3	132–136
5	10	135	4.0	1.8	133–137

SD = standard deviation, SE = standard error.
*Data presented to three significant figures

The data set on the far right is the mean value plus or minus one standard error, and provides a simple approximation of the accuracy of each mean value. For example, although the mean haemoglobin level at a dose of 2 mg/kg is 127 g/L, a view of the precision of this data is given by the error range of 125 to 129 g/L. Similarly, the error range of the data at 7 mg/kg is 132 to 136 g/L.

It is clear that the level of haemoglobin rises with the dose of the experimental drug. But is this increase steady, with equal increments, or is it staggered, with unequal steps between the doses? One way of part-answering these questions is to present the data visually, in the form of a graph, as is shown in Figure 4.16.

At first glance it is clear that the mean values rise in a way that is proportional to the dose of the drug. It is also clear that levels of haemoglobin are tending to level out when higher drug doses are administered. Another consideration is that there is a great deal of overlap in the error bars between adjacent drug doses.

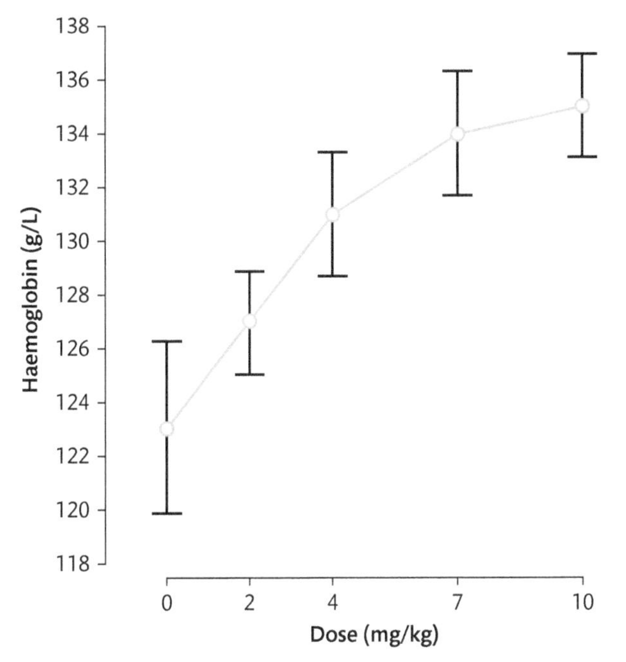

FIGURE 4.16
The response of haemoglobin to an experimental drug. The graph shows the mean level of haemoglobin in groups of animals given different doses of a drug, and the standard error associated with each mean value.

An important point about the use of error bars is that they give us absolutely no formal statistical information. Error bars give us a simple visual regarding the reliability of the mean value, but nothing more. Merely because the error bars of the data at 2 mg/kg and at 7 mg/kg do not overlap does not suggest that there is a genuine statistical difference between the data sets. This conclusion should only be drawn from formal statistical analyses, some of which are presented in Chapters 7 and 8.

4.8 Which type of visual presentation should I choose?

Having now considered the main options for presenting data in visual form, one important question remains: when is it most appropriate to use each of these options?

The choice of visual presentation can be difficult, but there are conventions. Although the distribution of continuously variable data can be assumed by comparing the mean with the median, and looking at the standard deviation compared with the mean, distribution is best demonstrated by the kinds of graphs depicted in Figures 4.1–4.4.

In demonstrating the relationship between two different sets of data that are linked in some way (such as the height and weight of a group of people), a scatterplot is appropriate (as in Figures 4.5–4.8).

Histograms (such as in Figures 4.9–4.11) are useful in showing the actual number or frequency of categorical data, but may also be used to summarize data with a continuous variation. An alternative presentation of the frequency of categorical data is with a pie chart (Figure 4.12), which in some cases is preferable to a histogram in that it provides a rapid perspective of the proportions of data points in different groups as a whole.

If showing continuously variable data from different groups, a dot plot is preferred as it shows each individual point (as in Figure 4.13). A less acceptable alternative is a box and whisker plot, such as in Figure 4.14. Here, the individual data points themselves are generally not shown but are summarized, often with mean and standard deviation.

Data collected over a series of time points, or in response to changes in a second index (such as the effects of a drug) are best plotted as a graph. This can be in a single index (as in Figure 4.15), or where individual data sets are combined into groups (as in Figure 4.16). A common feature of the latter is the use of error bars, which allow a rapid assessment of the quality of the data sets.

There are alternative ways of visualizing sets of data collected over time or in response to different concentrations of a drug (as in the haemoglobin data in Figure 4.16). In principle, it is possible to present this data as a correlation scatterplot, showing the relationship between haemoglobin and the dose of the drug (Figure 4.17). However, in this instance, such a plot is inappropriate as the haemoglobin data is the mean of several individual points, and the dose of the drug ranges over only five data points. A correlation plot should only be used when both indices can take any value and range freely (as in height and weight).

Similarly, the same data could be presented as a series of five histograms, each with its own error bar (Figure 4.18). But this too is inappropriate as it fails to show the full extent of the error bars, and implies that the five sets of data are independent, whereas it is clear that they are linked by increasing doses of the drug.

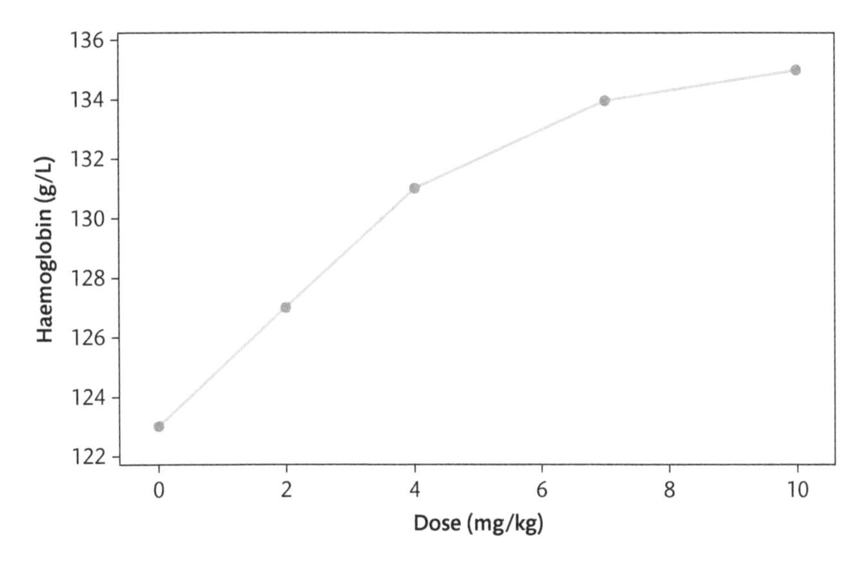

FIGURE 4.17

Scatterplot of haemoglobin and drug dose. Scatterplot of the relationship between haemoglobin and the dose of the drug.

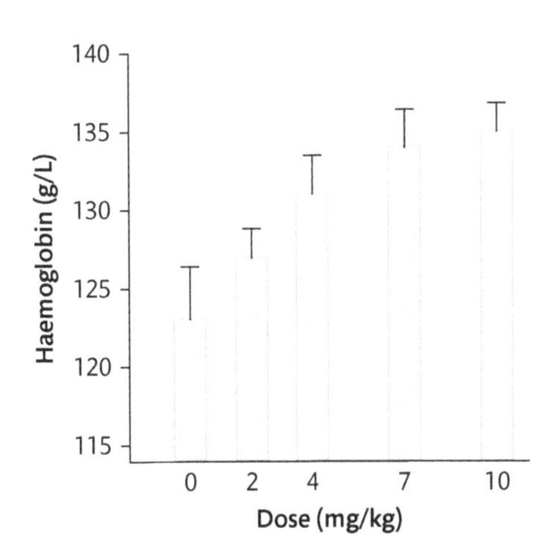

FIGURE 4.18

Histograms of haemoglobin and drug dose. Histograms of the relationship between haemoglobin and the dose of the drug. The lines rising from the top of each histogram represent the standard error, that is, they are error bars.

Chapter summary

- Continuously variable data with a normal distribution is represented by a bell-shaped curve. Data with a non-normal distribution is skewed to the left (by far the most common form) or to the right (rarely encountered).
- Scatterplots are used to show the degree of correlation between a pair of indices. The strength of this association is given by the correlation coefficient, which varies between 0 and 1; and any such relationship can be positive or negative.
- Data with a categorical distribution is often presented as a histogram. However, histograms may also summarize data that has a continuous variation.
- A pie chart expresses the frequency or proportion of different groups of data in terms of the 360° of a circle.

- A dot plot shows individual points of data in a scale, and provides a quick visual assessment of the distribution of the data. A box and whisker plot summarizes the major numerical aspects of the data set, such as the median and inter-quartile ranges.
- The relationship between pairs of data can be shown in a line plot, so-called because a line links each individual pair.
- Data showing changes over a time period, or the effect of increasing concentrations of a drug, can be represented with a graph.

Questions

4.1 What do you understand by the expression 'skewness'?

4.2 What is the major difference between a dot plot and a box and whisker plot?

4.3 What is the practical difference between the standard deviation and the standard error?

4.4 Determine the mean, standard deviation, median, and inter-quartile range of the data set I in self-check 4.2. You may have already plotted the data as part of self-check 4.2. What is your deduction regarding the distribution of this data set?

5

Quality, audit, and good laboratory practice

Learning objectives

After studying this chapter, you should be able to ...

- Understand the purpose of quality, audit, and good laboratory practice
- Appreciate the need for formal data analysis
- Explain why audit is a continuous process
- Describe key features of horizontal audit
- Recall the purpose of vertical audit
- Outline the need for self-audit

This chapter will address three important aspects of the laboratory. Having introduced these topics in Section 5.1, Section 5.2 describes in more detail how we ensure that the quality of the product of the laboratory, that is, its results, are correct. The second considers, in Section 5.3, why laboratory procedures are constantly being checked as being fit for purpose and, if needed, improved–the audit. Both these topics, and others, come together in Section 5.4 with a discussion of good laboratory practice.

5.1 Introduction

The laboratory may be likened to a processing plant, where a series of routine methods and procedures convert raw materials into product, which in the case of a laboratory are the analyses of blood, tissues, and other material, producing a result to be used in diagnosis and in the management of ill-health. However, in any situation there is the potential for error, and for routine processes to break down and become inoperable, giving a false result or product. Accordingly, any such system (be it in a clinical laboratory or in an industrial or manufacturing setting) must have a series of checks to ensure the product is acceptable to the end user, or customer. This is embodied by the term **quality**, which may be used to describe the accuracy

Quality
the degree to which a product is acceptable to the end user

Quality control
a procedure or set of procedures intended to ensure that the process of analysis adheres to a defined set of quality criteria or that meets the requirements of good laboratory practice

Quality assurance
a procedure or set of procedures that provide independent confidence that levels of accuracy and precision are being met

Quality management
A method for ensuring optimum design, development, and implementation of QA and QC

Audit
a systematic and independent examination to determine whether quality activities and related results comply with planned arrangements and whether these arrangements are implemented efficiently and are suitable to achieve objectives

Good laboratory practice
a set of criteria that ensure that a laboratory is a safe and efficient workplace

Standard operating procedure
the written instructions to be followed when performing an analysis

and precision of that product. One dictionary defines quality as "a high degree of excellence", another as "of a high standard".

If the product is of insufficient quality, then (in the biomedical science setting) an incorrect result may have serious repercussions for the individual providing the sample. Laboratory staff must ensure that their product is of sufficient quality. One of the mechanisms to confirm that a process is working well is to perform **quality control** (QC) and **quality assurance** (QA), as will be discussed in Section 5.2. A further process is that of **quality management** (QM), the administrative steps required to ensure successful QA and QC, and will be undertaken by a Quality Manager, whose potential remit may be the entire Pathology Department.

If a process 'fails' its QC or QA (that is, it is not of a sufficiently high degree of excellence or a high standard), steps must be initiated to identify and correct the error. In seeking to remedy a particular problem, a series of steps are reasonably easy to set in motion, as are depicted in Figure 5.1. However, the first three steps alone fail to assess whether the solution has actually worked: there is no check to determine the success or failure of the solution. The final step (Step 4) is required in order to assess the effect of the solution on the process.

This procedure is all very well, but it would of course have been better if the problem had not arisen in the first place. One the most effective ways in which this retrospective problem-solving process can be advanced upon is **audit**. Conceptually, audit is the equivalent of drawing an arrow from Step 4 to Step 1 in Figure 5.1, whereby our solution to the problem informs the future development of a given process such that the original problem will not return. Audits apply not only to the laboratory, but also to its staff, as is explained in Sections 5.3 and 5.4.

All of these factors (QC, QA, and audit) come together as key factors of **good laboratory practice** (GLP), described in Section 5.5. The latter is a set of criteria that ensure the efficient performance of a laboratory, and its staff, that it (and therefore its product) runs at a particularly high degree of quality. A crucial part of GLP is the **standard operating procedure** (SOP)—the written instructions to ensure staff carry out complex and routine operations to ensure the quality of the result. All laboratory analyses will have their own SOP, and some complex procedures may have a series of SOPs (such as booking in a specimen, preparing a work batch, and approving/reporting the result).

Cross reference
An example of an SOP is presented in Chapter 9

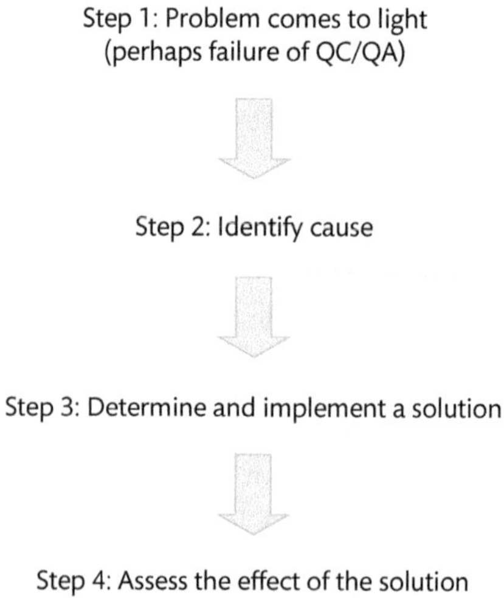

Step 1: Problem comes to light (perhaps failure of QC/QA)

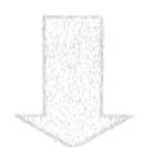

Step 2: Identify cause

Step 3: Determine and implement a solution

Step 4: Assess the effect of the solution

FIGURE 5.1
Problem solving

5.2 Quality control and quality assurance

Key aspects of GLP are the processes of quality control (QC) and quality assurance (QA), which are the responsibility of the Quality Manager. These are undertaken to ensure the process of analysis provides, and continues to provide, the correct result. The key difference between the two is that QC is done internally (and so may be described as IQC), whilst QA is performed in collaboration with a body outside the laboratory, so may be described as external quality assurance (EQA). QA and QC are not simply about laboratory results: they are also relevant in ensuring staff are appropriately trained and remain up to date with developments in their chosen field.

Quality control

In order to be confident that an analytical process has acceptable reproducibility, accuracy, and precision, the formal process of QC is performed. If the analysis delivers a continuously variable number (as in most biochemistry [such as serum urea] and haematology [such as a red blood cell count]) the same sample is analysed several times, and the variability of the results determined. In disciplines where the result is categorical (as is often present in cell pathology [such as the intensity of staining]), the same principle of repeat testing is applied, but the outcome measure is different (such as light, moderate, and heavy staining). The test sample may be one of the laboratory's own, or provided by an external organization, perhaps (if appropriate) the manufacturer of the analyser. However, the key aspect of QC is that the laboratory has a firm idea of what result is expected, and that the analysis is performed by the laboratory's own staff.

For continuously variable indices, the key index is the coefficient of variation (CV), and with it the concepts of intra-assay (generally, all assayed in a single batch, perhaps on the same day) and inter-assay (measured on different days) coefficients. These are important in establishing the value of a new method, and also in QC of an established technique.

Cross reference

Chapter 3 has details of intra- and inter-assay coefficients of variation. Both are obtained by dividing the SD by the mean

Consider the following set of consecutive results on the same sample of tissue (sample X), which are being used to test the QC of a particular process over several weeks with an inter-assay CV ...

50, 47, 48, 51, 49, 49, 51, 49, 47, 50, 50, 52, 51, 54, 55, 53, 50, 49, 55, 53

Overall, the mean result of this data set is 50.6 units, and the CV is 4.7%, which at first sight may be acceptable. However, if we group this data set into two sub-groups—the first ten points and the second ten points, we find that the mean of the first ten is 49.1 units, whilst that of the second set is 52.2 units, a difference of 6.3%. Note that this difference (6.3%) is larger than the CV (4.7%), and accordingly calls into question the likelihood that there is something different in the two sets of data, and so may prompt an investigation.

Closer analysis of these two groups shows that the CVs are actually even smaller (2.9% in the first and 4.1% in the second) than the CV of the entire group (4.7%). This is further evidence that there has been a change in the technique somewhere in the middle of the data set. Therefore, these figures will generate activities by senior staff to discover the fault and then rectify it. Presenting these data graphically produces a **Levey-Jennings plot** (Figure 5.2), and allows a quick and simple assessment of the performance of the method.

Levey-Jennings plot

a graphical representation of the variability of a technique over a period of time

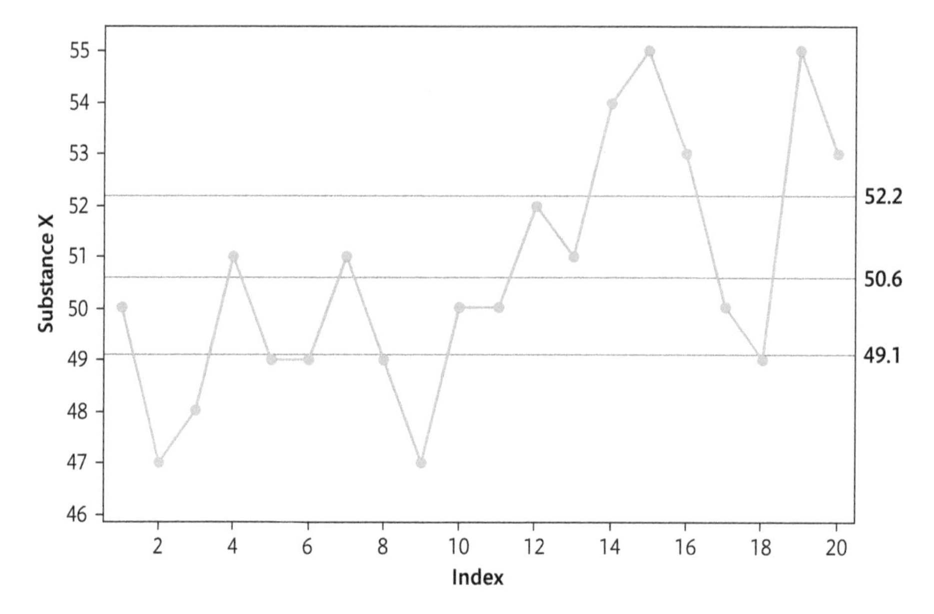

FIGURE 5.2

A Levey-Jennings plot. Variation in the levels of Substance X over a seven-week period. Over the whole seven weeks, the mean level is 50.6 units/mL (as indicated by the middle horizontal line) and the CV is 4.7%. However, the mean of the first ten samples (49.1, shown by the lower horizontal line) is lower than the second batch of ten samples (52.2, shown by the upper horizontal line), and the graphic shows more variation in the latter samples compared with the former. We can interpret this data as suggestive of a difference at around the 10th or 11th sample point. The nature of this difference cannot be determined by the analysis, but it is likely to set in train an investigation to determine the cause of the change.

Quality control of analyses that are categorical have much in common with 'numerical' analyses. In cell pathology, issues include the correct intensity of staining, especially in immunocytochemistry, where the quality of the antibody is crucial. In immunocytochemistry and standard (e.g. haematoxylin and eosin) processes, positive and negative control tissues must be set up alongside the test sample to ensure staining is appropriate, and these will be written into the SOP. Tissue sections can act as their own internal control if they contain both normal and abnormal (malignant, inflammatory) tissues. In microbiology, a control slide known to contain tubercle bacilli should be included with every batch of slides stained with a Ziehl-Neelsen stain. In molecular genetics, positive and negative controls are also important, and in gel electrophoresis, these are mandatory. In cytology, QC must address the sample rejection criteria as a false negative may lead to cervical cancer. Automatic staining and screening devices must address which samples are not fit for imaging due to technical reasons or poor cellularity, resulting in rejection. In semen analysis, the Association of British Andrologists publishes guidelines of best practice, QA, and QC. These comments are brief, and the reader is directed to their own Quality Manager.

Quality assurance

An extension of this principle is to provide quality assurance. This is effectively where the host laboratory assesses a tissue section, or measures levels of the molecule, antibody, or metabolite in question not in a sample of its own, but in a sample provided by another (independent) laboratory. There are several quality assurance schemes, one of the most common being the National External Quality Assurance Scheme (NEQAS). This commercial laboratory provides samples of blood and tissues to subscribing laboratories which perform certain tests and then return their result to NEQAS. The NEQAS can then compare the result from one particular laboratory with that from all the other subscribing laboratories that have measured an aliquot of the same sample on their own analyser.

The NEQAS offers QA in specific areas such as haematology, blood transfusion, microbiology, cellular pathology, *in situ* hybridization, clinical chemistry (including cerebrospinal fluid), leukocyte immunophenotyping, coagulation, molecular genetics, parasitology, immunology, andrology, histocompatibility and immunogenetics, immunocytochemistry, and allergy. However, in other settings, such as gynaecological cytology, standards may be monitored by a regional **Quality Assurance Reference Centre**.

Quality Assurance Reference Centre

A regional centre which oversees the quality of cervical screening and colposcopy services on behalf of the NHS Cervical Screening Programme

Worked example

Let us suppose your laboratory subscribes to a QA provider that supplies three blood samples for you to test. Ten other laboratories using the same analyser also subscribe to this process. The return from the QA provider may be a list of results on these three different tests from the eleven laboratories:

Laboratory number	Test 1	Test 2	Test 3
1	100	5.6	46
2	105	6.1	48
3	126	5.7	46
4	104	5.8	48
5	99	5.8	50
6	102	6.0	46
7	104	5.9	46
8	102	5.7	48
9	101	6.1	40
10	103	5.8	54
11	102	6.0	48
Average	104	5.9	47

Several features are noteworthy. In test one, the difference between the highest (126) and the lowest (99) is 27%, which is large and likely to be due to at least one unusual result. However, ten laboratories give a result between 99 and 105 (a difference of only 6%), but one value (126, from laboratory 3) is far removed—being 21% higher than the mean value. This outlier is obviously suspicious. In test two, all eleven laboratories return a result between 5.6 and 6.1 (a difference of 9% between highest and lowest) with no obvious outlier, so the data seems sound.

The difference between the highest and the lowest for the third test is a huge 35% (that is 54/40) and this also implies an outlier–obviously the result of 40, from laboratory 9. Removal of this point brings down the highest/lowest difference by a half–to 17%. Perhaps laboratory 9 needs to look closely at the performance of its analyser and/or linked procedures. A single result is difficult to interpret, but over a series of several QA assessments, if laboratory 9 consistently returns the lowest result, it seems likely that a systematic error is present, and corrective action may be necessary.

In a cell pathology setting, if 17 of 18 laboratories consistently find the intensity of a certain antigen in a section of liver to be 'moderate', whilst the 18th find the intensity to be 'low', then perhaps questions need to be asked. Another quantifiable result may be the infiltration of a section of tissue with leukocytes (absent, mild, moderate, marked). In some cases, the titre of an effect may be an end point, an example being the degree to which a sample of serum containing an autoantibody is detectable. Perhaps 25 from 30 laboratories report a titre of an anti-nuclear antibody to be between 1/160 and 1/320, and the remaining 5 laboratories report titres between 1/20 and 1/40. If so, it could be argued that these 5 laboratories are out of consensus and may consider checking their procedures.

Subscription to an independent quality assurance organization such as NEQAS is mandatory for all NHS and other laboratories that provide data upon which clinical decisions are made. Indeed, paperwork provides proof of adherence to the system and good quality assurance, and may be demanded by inspectors. QA, like QC, can be easily represented graphically with a Levey-Jennings plot. Indeed, this means that trends can rapidly be noted and acted upon. The best laboratories will place these plots in a prominent position so the QA/QC performance can be seen by all their staff.

5.3 Audit

Audit is about prospective evaluation and improvement. In theory, this evaluation can be of a person or a procedure, and can be applied to virtually all forms of human activity: for example, in teaching, engineering, banking, and finance. The process of audit has a common structure regardless of the particular industry or activity in which it is being used. In the laboratory, audit covers the entire process of analysis, and so includes staff and equipment. In biomedical science, the process of audit generally examines everything from the receipt of the specimen to the release of the data, but it can in some circumstances be extended to include the reason for requiring the analysis through to the consequences of the result–often clinical management.

Audit can be carried out in almost any situation, and there are common steps regardless of the profession or occupation concerned. Accordingly, the International Organization for Standardization (the ISO) publishes a series of guidelines and recommendations. Those pertaining to biomedical science include ISO27789 (describing a common framework for audit trails for electronic health records), ISO11073-90101 (point of care testing), and ISO20776 (microbiology testing). However, perhaps the best known are ISO9000 and its successor, ISO9001, which define audit as:

> **a systematic, independent, and documented process for obtaining evidence and evaluating objectively the extent to which audit criteria are fulfilled**

ISO9001 is part of a developing series of standards reaching back decades, and which deals with quality management systems. One of the key features is the awarding of a certificate that gives assurance and confidence in that organization in that a minimum level of competence and administration has been attained. Updated regularly every few years, ISO9001:2008 is a document of some 30 pages that sets the required standards of quality in management, but also of external audit.

Although the ISO documents are aimed at management strategies to improve business processes, and so increase profits, many principles are nonetheless applicable to laboratory management. A good example of this is good record keeping with well-formed SOPs.

The ISO system is not limited to laboratory management. For example, ISO13485:2012 is specifically dedicated to medical device manufacture, which includes any device used to diagnose, prevent, or treat disease, and so embraces all laboratory equipment. High-quality audit is a requirement for the formal accreditation of the laboratory, and systems such as those of the ISO are an essential requirement. A further demand of good laboratory practice is to be able to demonstrate that any process operates to defined criteria.

Establishing an audit

A key question is 'why audit?' A scientific answer is that there is always change, and so staff and processes must be constantly re-evaluated; a more philosophical answer is that failure to audit leads to stasis, complacency, and so a possible loss of focus that may potentially lead to error. Indeed, there is a school of thought that promotes the idea that every single process or procedure should be regularly audited, even if it appears to be working effectively. SOPs are certainly part of this process, with a limited lifespan, and each will carry a date for its review. We can summarize the process of audit in three main aspects as follows:

1. What is it we are trying to do?

Audit must be focused on a number of key process indicators. Every NHS laboratory will have one or a team of staff who will be experienced in, and responsible for, defining these questions. Examples of indicators include:

- the turn-around time for the analysis of a batch of samples
- the number of samples that are rejected in sample reception
- the number of analyses that must be repeated.

But whatever the topics may be, they must be clinically relevant, clearly defined, and easily measured. The overview is that audit is a quality improvement process, with the planned review and examination of records and activities to assess. The objectives of the audit may include:

- testing adequacy of control systems
- ensuring compliance with existing policies and procedures, and so ...
- recommending changes, should any be necessary.

As regards the latter point, many are unprepared for the possibility that an audit will fail to find any areas where a clear improvement can be made. However, this does not invalidate the purpose of an audit so long as it is correctly conducted.

2. What data should we collect and what do we do with it?

The type of data to be collected must be pre-determined, as must the period of time over which audit will run. However, these must also be formally defined, ideally by a statistician, because conducting data over too short a time period, or collecting too few data points (called the sample size), are both common causes of false-negative results.

An example of this may be the turn-around time of a batch of samples. The data to be collected is the time in minutes from the samples being delivered to the laboratory to the results being presented to senior staff for approval. Let us suppose that on certain days, the average turn-around is 45 minutes (fluctuating between 40 and 50 minutes), whilst on other days the same process often takes over 60 minutes, a marked increase of a third or more. This difference is large and implies a change in the procedure (of which the formal analytical SOP is only a part). One way to find if this is indeed the case is to break down the analysis to defined steps (as in an SOP), and record each step in order to pinpoint that step that has the greatest impact on the overall turn-around time. One issue may be that certain key staff are out of the laboratory and so are unable to take part in, or oversee, the procedure. The solution is therefore to re-arrange work schedules so that the correct staff mix is present at each stage. The final part of the audit is to collect turn-around times after the review of staff rostering, which should iron out the variability.

Once collected, the data is likely to be entered into a spreadsheet such as Excel, a software package available on many computers. One of the problems with Excel is that it has a limited capacity to analyse data. Excel is perfectly capable of summarizing a set of data, providing indices such as the mean and standard deviation, but for more complex analyses a dedicated statistical package is required, examples of which include Minitab and SPSS.

Cross reference

We discuss the statistical analysis of data in more detail in Chapters 1, 3, 6, 7, and 8

This formal analysis is crucial, as major decisions are likely to follow from the data collected. Accordingly, data should only be analysed by trained staff. It is likely that each NHS hospital in the UK will have a Research and Development Department who can advise on training in analytical technique from a statistician. Indeed, many Hospitals have units with names such as 'Clinical Improvement', often with hospital-wide responsibility for audit, and may well have an in-house hospital statistician.

3. What can we do with all this information?

Once formal analysis is complete, assessment can begin. It is likely that data can be compared with that from a previous audit. However, once more, formal statistical analysis must take place to ensure that the correct interpretation is made. There are a large number of statistical tests that be employed, and an invalid answer may be obtained if the wrong test is applied. Chapter 7 has details of these tests. After the data is verified, decisions can be made and the analytical procedure altered (or not). However, this is not the end of the process: the final step is to define the parameters for re-analysing the process at some time (perhaps weeks or months) in the future.

Audit is a continuous process

Audit is not simply a procedure to establish the cause of a particular problem—it is a continuous and rolling programme. Staff responsible for quality and audit will monitor any change in the process or procedure to ensure that such changes are correctly implemented, are effective, and are maintained. It follows that staff may need additional training and that SOPs will need to be rewritten.

Regardless of the situation, audit can be simplified to a series of steps that form a continuous cycle (Figure 5.3). Hence audit is not an end in itself: if it does identify areas in need of improvement, steps must be taken to ensure any changes made are sustainable—for example, the redistribution of resources, including staff.

SELF-CHECK 5.1

What body sets the required criteria for audit?

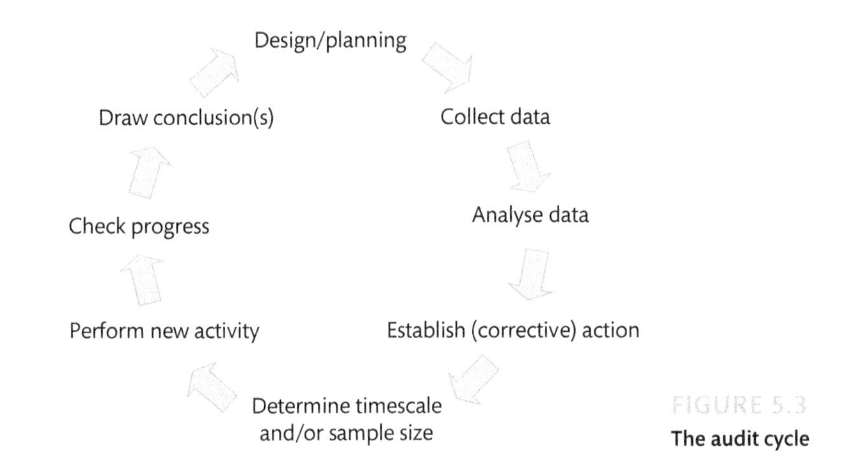

FIGURE 5.3
The audit cycle

5.4 Types of audit

There are many types of audit. Some refer to the practitioner themselves, who can address their own personal professional standing in self-audit. As regards the physical aspects of the laboratory, audit may be characterized as being internal and external. The former involves only staff within the laboratory, and perhaps addressing issues of QC and QA. External audit involves other independent agencies such as the MHRA and the Health and Safety Executive. These bodies, whose primary interests are the patients, have powers of investigation and also to impose directives on the laboratory, such as requiring the replacement of obsolete equipment and ensuring the correct training of staff. In extreme cases they may close the laboratory.

In the UK, hospital laboratories are accredited by the United Kingdom Accreditation Service (UKAS), which itself certifies 'proficiency testing providers', and in doing so acquired the organization 'Clinical Pathology Accreditation' (CPA). The UKAS now provides accreditation for all NHS pathology laboratory disciplines to the internationally recognized standard ISO15189:2012 in place of the CPA. Although no longer directly operating, the CPA set a standard for good laboratory practice that continues to be relevant. For example, the CPA's mission statement was structured in a similar manner to that of the ISO with much common vocabulary.

One of the CPA guidelines stated that process audits shall:

- be planned and scheduled
- be conducted against agreed criteria
- be carried out by trained personnel
- record those activities, areas, or items to be audited
- report non-conformities and deficiencies
- recommend a timescale for corrective action
- set the date for the next audit.

Furthermore, the CPA recommended a number of other steps in the process of audit, such as that:

- departments establish an audit calendar
- there is a plan for a series of internal audits over a period of a year (or perhaps each quarter)
- all staff are aware of the outcome of past audits (perhaps on a notice board) and when they will be refreshed

- all pre-examination, examination, and post-examination procedures are checked for concordance
- there is contingency for unscheduled audits should they be called for, perhaps as the result of an unusual batch of results
- audit will be overseen by a formal group consisting of several different grades of staff
- a senior member of staff is in overall control/has responsibility for audit.

SELF-CHECK 5.2

Suggest four key steps in the process of audit.

Vertical audit

Vertical audit often follows a particular sample as it makes its journey through a particular laboratory (Box 5.1). This is likely to include issues in sample reception, processing, and reporting. There may well be dozens of small but significant processes involved in this journey, and each must be addressed. Vertical audit may also occur at different times of day. It is possible, for example, that at busy times, samples are placed temporarily in a refrigerator, which at less busy times does not need to happen. This may have an impact on the results of the analysis as samples may not have the time to reach the room temperature generally demanded by analysers.

Comparisons of audit practice between laboratories may also be enlightening: haematology and biochemistry practice may differ for purely historical reasons, leading to the possibility that one can be optimized. Similarly, colleagues in microbiology and histology may be invited to provide fresh perspectives on the working of other departments, and vice versa.

BOX 5.1 Examples of potential steps in a vertical audit

1. The sample arrives in specimen reception
2. It is booked in and checked for integrity and safety (blood spillage etc.)
3. The request slip has a minimum of information (patient's details, justification for the analysis, time of collection, details of the requesting practitioner)
4. It is formally accepted in the analytic process and logged by the pathology computer
5. The sample is stored locally in specimen reception
6. When a sufficient number of samples for a specific test or tests have been received, they are collected into a work sheet
7. The samples move from specimen reception to the analysing laboratory
8. The samples are analysed
9. The samples are allotted to post-analysis storage (perhaps at 4°C)
10. The results are checked by a Scientist and then authorised as being ready for dispatch
11. The result is sent out to the requesting practitioner
12. The sample is stored for a defined time period, and then disposed of

Naturally, different laboratories will have their own system: this pathway is not intended to be exhaustive. Some steps may be merged and/or occur in different places; the use of checklists ensures that steps are not missed.

FIGURE 5.4

Horizontal audit

Horizontal audit compares different processes or equipment, and may be specific for one element of a procedure or policy. In many cases these audits can be across the entire pathology section. An example of this may be the regular audit of the electrical safety of all analysers, or the frequency of servicing by the manufacturer. Another may be to regularly review the expiry dates of chemical reagents, buffers, etc. This would include those produced by the laboratory and those bought in from an outside source. However, in many cases, reagents that are dedicated to a particular analyser and that are provided by the manufacturer have a safety system that rejects a buffer if it is outside its 'sell-by date'. When was the last time the laboratory had a thorough 'spring clean'?

Horizontal audit extends to staff training and development. For example, have all been 'signed off' on particular SOPs, and have they all read the appropriate Departmental Manual? Are they fully aware of their roles, responsibilities, and limitations?

Self-audit

In the UK, all staff employed in NHS laboratories must be registered with the Health and Care Professions Council (HCPC). This body sets and maintains standards of proficiency and conduct, one of which is that practitioners are required to show evidence that they are keeping up to date with new issues. This is often described as '**Continuing Professional Development**', or CPD, and the assessment of the CPD record of all staff in a laboratory is an example of horizontal audit. Certainly, each staff member must be signed off as competent in a particular process (probably an SOP), but it may be that their professional standing will be improved by becoming trained to competency on processes not yet part of their job description.

Continuing Professional Development
the process by which practitioners maintain current knowledge

The concept of self-audit may well be part of their annual review, and so become part of their personal and professional development. Some aspects of professional development can be addressed within the scientist's local setting, such as participation in a lunchtime journal club. Others may be attending an external conference or completing an on-line training module.

The primary organization providing CPD for biomedical scientists is the Institute of Biomedical Sciences (IBMS). Those clinical scientists not part of the IBMS may subscribe to the CPD process of the Royal College of Pathologists (RCPath), should they be affiliated. An important part of CPD (and, indeed, there are many who consider this should be part of personal development) is reflective practice, or reflective learning. This is the personal view of the scientist's own practice, where he or she considers all activities pertinent to their work, many of which are questions. For example, upon returning from a conference, the scientist may reflect 'What did I learn at this conference that will make me a better scientist?'. Should answers be informed, critical, and measured, they may well satisfy CPD criteria. According to Davies (2012) benefits of reflective practice could include:

- Acquisition of new knowledge and skills and a source of feedback.
- Encouragement of self-motivation and self-directed learning.
- Further understanding of own beliefs, attitudes, and values.
- Increased learning from an experience or situation.
- Identification of personal and professional strengths and areas for improvement.
- Identification of educational needs and promotion of deep learning.
- Possible improvements of personal and clinical confidence.

SELF-CHECK 5.3

Identify an aspect of horizontal audit when you are next in a laboratory.

Clinical audit

Audit is also important in clinical practice. A good example of this is in cervical cancer screening, where women are invited to attend to determine the presence of this malignancy, and, if present, to ensure the correct pathway is followed. The NHS Cervical Screening Programme publication 28 describes a protocol for auditing the screening history of women with cervical cancer. Relevant steps are:

- Review of invitations sent to the women, and those taken up.
- Review of all cytology and histology samples.
- Review of clinic attendance, colposcopic findings, and management.

A similar clinical audit of patients attending an oral-anticoagulant clinic could provide better service in two ways. The first being steps involved in generating an international normalized ratio (INR), and secondly, the step taken by a 'dosing-officer' (who may be a scientist, pharmacist, or other health care professional) to ensure the dose of the anticoagulant is correct. Other audit steps may determine those factors leading to failure of patients to attend for subsequent appointments.

Cross reference

Further details of cervical cancer audit are present in the *Cytopathology* textbook of this Fundamentals of Biomedical Science series, edited by B. Shambayati

5.5 Good laboratory practice (GLP)

Although the concept of GLP was developed in a formal setting several decades ago, and in respect of non-clinical research and other laboratories, many aspects are directly relevant to clinical laboratories. An important non-clinical body is the Organisation for Economic Co-operation and Development (OECD), who set the criteria, defining GLP as *a quality system concerned with the organizational process and conditions under which non-clinical health and environmental safety studies are planned, performed, monitored, recorded, archived, and reported.* In the UK, GLP regulations were published in 1999 and updated in 2004, and effectively mirror those of the OECD. Box 5.2 summarizes certain OECD principles of GLP, and is clear that almost all apply to how biomedical and clinical science is performed, an exception being 'performance of the study'.

BOX 5.2 Key OECD principles of GLP

1. **Organization and personnel**
2. **Quality assurance program**
3. **Facilities**
4. **Equipment, reagents and materials**
5. **Test systems**
6. **Test and reference items**
7. **Standard operating procedures**
8. **Performance of study**
9. **Reporting of results**
10. **Archival–storage of records and reports**

Training, education, and staff development will be part of Section 1 (organization and personnel). Indeed, GLP demands that the department has a named training officer (much like a named safety officer and a named quality manager).

In the clinical laboratory setting, QC, QA, and audit (as described earlier in this chapter) are key factors of GLP. The latter is a set of criteria that ensure the efficient performance of a laboratory, and that it (and therefore its product) runs at a particularly high degree of quality, and which include features such as staff training and health and safety issues. A key organization setting standards for GLP in the UK is the Government's Medicines and Healthcare Regulatory Authority (the MHRA), which itself refers to documents from the European Community, where the European Medicines Agency is the key body. Following Brexit in March 2019, it may be speculated that powers will return to the MHRA. However, although the MHRA definition of GLP often refers to agrochemicals, veterinary medicine, and cosmetics, many of its principles are fully applicable to biomedical science as is practised by NHS pathology laboratories.

Chapter summary

- Quality control and quality assurance provide confidence that a result is accurate, precise, and reliable.
- Audit is a process whereby improvement can be made. It is a controlled and regulated activity that is vital for the scientific health of the laboratory and the professional health of the practitioner.
- Data collected by audit must be formally analysed: in many cases a statistician is involved.
- Audit is not merely a one-off assessment: it is a cyclical process where the effects of changes are monitored and new opportunities for improvement are identified.
- Vertical audit often considers the fate of an individual sample from its arrival in the pathology department to the release of the result. Horizontal audit considers a common theme within a single laboratory or across different laboratories.
- Good professional practice by an individual demands self-audit to ensure they remain effective practitioners. This inevitably requires subscribing to a system that offers continuing professional development.
- QA, QC, and audit are key aspects of good laboratory practice, a set of criteria that ensure that a laboratory is a safe and efficient workplace.

Suggested reading

- **Moore, E. Quality assurance and management. In *Biomedical Science Practice*, Eds Glencross, H., Ahmed, N., Wang Q. Oxford, Oxford University Press, 2011.**
- **Theodorsson, E. Quality Assurance in Clinical Chemistry: A Touch of Statistics and A Lot of Common Sense. *J Med Biochem*. 2016; 35:103–112.**
- **Yu, S., Cui, M., He, X., Jing, R., Wang, H. A review of the challenge in measuring and standardizing BCR-ABL1. *Clin Chem Lab Med*. 2017; 55:1465–1473.**

- Pavlova, S., D'Alessio, F., Houard, S., Remarque, E.J., Stockhofe, N., Engelhardt, O.G. Workshop report: Immunoassay standardisation for 'universal' influenza vaccines. *Influenza Other Respir Viruses*. 2017; 11:194–201.
- Endrullat, C., Glökler, J., Franke, P., Frohme, M. Standardization and quality management in next-generation sequencing. *Appl Transl Genom*. 2016; 10:2–9.
- Gargis, A.S., Kalman, L., Lubin, I.M. Assuring the quality of next-generation sequencing in clinical microbiology and public health laboratories. *J Clin Microbiol*. 2016; 54:2857–2865.
- Torlakovic, E.E., Nielsen, S., Vyberg, M., Taylor, C.R. Getting controls under control: the time is now for immunohistochemistry. *J Clin Pathol*. 2015; 68:879–82.
- Cree, I.A., Deans, Z., Ligtenberg, M.J., et al. European Society of Pathology Task Force on Quality Assurance in Molecular Pathology; Royal College of Pathologists. Guidance for laboratories performing molecular pathology for cancer patients. *J Clin Pathol*. 2014; 67:923–31.
- NHS Cervical Screening Programme http://www.cancerscreening.nhs.uk/cervical
- Johns, C., Burnie, S. *Becoming a reflective practitioner* (4th ed.). Chichester, UK; Ames, Iowa: Wiley-Blackwell, 2013.
- Davies, S. Embracing reflective practice. *Education for Primary Care*. 2012; 23(1): 9–12.

Useful websites

- The International Organization for Standardization. http://www.iso.org
- The United Kingdom Accreditation Service. http://ukas.com
- The European Medicines Agency: http://www.ema.europa.eu/ema/
- The Medicines and Healthcare Regulatory Authority (MHRA). http://www.mhra.gov.uk/Howweregulate/Medicines/Inspectionandstandards/GoodLaboratoryPractice/Structure/
- The Good Laboratory Practice Regulations, Statutory Instrument 1999 No 3106. http://www.legislation.gov.uk/uksi/1999/3106/contents/made
- Statutory Instrument 2004 No. 994: The Good Laboratory Practice (Codification Amendments Etc.) Regulations 2004. http://www.legislation.gov.uk/uksi/2004/994/contents/made
- The Health and Care Professions Council. http://www.hcpc-uk.org/aboutus/
- The Institute of Biomedical Science. www.ibms.org
- The Royal College of Pathologists. www.rcpath.org
- The Organisation for Economic Co-operation and Development: OECD. ENV/MC/CHEM(98)17 part two
- UK National External Quality Assurance scheme. https://ukneqas.org.uk/
- The Association of British Andrologists. https://www.aba.uk.net/
- The WHO Handbook on GLP. https://assets.publishing.service.gov.uk/media/57a08b53ed915d3cfd000c6c/glp-handbook.pdf

Questions

5.1 What is the key difference between QA and QC?

5.2 The EQA report of results from different laboratories is as follows:

Laboratory	Result
1	165
2	160
3	158
4	166
5	161
6	157
7	162
8	160
9	145
10	158
11	165
12	160
13	162
14	159

Using standard graph paper, construct and comment upon a Levey-Jennings type plot, placing the laboratory number on the horizontal axis.

5.3 Place the bullet points of the benefits of reflective practice (Section 5.4) in rank order of importance.

6

Research 1: Setting the scene

Learning objectives

After studying this chapter, you should be able to ...

- Understand the purpose of research
- Explain the concept of the hypothesis
- Understand the importance of the sample size determination
- Describe the importance of probability and confidence
- Outline common types of research, how these are designed, and those permissions/approvals that are required
- Appreciate the need for high standards of care in data collection and transfer to a spreadsheet

In our first look at research we will be setting the scene by asking "What is research?" Many will consider research, in general terms, to be the discovery of new facts. In a biomedical science context, however, we see two types of research:

- **that which extends our knowledge of the pathophysiology of human disease (often called basic laboratory research), and**
- **that which improves patient outcome—typically called clinical research.**

The two are linked, in that research in human disease is of little value unless it leads to an improvement in the health of the individual, and so the nation. A good example of this is the basic science research that established the metabolic pathways leading to the synthesis of cholesterol. The key enzyme in this metabolism is hydroxyl-3-methylglutaryl coenzyme-A (HMG-CoA) reductase. Identification of this enzyme led to the development of the class of drugs that interfere with it and so inhibit the production of cholesterol—hence their name: HMG-CoA reductase inhibitors (commonly known as statins). These drugs have been used in clinical research to demonstrate unequivocally that reducing serum cholesterol with statins leads to a reduction in the risk of cardiovascular disease such as heart attack and stroke.

Both types of research start with an important concept, that of the hypothesis (Section 6.1). We will begin our exploration of research by looking at the different forms a hypothesis may take, and how to determine the sample size, which is how many people will need to be recruited, or how many experiments or tests will need to be carried out to validate a study. A major (some would say the major) feature of statistics—probability—will be explained in Section 6.2. Basic research design will be described in Section 6.3, followed by more focused discussion on clinical (Section 6.4) and laboratory (Section 6.5) research design. The chapter concludes in Section 6.6 with advice on data collection and spreadsheets.

6.1 The hypothesis

All serious research must address a **hypothesis,** which may be defined by the Cambridge English Dictionary as an idea or explanation for something that is based on known facts but has not yet been proven. An alternative view is that a hypothesis sets out the particular research question that the researcher seeks to answer. However, the research itself must be novel—that is, it must be original; this leads us to the concept of an *original* hypothesis. Use of this qualifying adjective focuses the mind of the researcher(s) and the audience to ensure that the information that is being sought is unknown, is new, and is not merely supporting other findings.

Hypothesis
the supposed or proposed explanation, possibly made on the basis of limited evidence, for an observation

In some rare circumstances, findings may be so important that they need to be independently confirmed, in which case there is no original hypothesis. If so, those confirming the original data often design their research so as to find new aspects, in which case their work will 'confirm and extend' the work of others. The research process must be designed so that it tests the validity of the original hypothesis—hence the concept of hypothesis testing. Another way of saying this is that the hypothesis must be proved or disproved. Let us now consider the hypothesis in more depth.

The null hypothesis

In general terms, the null hypothesis is the equivalent of saying 'nothing happens unless you do something', itself a distant corollary of Newton's first law of motion (summarized as '*Every body persists in its state of being at rest or of moving uniformly straight forward, except insofar as it is compelled to change its state by force impressed*'). In research terms, the null hypothesis states that there is no difference between two sets of data drawn from a common population. The null hypothesis is abbreviated to H_0, or perhaps H_o.

For example, we may have a null hypothesis that men are not taller than women. In other words, and from a statistical point of view, this hypothesis says 'If we were to measure the heights of a selection of men and women drawn from a common population, we would see no difference in the measurements'. It is the task of the researcher to demonstrate the null hypothesis is not valid by producing data in favour of the proposition that there genuinely *is* a difference present. As just noted, when testing the hypothesis that men are not taller than women, the data collected would be measurements of the heights of men and women.

The alternative hypothesis

The alternative hypothesis (abbreviated to H_1, or perhaps H_A) is the reverse of the null hypothesis—that is, that there really is a difference between the two sets of data drawn from a common population. To extend our example, H_A might be that men are taller than women. Again, we can detect such a difference by using an appropriate research method, most obviously by measuring the heights of men and women, and then comparing the results.

If the data gathered supports the alternative hypothesis then it is accepted (or proven) and so by definition the null hypothesis is rejected (or disproven). The procedure by which we determine which of the hypotheses is true is by *testing*, so we say we would *test* the hypotheses. As we will see in this and the chapters that follow, there are many different tests we can apply.

The quantified hypothesis

It is generally not enough to have an original hypothesis; one must extend the hypothesis by proposing how big a difference you are likely to see between the two data sets (that is, how big an effect you think is likely to be present). Examples of this include the number of people with a certain disease, how big a change in the physiology of the body (such as blood pressure) may be caused by an intervention such as a drug, or how much of a difference in levels of a certain cell or molecule are present in the blood between two groups of subjects. Accordingly, the original hypothesis becomes the original *quantified* hypothesis. Returning to our earlier example, the original quantified hypothesis may be that men are 5 cm taller than women. Incidentally, in these examples we assume, unless directed to the contrary, that all men and women are healthy adults and are of roughly the same age. Notably, at ages 10–12 years, on average, girls are often taller than boys as they enter puberty a little earlier. Accordingly, we would describe age as a **confounding factor**, and we should specify in the research design that only those aged over 18 years are to be included.

Confounding factor

a factor that is likely to have an effect on the original hypothesis, and therefore which must be addressed in the analysis of the data

The same principles apply in basic laboratory research, as may be conducted in a university setting, or in a pharmaceutical company. It may be that researchers need to be sure what dose of their drug has the optimum effect on cells growing in tissue culture, or on the catalytic effect of an enzyme. The null hypothesis states the drug has no effect: the alternative hypothesis that the drug does indeed have an effect. The quantification of this alternative hypothesis will follow by asking how much of the drug (in terms of mg/mL, or ng) is effective. In the case of cell growth, an important co-hypothesis is that the dose of the drug does or does not have a cytotoxic effect on the cells.

Cross reference

A brief discussion of sample size and the power calculation is present in Chapter 3, Section 3.2

The sample size

A crucial element of testing a quantified hypothesis is to correctly identify the minimum number of experiments that must be performed for the test to be valid—for example, the minimum number of participants (in our example, the number of men and women) that have to be recruited, or the minimum number of observations that have to be made in order to robustly test the hypothesis. This crucial number is called the **sample size**, and is determined by the process of the **power calculation**. This process relies on the numerical differences embodied in the quantified hypothesis, and must be robust in order to minimize the possibility of errors.

Sample size

the number of subjects to be recruited, or number of observations to be made, in order to test the original hypothesis

Power calculation

the process by which the sample size is determined

The mathematics of the power calculation relies on one single index that is the basis of the hypothesis. This index is called the **test statistic**; in the example given above it is the height of the men and women in units of metres or centimetres. In a study of diabetes, the test statistic may be the level of blood glucose; in a study of renal disease the test statistic may be serum creatinine; in an outcome study it may be the number of deaths. In tissue culture it may be proportion of cells showing an effect, or the secretion of a soluble product.

Test statistic

the index upon which the power calculation is based

All statistical software packages are equipped with programs enabling a rapid and accurate power calculation, and there are also some calculators freely available on-line. These calculators demand a number of indices of the researcher. These include:

- The difference in the value of the test statistic as set out in the hypothesis
- The standard deviation of the test statistic.

This second point therefore expects the researcher to already have some data on the test statistic, more likely its mean and standard deviation. These are likely to be available from other sources, otherwise the researcher must generate this information themselves by measuring the test statistic in a representative number of people. This information is called **pilot data**, and is likely to yield not only the mean and standard deviation, but also tell us about the nature of the distribution of the data—for example, whether it has a normal or a non-normal distribution.

Pilot data
a small amount of information that is used solely to provide preliminary information. It cannot be used in formal analysis

Error

An important part of the power calculation is that it states the sample size needed to minimize the likelihood of error, of which there are two sorts: **false positive** and **false negative**. A false positive occurs when a difference is found when one ought not to have been found. An example from haematology is the finding that the size of the red blood cell (the mean cell volume) is larger in men than in women. It is an established fact that mean cell volume does not differ between the sexes. Therefore, finding a difference in the sizes of red cells between the sexes is incorrect. False positives are also described as type I errors.

Cross reference
The mean, standard deviation, and distribution of a data set are explained in Chapter 1

A false negative occurs when no difference is found when it should have been found. A false negative may be illustrated by the difference in height and weight between the sexes: it is an established fact that, on average, men are both taller and heavier than women. Therefore, failure to find such a difference must be because of an error, where the result has found no difference where a difference should indeed be found. False negatives are also described as type II errors.

False positive
finding a difference where one should not be present—also called a type I error

False negative
failure to find a difference where one should be present—also called a type II error

In most cases, the main reason for type I or type II errors in these examples is that not enough men or women have been studied. In the type I error it may be that a small but significant number of women who coincidentally have large red blood cells, and/or men with unusually small red blood cells have been recruited. Similarly, the type II error may be a consequence of some tall women and short men having been recruited who are collectively unrepresentative of their sex. Both these errors can be overcome by increasing the sample size.

SELF-CHECK 6.1

A study from a small town in the Midlands concludes that men live longer than women. Comment on this finding in terms of the two types of error.

SELF-CHECK 6.2

Construct a series of hypotheses on hair length and sex. What common demographic feature will need to be accounted for?

Worked examples (part 1)

Let us broadly hypothesize that men are heavier than women, a hypothesis that is far from original. Accordingly, the null hypothesis is that there is no difference in the weights of the sexes, and the alternative hypothesis is that men weigh more than women. We extend our broad hypothesis to a quantified hypothesis by suggesting that men are 5 kilograms heavier than women. Thus, the test statistic is weight.

In order to determine the number of men and women to be recruited (the sample size) we move to the power calculation. In measuring a mixed group of perhaps 20 men and women to generate some pilot data, we find that their mean weight is 80 kilograms with a standard deviation of 10 kilograms.

By inputting this data (a hypothesized difference of 5 units in a test statistic with a standard deviation of 10 units) into a power calculator, we find that the sample size is 64. This means we must weigh 32 men and 32 women in order to adequately test the hypothesis. However, to be prudent, it is common to recruit slightly more than the sample size (perhaps up to 35 or even 40 people in each group), merely to provide a little more confidence that any result is genuine.

In a pure laboratory setting, the statistical power of a study depends on the variability of the method for detecting the test statistics. Suppose we wish to determine the effect of a new anti-inflammatory drug on the release of a cytokine (such as IL6) by macrophages *in vitro*. We need to know how much IL6 is being secreted under standard conditions, and the approximate mean and standard deviation of this figure. We can them hypothesize that certain doses of the drug reduce the release of the IL6 a certain amount. In most cases the experiment will be conducted in the wells of a tissue culture plate: it is expected that the experiment is conducted in at least three wells for each dose of the drug. It is important to note that there is more to the power calculation than these simple examples, as will be explained in the section that follows.

6.2 Probability

A key aspect of hypothesis testing is that we very rarely make the universal assertion that, for example, a group of men are actually taller than a group of women. Instead, researchers take a more guarded position by saying that a group of men are *probably* taller than a group of women. This distinction may seem minor, but it is of crucial importance in statistics.

Probability
a term to describe the likelihood that a particular difference is due to random chance or is genuine

But '**probability**' needs quantifying to give it credence. Just saying that something is 'probable' is too subjective to be useful. Consider the toss of a coin: if the coin is unbiased (such that there is an equal chance of the coin landing on a head or a tail) the probability of getting a head or tail is ½ or 0.5 for each. Now suppose it lands heads or tails four times in a row—the likelihood of this happening randomly (purely by chance) is $0.5 \times 0.5 \times 0.5 \times 0.5 = 0.5^4 = 0.0625$. Statisticians are prepared to accept this as being within the bounds of randomness: it would still suggest the coin is unbiased. However, landing heads or tails one more time brings the probability of the coin being unbiased to 0.03125 (that is, 0.5^5), or 3.125%, a very low probability. Another way of looking at this is that there is now a 100−3.125 = 96.875% chance of the coin being biased. Could we really consider this as likely to be due to chance? A probability of 3.125% would lead us to think it *unlikely* that a truly unbiased coin would give five heads in a row—and so we would come to the conclusion that a coin that does give five heads in a row must, in fact, be biased. We use the lower case letter p as a shorthand for probability, so in the example above p=0.0135 that the coin is biased.

Similarly, if a standard six-sided die is cast, the probability of it landing on any one number is 1/6, or 0.1666. Expressing this as a percentage, we say there is a 16.7% chance of any one number being uppermost on the die. The probability of it landing on the same number a second time is therefore $0.1666 \times 0.1666, = 0.028$ (that is, p=0.028). As a percentage, therefore, we can say that the same number being uppermost will occur by chance 2.8% of the time.

In research terms we use probability as a benchmark to define whether a difference we have found between two sets of data is due to chance or is genuine (that is, is due to some real feature). Specifically, statisticians have defined a level of probability of <0.05 (5%) as being significant. If $p>0.05$, perhaps $p=0.06$ (6%), the difference is not sufficiently large enough for us to be confident that what we are seeing isn't simply due to chance. We could also say there is a 94% chance that the difference is genuine. Conversely, if we have found that $p<0.05$, perhaps $p=0.04$ (4%), then we consider this to be sufficient evidence that the difference we are observing is genuine. Indeed, we are 96% confident that the difference is unbiased. So in the example given above of heads or tails five times in a row, $p=0.03125$ is <0.05, so we can be 96.875% confident that the coin is biased.

In a fresh example, let us suppose that in the month of May there have been eight instances of fluid leaking from a specimen bottle, from a total of 318 specimens received. Thus, the leakage rate is 2.5%. Now let us suppose that in July there were 355 specimens received, but that 20 of these leaked, giving a rate of 5.6%. This increase in the rate of leakage seems to be large, but is it significantly large? By applying a statistical test (the Chi-squared test, detailed in Chapter 7) to this data we obtain $p=0.043$. This means that we can be 95.7% sure that this increase is significant, and only 4.3% sure that the increase is simply due to chance. This result is likely to lead to an investigation (perhaps an audit) to discover the cause of this increase, and then steps to minimize the rate of future leakage.

We shall return to this enormously important concept of probability time and time again in the remainder of this chapter and those that follow.

SELF-CHECK 6.3

A pack of 52 playing cards consists of 13 cards in each of four suits (clubs, diamonds, hearts, and spades). In each suit there is one ace, one king, one queen, and one jack.

(a) What is the probability that a card selected at random will be a queen?

(b) What is the probability that any three cards drawn at random will all be spades?

Power

Bringing together probability and error brings us to **power**, of which there are two types—alpha and beta. The alpha power of a study is such that it minimizes the risk of type I error—the risk of a false positive (such as finding men to have longer hair than women). The minimum value of alpha that must be attained is such that the probability value must be less than 5%, that is, $p<0.05$.

Power
the ability of a data set to adequately address the risk of types I and II error

Similarly, the beta power of a study must be sufficiently large so as to minimize the risk of type II error—a false negative (such as failing to find that, on the whole, women have longer hair than men). Again, in a parallel with the alpha value, the minimum value of beta must be that the probability is less than 20%. However, because we are seeking to deny a negative, we describe the beta power in terms of 1 - beta, so that 1 - beta = 0.8 is generally cited.

A key aspect of the power calculation in basic laboratory research is the number of experiments being undertaken; in clinical research, the number of subjects to be recruited.

Worked examples (part 2)

We can now review our original hypotheses, the first being that men are 5 kilograms heavier than women, and that the power calculation states that 32 men and 32 women need to be recruited. A statistician would say that we are actually testing the hypothesis that men are, on average, *probably* or are *likely to be* 5 kilograms heavier than women. The standard level of probability of $p<0.05$ (that is, that we need to be over 95% sure that men are heavier than women) is needed to minimize a degree of type I (false positive) error.

In the same way the standard level of power needed in minimizing type II (false negative) error is 0.8. But is this minimizing of the risk of error sufficient? We can adjust the power calculation to reduce this risk even further, and so make the outcome of the hypothesis more robust, and so believable. More power is obtained by increasing the sample size (Table 6.1).

The model we have been developing is that we take a level of probability of <0.05 to be significant. If so, we are 95% sure the difference is genuine. But suppose we want to be even more

TABLE 6.1 **The relationship between probability, sample size, and power**
Reference point: the test statistic has a mean (standard deviation) of 100 (10) units in a control population. The quantified hypothesis states that levels are different in a case population. The combined number of cases and controls (being the sample size) for differences in levels of the test statistic at different levels of alpha (p<0.05, p<0.02, and p<0.01) and different levels of beta (0.8, 0.9, and 0.95) are as follows:

Power level of beta = 0.8 (minimizing type II error)	**Sample size required**		
Level of test statistic in the case population	p<0.05	p<0.02	p<0.01
102	394	504	586
105	64	82	96
110	17	22	26
Power level of beta = 0.9 (minimizing type II error)	**Sample size required**		
Level of test statistic in the case population	p<0.05	p<0.02	p<0.01
102	450	567	655
105	73	92	107
110	19	25	28
Power level of beta = 0.95 (minimizing type II error)	**Sample size required**		
Level of test statistic in the case population	p<0.05	p<0.02	p<0.01
102	527	653	746
105	86	106	121
110	23	28	32

The key point to note is that as the need (or desire) to reduce the risk of type I error (with an increasingly small p value) and type II error (with an increasingly large beta power) increases, then the sample size increases accordingly.

convinced that the difference is genuine, perhaps to a level of 98%. This translates to a likelihood that a difference is present by chance to be only 2% (that is $p=0.02$). The power calculation in Table 6.1 tells us we need to increase our sample size to 82 people (that is, 41 men and 41 women). If we continue in this vein, and demand 99% confidence that men are 5 kg heavier than women, we need good data from 96 people, and then if we find such a difference we can safely say the probability that the difference is genuine is $p<0.01$.

The changes in the sample size we have just discussed help us to be increasingly sure of reducing the risk of type I error, that of a false positive. We can perform the same calculation for reducing the risk of type II error, that of a false negative, which is given by the power of beta being 0.8. In the same way that probability is higher with a larger sample size, we must increase the sample size if we wish to reduce the risk of a false negative; this concept is also illustrated in Table 6.1. If we want to reduce the risk of a false negative we must increase our value of beta from 0.8 to 0.9, and we need to increase the sample size from 64 people to 73 people accordingly.

Discussing power in a basic laboratory setting, as in the second example (the effect of a drug on a cell or an enzyme) is more complex. Experiments of this nature often look at a range of time points, and perhaps at a range of cell types, and at different doses of the drug (that is, a dose-response analysis). Certainly, the need for power arises directly from the coefficient of variation (CV) of the method for determining the test statistic (the cytokine IL6, or activity of an enzyme). If the assay is very precise, with a small CV, then a small number of wells or experiments will be needed for statistical significance. But if the assay has a large CV, a correspondingly large number of replicates will be necessary. Broadly speaking, methods based on cell biology have a larger CV than immunoassays, which in turn have larger CVs than standard biochemistry methods. Inevitably, researchers will have built up a depth of experience to determine the best research design.

Confidence

The previous section has demonstrated a relationship between sample size and power. If we want to be increasingly confident that a statistic (such as the difference in weight between the sexes) we have found is genuine, then we have to increase the number of subjects. This is embodied in a **confidence interval (CI)**, which tells us the limits to which we can be confident about the accuracy of the data we are discussing. The mathematics of this are relatively simple, according to the following equation ...

Confidence interval
an upper and lower figure that together define the degree of confidence for a data set

$$95\% \text{ confidence interval} = \text{mean} \pm 1.96 \times (\text{standard deviation/square root of n})$$

... where n is the sample size. Consider the height of healthy adult men. If we measure the height of 8 men, and the result is a mean of 1.72 metres with a standard deviation of 0.06 metres, then by inputting these figures in the equation given above, we can be 95% confident that the true height of men is somewhere between ...

$$1.72 \pm 1.96 \times (0.06/\text{square root of } 8)$$
$$= 1.72 \pm 1.96 \times (0.06/2.83)$$
$$= 1.72 \pm 1.96 \times (0.02)$$
$$= 1.72 \pm 0.04$$
$$= 1.68 \text{ metres to } 1.76 \text{ metres}$$

For simplicity, all numbers are taken to two decimal places. This interval means that we have the same confidence that the true height of a man is 1.69 metres as we do that his height is 1.75 metres. The value 1.96 comes from fact that 95% of a completely random set of data points that have a perfect normal distribution can be found within a range of the mean plus 1.96 × the standard deviation and the mean minus 1.96 × the standard deviation. Incidentally, this is where we get the expression mean ± two SDs, where 1.96 gets rounded up to 2.

For many, this confidence interval may be too wide, especially in a clinical laboratory setting. Any 95% CI can be improved by increasing the sample size. With data from twice as many men of the same mean height (1.72) and standard deviation (0.06), the equation becomes ...

$$1.72 \pm 1.96 \times (0.06/\text{square root of } 16)$$

$$= 1.72 \pm 1.96 \times (0.06/4)$$

$$= 1.72 \pm 1.96 \times (0.015)$$

$$= 1.72 \pm 0.03$$

$$= 1.69 \text{ metres to } 1.75 \text{ metres}$$

So, by doubling the sample size we have reduced the confidence interval by 0.01 metres at each of the two endpoints. With a much large sample size of 120 men, the 95% CI becomes much tighter at 1.71–1.73 metres.

In some circumstances, we may need even more confidence about our data than the standard 95%. A more stringent 99% CI can be calculated by substituting 2.57 for 1.96 in the equation. If so, then the 99% CI for the height of the men, based on a sample size of 8, becomes 1.67–1.77 metres, and for a sample size of 16 it becomes 1.68–1.76. Note that the confidence interval gets smaller as the percentage of confidence get larger. As with the value 1.96 described above, the value 2.57 comes from the fact that 99% of a completely random set of data lies with a perfect normal distribution can be found within a range of the mean plus 2.57 × the standard deviation and the mean minus 2.57 × the standard deviation.

SELF-CHECK 6.4

The mean weight of a group of 36 women is 70.9 kg, and the standard deviation is 9 kg. Calculate the 95% CI for the weight of these women.

Now that we have introduced a number of important key terms, we can apply these to different types of research.

6.3 Research design: general concepts

The original hypothesis must be rigorously tested, and the method for the test must be precisely designed to enable this to occur. Once the researcher(s) has settled on an original hypothesis and a sample size or number of experiments sufficiently large as to be able to test the hypothesis with confidence, the next step is to seek permissions.

Permissions

The route to research may appear clear, but there are in fact a series of hurdles that researchers must address. Although seemingly irritating, these hurdles are designed to ensure the research is done well, and safely.

- The first step is undoubtedly the protocol. This key document contains all the facts and figures needed to conduct the research, outlining why the research is needed, the original hypothesis, the power calculation (inevitably provided by a card-carrying statistician), and the method. If appropriate, the latter may be how the research material is to be collected and/or how the subjects recruited. There will be a section on how the data will be analysed, and another on costings—of the reagents and equipment, and perhaps the salaries of the researchers. The protocol will be fully referenced, and may have supplements such as the curriculum vitae of the researchers and letters of support from colleagues within the establishment, and others based elsewhere. With all this detail required, some protocols may run to thousands of words.
- Much research work has been wasted because the workers, perhaps over-enthusiastically, jumped straight from their idea to sample collection. A crucial part of research is to discuss ideas and methods with colleagues, some of whom will be experienced in research and so may spot problems—a process call **peer review**. Some researchers will show their protocol to colleagues within their hospital or university at an early stage, in the hope that errors may be spotted and corrected, and suggestions made. It may be these colleagues who provide a letter of support.

Peer review
the processes where a group of the researcher's peers will comment upon a protocol or other document

- No research in the NHS can be conducted without the permission of the Trust's Research and Development team. Similarly, Universities will also have a process for approving research. They may have their own set of paperwork, but the researchers will already have most of the required information in the protocol. Should NHS researchers be linked to a University, a second layer of paperwork may be required of the latter, especially if the research is part of a higher degree. The purpose of this layer of administration is to ensure the work is done well and safely, as the Trust and/or university effectively give their permission, and so will be held responsible if something goes wrong. It is also likely that 'intellectual property' will need to be addressed, as this may generate future income. If so, then a patent may be sought. These bodies may well seek external peer review opinions as part of their own procedure (hence the crucial importance of the quality of the protocol). Researchers must prove that they are capable of conducting the research by having completed 'Good Clinical Practice' training, a course often offered by the Trust or a local research collaborative.
- Almost all large research projects will require funding. This may come from the Government (Medical Research Council, etc.), charities (British Heart Foundation, Cancer Research UK etc.), industry (such as a manufacturer of a device or of a pharmaceutical) or elsewhere. These organizations will inevitably have their own procedure to justify the financial aspects of the research, but in any case these issues should be part of the protocol.
- Research on clinical samples (blood, urine, cerebrospinal fluid, tissue etc.) will inevitably require approval of the Local Research Ethics Committee (LREC). One of the major concerns of this body will be for the safety of the patients, whose written consent will be required. Should the samples already have been collected, perhaps elsewhere, or patient's permission not possible, then the Human Tissue Act will be relevant. The relationship between patient and health care professional is of course not equal, and the LRECs are rightly concerned that patients are not coerced into taking part. Informed consent must be obtained, after the subject has read a document called the 'Patient Information Sheet', which describes in detail what the subject can expect if they participate.
- Should the research involve determining the effect of a new medication or clinical procedure, then the Medicine and Healthcare products Regulatory Authority (MHRA) will need to give their approval. This will be relevant in a drug trial and will closely involve the particular industry partner that markets the new drug. The key point here is that the research tests a new process against normal clinical practice. If the latter is not the case, and there is no change to clinical practice, then the MHRA is unlikely to be involved.

- Much of the official paperwork is now on-line, such as the LREC documents and the Integrated Research Application System (IRAS).

Further information of this nature is present in sections that follow. In many cases these bodies will have dedicated websites where guidance and the relevant forms can be downloaded. Although there are dozens of different types of research, and each is designed to test a particular hypothesis, we have space only for those that are most relevant to biomedical science. We will first consider clinical research, and then look at basic laboratory research.

6.4 Clinical research

Broadly speaking, we can classify clinical research (mostly performed in a hospital setting) as being observational and/or interventional. In the former, we simply obtain data from a group of people (such as sex, age, body mass index, the frequency of heart disease and diabetes, or the levels of a particular molecule in their blood), and then draw conclusions. In some instances, we compare one group of subjects with another—these are **case-control studies**. Research looking within one single group of people is a **cohort study**. In an **interventional study** we subject the participants (probably patients with a particular condition) to a change (such as may be due to a drug) and then record the effects of the change, which may be months or years later. Regardless of the type of study, an important aspect is confidentiality. The regulatory authorities rightly place great emphasis on preventing the release of personal and sensitive information about the subjects being investigated. Accordingly, in many spreadsheets, information is anonymous, but coded.

Case-control studies

Case-control studies are the most common type of clinical research in biomedical science: the example of the height of men and women we have been working on is an observational case-control study. An absolute requirement in case-control studies is that these two groups are matched for certain factors likely to have a bearing on the results, such as age and sex, or are at least addressed (as in the height of those aged 10–15). This principle is applicable in studies of, for example, lung cancer or osteoarthritis, but also applies widely in other conditions, such as matching for menopausal status and parity in studies of reproductive or breast disease. In studies on diabetes, it is advisable to match cases and controls for body mass index (if possible), as this is an important pathophysiological feature of this risk factor. Failure to match for these confounders may lead to bias and so a false positive or negative.

Let us suppose a new molecule has been discovered that we feel may have a role in a certain disease. Our hypothesis may be that the expression of a molecule at the surface of a white blood cell, the frequency of a mutation in a certain gene, or blood levels of this molecule (let us call it Substance X) is/are higher in subjects with a particular disease compared to people free of the disease. In this setting those with the disease are called 'cases', whilst those free of the disease are called 'controls'. This brings us to the null and alternative hypotheses:

- The null hypothesis states that levels of Substance X are no difference between the cases and the controls.
- The alternative hypothesis states that Substance X is higher in the cases compared to the controls.

Cross reference

We have already introduced the concept of a mythical 'Substance X' in Chapter 3, where Figure 3.2 is relevant

Since this molecule is newly discovered, the hypothesis is original. But we must now extend our hypothesis by proposing exactly by how much more are levels of Substance X expressed on the white cells, or are at higher blood levels in the cases than in the controls—that is, we must have a quantified hypothesis.

Let us suppose that there is preliminary data from a group of healthy subjects indicating that the average level of Substance X (its mean value) in the plasma is perhaps 100 mmol/L. However, there will of course be variation in levels of Substance X in these people: some may have a result of 92 mmol/L, others may have a result of 112 mmol/L. Combining all these differences (or deviations) from the mean value gives us the standard deviation (SD). So let us conclude that the mean (SD) of Substance X in health is 100 (25) mmol/L.

However, how do we define healthy, and where do 'healthy' controls come from? Ideally, controls would be drawn from the same community as has provided the cases. Quite often, a case will present to hospital with his/her spouse, providing the opportunity to recruit the presumably healthy spouse as a control. Faults with this approach include diseases where one sex is predominant (breast and prostate cancer), so recruitment of the spouse is pointless. An alternative is to 'invite' the case to attend with an age and sex matched friend. An advantage to the potential control is a free health check-up, although this can, of course, find unwelcomed asymptomatic disease. Many studies recruit hospital staff to be controls. A problem with this is that, unsurprisingly, workers in the health care industry are more healthy than the general population. This also has repercussions for the definition of a reference range.

Cross reference

Mean and standard deviation have been introduced in Chapter 1

Once we have an idea of levels of Substance X in a healthy population, we can form a hypothesis regarding levels in a population with a particular disease. Let us hypothesize that the average level of Substance X in the cases with a particular disease is 125 mmol/L. Let us also assume that the standard deviation of levels of Substance X in the cases is the same as in the controls, that is, 25 mmol/L.

So, in summary, we hypothesize that levels of Substance X are 25% higher in the cases (that being mean 125 (standard deviation 25) mmol/L in patients with a particular disease) than in the controls (that being 100 (25) mmol/L in healthy subjects free of the disease). So the increase being hypothesized is of one complete standard deviation (25 mmol/L).

The next step is to determine the correct sample size (the number of cases and controls to be recruited), for which a calculation is required—the power calculation. Fortunately, all major forms of statistical software packages include such a calculator, and one software package tells us the sample size is 18, so we need 9 cases and 9 controls. However to be prudent, it is often wise to recruit slightly in excess of the sample size, so that 10 or 12 subjects in each group are needed. It is then simply matter of measuring the height of this number of men and women, and comparing the two data sets with an appropriate statistical test. We explain this process in Chapter 7.

The number of patients to be recruited is closely related to the hypothesized difference. If a smaller difference is hypothesized (such as half of a standard deviation), then perhaps 30 cases and 30 controls will be needed. It follows that the smaller the hypothesized difference, the greater the sample size. The problem here then is that researchers may hypothesize an impossibly large difference that will require a small sample size, and that in turn will be easy to recruit. A further requirement is that the hypothesis be clinically feasible. It is all very well hypothesizing difference in a new measure of iron metabolism in 30 patients with a haemoglobin result between 170 and 200 g/L, but how easy will it be to recruit these patients?

Once the design has been established, sample collection can begin, perhaps by approaching patients on the ward or in out-patient clinics. The bench work then proceeds, and once the recruitment target has been reached and analysis completed, analysis can begin. Chapter 6 explains key steps in the analysis of data obtained from cross-sectional studies.

Cohort studies

Laboratory scientists are less likely to be directly involved in work of this nature. Cohort studies look at a single group of subjects, and generally determine relationships within this group. Consequently, most cohort studies are observational, but nonetheless these relationships can be very important.

One of the best-known cohort studies is that known as the 'Framingham' study, named after a town in Massachusetts, USA. This study recruited over 5000 people, and searched for relationships between factors such as hypertension, smoking, blood cholesterol, and diabetes with heart disease such as myocardial infarction. A study of this nature is often described as cross-sectional, because it is believed that the subjects represent a true cross-section of the population from which they are recruited.

A second feature of cohort studies is that they can be used to test additional hypotheses. In the case of the Framingham study, returning to the subjects at regular intervals over the years enabled researchers to demonstrate the efficacy of a number of important health initiatives. One such initiative was that regular physical exercise reduces the risk of developing heart disease, whilst obesity increases the risk of heart disease. In this case the study may be described as 'follow-up'.

In the UK, in the early 1950s, Doll and Bradford Hill sent a questionnaire to all British doctors, and asked if they smoked tobacco. Some 29 months later, they followed up the doctors, and collected information regarding major health issues, demonstrating that smoking brings an increased risk of lung cancer. This study is not simply a cohort study, but is also prospective, in that the researchers specified in advance those factors that they believed to be significant.

Cohort studies often fail to pre-specify as exact an original hypothesis as a case-control study. Certainly, cohort studies will have aims and objectives, but these are frequently quite broad, and this in turn is often because researchers seek to determine which of several different measures are related to each other, or are most strongly related to an outcome such as the development of cancer or heart disease.

However, large studies such as those of Framingham and the British doctors enrolled thousands of subjects, so have a considerable degree of statistical power. This means that they are able to ask several hypotheses at the same time, with a low risk of false positives and false negatives. These studies are described as '**epidemiological**', deriving from epidemics and public health, and are concerned with the patterns and causes of diseases. These studies have such power that very complicated analyses can be performed, and their conclusions often have considerable implications for the health of the wider population.

Epidemiology
literally, the study of epidemics, but now taken to be the study of patterns of disease in large populations. It is the subject of Chapter 8

Intervention

Prevention is indeed preferable to cause, but once disease is present (such as an infection, cancer, diabetes, or hypertension), it may be treatable, perhaps with drugs and/or surgery. An intervention study is one where researchers introduce a particular form of therapy, such as a drug designed to treat a particular problem, and then observe the effect. This is often done in the setting of a clinical trial, where subjects may be randomly selected to receive an active intervention (such as a drug) or to receive usual care. Laboratory scientists may also be able to participate in studies of this nature, which will be in close cooperation with medical colleagues.

One of the problems of an intervention study is the possibility that the test statistic (such as blood pressure in a study of hypertension, or serum creatinine in a study of renal disease) will improve even in those not taking the new drug. A way of getting around this is to 'randomize' the patients: to give them the active drug or a tablet with no active ingredient (a placebo), with each patient being assigned an intervention (drug or placebo) completely at random.

Such a study can be further improved by making the patient unaware of which tablet they are taking—this is called being 'blind'. Blinding can be extended by ensuring that the complete health care team (doctors, nurses, physiotherapists etc.) are also unaware of what the patient is taking, giving us the concept of a 'double blind' trial. In some cases, a particular study centre may be unusually successful in completing this research, leading to a possible bias. This can be reduced by recruiting several different research centres.

The gold standard study of a new drug is a randomized, double blind, multi-centre, placebo-controlled trial. The null hypothesis states that there is no difference in the test index (blood pressure, creatinine, cholesterol) in those taking the drug compared to those taking the placebo. The alternative hypothesis states that the drug has a different effect on those taking it compared to those taking a placebo. Incidentally, some placebos can have powerful effects: in one study of hypertension, the placebo lowered blood pressure in those taking it (possibly because the patients and research team were aware of the trial, and so were more active in addressing general overall health issues). Fortunately, the new drug was far more effective in reducing blood pressure.

Regulation of clinical trials

The history of clinical trials is scarred by instances where serious problems have arisen in patients taking drugs thought to be safe (thalidomide comes to mind). In the UK, bodies such as the MHRA are concerned with the safety and efficacy of medicinal drugs and devices (an example of the latter being heart pacemakers). Any study subjecting people (be they patients or healthy individuals) to new drugs (inevitably an intervention study) must be authorized by the MHRA. Within Europe, the European Medicines Agency is a higher regulatory authority, whilst in the USA the Food and Drug Administration (FDA) must give consent to a trial. This will change after Brexit in March 2019.

As described above, even studies simply taking blood from subjects without a change in their medications (that is, those that are not interventionist) must gain the approval of LRECs, who consider the subject's perspective. The 'Declaration of Helsinki', administered by the World Medical Association, sets the standard of care for investigations on human subjects, and must be adhered to. In the UK, each NHS Trust has a Research and Development unit that can advise.

6.5 Basic laboratory research

The design of most basic laboratory research, often away from a hospital pathology setting, is reasonably straightforward and has much in common with the design of clinical research, such as the testing of a hypothesis, sample size determination, and choice of analytes. For example, in hypothesizing a different expression of a certain molecule by cells or tissues of a tumour, an important comparator group comprises cells of tissues that are not malignant. Here, the tumour cells to be examined histologically are the 'cases', and the normal cells are the 'controls'. The parallel extends to the need for a power calculation to ensure enough sections of tumour and normal tissues have been examined, or the appropriate number of experiments performed. Again, the problem arises of the consent from the patient to perform research on their tissues, and so collaboration with researchers within a hospital, and who have access to malignant and normal tissues, is essential. However, some research, such as drug discovery by the pharmaceutical industry, can be extremely complex and technical. Leaving these (and other) aspects aside, we can focus on three strands in basic laboratory research most relevant to biomedical science.

Biochemistry

This strand involves research on metabolic and physiological pathways and their component atoms, ions, and molecules. Research may be on 'pure' chemicals and reagents that can be purchased from reagent suppliers (an enzyme, its substrate, and potential inhibitors), or that which uses plasma or serum, in which instance the consent of the cases, and controls (if there are to be controls), and LREC approval will have to be obtained. Typical hypotheses include those of the effects of drugs on a particular process. For example, the effects of new anticoagulants on clot formation and/or dissolution can be tested in plasma and/or serum before their effect in animals and then healthy persons is determined. Central to research of this nature is the dose-response curve, which tests the expectancy of a linear relationship between the dose of the drug used and its effect on the system.

Tissue culture

Work of this nature is often described as '*in vitro*', following the historical view that it was performed in glass dishes, now superseded by plastic tissue culture plates of 6–24 wells. In testing a hypothesis that a particular drug influences a certain metabolic pathway, one design may be to use tissue culture. A common approach is to add the drug to special fluid in which a transformed (cancer) cell line is being grown, and then periodically harvest the cells and/or the tissue culture supernatant. Some cells will be cultured in wells of a tissue culture plate free of the drug (the negative control). Alternatively, cells may be obtained from animals, or from 'left-over' human tissues, such as those discarded after surgery (the uterus after hysterectomy) or after childbirth (the placenta and umbilical cord, the latter a source of endothelial cells). Dose-response curves can also be established in tissue culture, and an important further aspect is the potential for a cytotoxic effect. The latter data will directly feed animal studies.

Techniques such as immunofluorescence, molecular genetics, and enzyme linked immunosorbent assays may be applied to cells, cell lysates, and the supernatant respectively to determine changes in, for example, gene expression or levels of metabolites such as an enzyme, growth factor, or cytokine. When using cell lines or animal tissues, there is no direct human provider of the research material (i.e. a patient's serum sample), so LREC issues will not arise. It is also possible to avoid these issues by ensuring that material from patients and controls that has already been taken for normal clinical practice is anonymous.

Although not 'tissue', the culture of microbes is a further useful tool for research into new drugs and therapeutics, such as antibiotics. Much of the research design using cell lines is applicable to microbiology culture.

Animal models

Other hypotheses may be tested in animal models, some of which may be genetically modified (such as knock-outs). Animal experimentation has several advantages, such as that the hypothesis can be tested in a homogenous system (that is, assuming an inbred strain where all animals are genetically identical) and that many variables that confound clinical research (such as diet and lifestyle) can be controlled. Such experiments must be approved by local bodies and by the Home Office, from whom a licence must be obtained. Once more, the number of times an experiment needs to be repeated or the number of animals to be tested (perhaps some with a new drug, others with a placebo, mimicking a clinical intervention study) must be defined statistically with a formal power calculation before the work begins. Animal work is essential to determine the safety profile of a new drug (that is, toxicology), one method being to determine the dose at which 50% of the animals die (the lethal dose 50, or LD50), and so which will guide clinical work. A linked statistic is the dose at which the drug is effective in half of the animals concerned—hence effective concentration 50, or EC50. The precise definition of *effective* is, of course, crucial. A third statistic is the concentration at which an inhibitor suppresses 50% of a certain reaction (such as that of an enzyme/substrate), hence IC50. However, as all research must benefit patient outcome, all data from animal models must be rigorously tested in pre-clinical studies.

Cross reference

Section 7.6 has worked examples of the analysis of dose-response data

6.6 What data do I need to collect and what do I do with it?

The precise data to be collected follows directly from the original hypothesis, the protocol, and the design of the research. Although not essential, there is still a place for information from the patient or control to be collected onto paper as a pro forma.

The pro forma

In a purely laboratory project, there may be research data on up to half a dozen biochemicals, some of which may be the products of an enzyme-mediated reaction, or perhaps the intensity of staining of a section of a tumour. As many indices in clinical research are influenced by the subject's age and sex, and also often by other factors such as body mass index, blood pressure, past medical history, and concurrent disease and drug therapy, all these details must also be collected. These 'confounders' may all influence the test statistic and so the original hypothesis, and may require adjusting for in analysis. Accordingly, a great deal of other information on each participant is often collected, generally on a pro forma (Figure 6.1). Quite often, this is a sheet of A4 paper, but data may also be collected in electronic form on to a laptop or tablet. Figure 6.1 also has a brief note about the design and purpose of the study, and other abbreviated information relevant to the particular project.

Only data that is part of the protocol, and which has a bearing on the hypothesis, is to be collected. For example, collecting marital status is pointless unless there is a plausible reason why being married or not influences the test statistic or an outcome. Similarly, collecting data on number of siblings or education must be scientifically justified. More pertinent to clinical research is race and ethnicity, as, for example, Indo-Asians are at greater risk of type 2 diabetes than are white Europeans, whilst Afro-Caribbean men are at greater risk of prostate cancer than are Caucasian men. It is important that as this sheet contains personal information, it must be kept strictly confidential to the research team, the members of which will be stated in the permission and approvals documents.

The spreadsheet

Once collected, the raw data needs to be entered into a spreadsheet to enable analysis. However, for reasons stated just above, information stating or giving clues as to the identity of the patient must be excluded or at least minimized—the person doing the analysis has no interest in the patient's name or address. Usually, a single reference number links the pro forma to the spreadsheet.

The Eye in diabetes and CVD **Clinic date**....................................

Inclusions: Type 2 diabetes with/without MACE, being ACS, CVA or arterial procedure to be stated in patient's notes.
Exclusions: Within 3 months of the above, cancer, inflammatory disease (RA, SLE etc), on antibiotics, steroids, cytotoxics (implies inflammation), on medications for psychiatric indication.

Patient's Sticky Label

To be stuck here

GP detail if sticky label available

Clinical and Demographic

1. Unique Patient Identifier... 2. Ethnicity ....

3. Telephone number...

4. Age... 5. Disease duration ....

6. Blood pressure... 7. BMI..

8. HbA_{1c}.. mmol/mol

9. Creatinine................................ µmol/L = eGFR.............................. mL/min/1.73m^2

10. If diabetic: uACR...

11. Medical/surgical history e.g. MACE (what/when)..

Medications: name and dose

12. Anti-hypertensive 1 (ACEI, ARB, CCB)...

13. Anti-hypertensive 2 (BB, nitrate, diuretic)..

14. Hypo-glycaemic...

15. Lipid-lowering...

16. Anti-thrombotic (aspirin/clopidogrel)...

17. Anti-coagulant...

18. Other drugs (T4, analgesic)...

FIGURE 6.1
Example of a pro forma

Although imperfect, the most convenient spreadsheet is Microsoft's Excel, not merely because it is present on virtually all personal computers, whether desktop or laptop. An example of a small amount of raw data in an Excel-style spreadsheet is presented in Figure 6.2. This data is a mixture of identification (columns A and B), demographic (columns C–F) and clinical (G and H) material. The research result is in column I. The information in the patient identification

A	B	C	D	E	F	G	H	I
0	Patient ID	Initials	Sex	Age	BMI	Marital Status	Diabetes	Hb
1	2112	SG	M	45	26.4	M	N	150
2	4713	NT	F	46	35.3	M	Y	126
3	8314	HV	M	74	22.7	M	N	134
4	3917	FR	F	36	27.4	DW	N	117
5	6320	HY	F	58	18.9	M	N	129
6	5022	UM	M	65	26.4	M	N	136
7	7127	KP	F	36	23.9	S	N	139
8	3828	LB	M	48	23.6	M	N	145
9	6230	ED	M	54	25.7	M	N	139
10	5532	BF	F	58	28.3	M	N	130
11	4833	BW	M	52	26.4	DW	Y	142
12	6035	RV	F	43	25.1	S	N	135
13	7436	NY	F	57	28.6	DW	N	128
14	5837	AV	M	45	28.5	M	N	156
15	3640	HL	F	63	31.6	M	Y	131
16	7442	ON	M	54	26.6	M	N	138
17	5843	MR	M	68	24.1	DW	N	145
18	7846	LB	F	69	37.8	M	Y	130
19	5147	BT	F	56	24.4	M	N	125
20	2949	EC	M	41	29.5	S	N	149

Key: ID = identifier, M = male, F = female, BMI = body mass index, Hb = haemoglobin, S = single, M = married, DW = divorced, widow or widower, N = no, Y = yes.

FIGURE 6.2
An example of a spreadsheet

columns A and B is not for analysis. Of course, all this information is highly confidential, and the LREC will insist on it being seen by the research team only, and kept stored in a locked cupboard/filing cabinet in a room that is locked at night.

As we have discussed, the spreadsheet must be validated, as the greatest source of error in research is that of the analyst who makes mistakes in the transfer of the data from the pro forma (or another source) to the spreadsheet. Some such errors are easily spotted, such as missing the decimal point in age: 45.5 years becoming 455 years. However, others may be almost impossible to detect, such as age 54.5 years instead of 45.5 years. The data in this spreadsheet is not directly analysable, but fortunately it can be cut and pasted into a formal statistical package. These issues are discussed in Chapter 7.

Note that some of the data is categorical (sex, marital status, presence or absence of diabetes) and some continuously variable (age, BMI, haemoglobin, research result). Furthermore, marital status has three categories, whilst sex and diabetes have only two. Generally, data is carefully collected and quality control measures are undertaken to verify the accuracy of this data.

Statistical packages

Analysis call, for a statistical package, of which several are available. Microsoft's software Excel, found on almost all computers and laptops, is very good for hosting a basic spreadsheet, but although it can summarize data, its analytical powers are poor. Fortunately, all dedicated software packages can import spreadsheets from Excel.

Probably the most user-friendly package is Minitab, which is perfect for the small to moderate-sized studies often undertaken by laboratory scientists, with less than perhaps one or two hundred sets of data, and is therefore ideal for those whose research is not extensive; perhaps being undertaken to satisfy the demands of an undergraduate dissertation, or a higher degree.

Data from thousands of participants, as may be collected by teams of researchers, is better analysed by a more powerful package such as SPSS, R, SAS, or Stata. The complexity of these latter packages is such that they are generally used by full-time epidemiologists or statisticians.

Chapter summary

- Practical biomedical research is directed towards a better understanding of physiology and/or pathology, or to improving patient outcome.
- Researchers must formulate an original hypothesis: an idea or explanation for something that is based on known facts but has not yet been proven.
- The power calculation tells us how many experiments must be done, or how many subjects must be recruited, in order to test the hypothesis. This number is called the sample size.
- If the sample size is too small, results are at risk from false positive and false negative conclusions.
- The standard level of significance is taken to be a probability (p) of <0.05. If we attain the p value, we can be over 95% sure that the result is genuine.
- Researchers must obtain permission or approval for their research from bodies such as their Trust/University Research and Development Department, a Local Research Ethics Committee, and the MHRA.
- Different designs for clinical research include case-control, cohort, and intervention. These are often subject to strict regulation.
- Data is generally collected onto a pro forma, and is then transferred to a spreadsheet for analysis using statistical software.

Suggested reading

- **Altman, D.G. *Practical Statistics for Medical Research*. Chapman & Hall, London, 1991.**
- **Blann, A.D., Mather, H., Miller, J.P., McCollum, C.N. (1995). Atherosclerosis risk factors: variation in healthy hospital workers and members of local communities asymptomatic for vascular disease. Implications for normal controls. *Br J Biomed Sci* 52, 31–34.**

- Blann, A.D., and Nation, B.R. (2008). Good analytical practice: statistics and handling data in biomedical science. A primer and directions for authors. Part 1: Introduction. Data within and between one or two sets of individuals. *Br J Biomed Sci* 65, 209–217.
- Blann, A.D., and Nation, B.R. (2009). Good analytical practice: statistics and handling data in biomedical science. A primer and directions for authors. Part 2: Analysis of data from three or more groups, and instructions for authors. *Br J Biomed Sci* 66, 1–5.
- Colquhoun, D. An investigation of the false discovery rate and the misinterpretation of p-values. http://rsos.royalsocietypublishing.org/content/1/3/140216
- Colquhoun, D. The reproducibility of research and the misinterpretation of p values. http://www.biorxiv.org/content/early/2017/07/24/144337.
- Holmes, D., Moody, P., and Dine, D. *Research Methods for the Biosciences*. Oxford University Press, Oxford, 2006.
- Peacock, J.L., and Peacock, P.J. *Oxford Handbook of Medical Statistics*. Oxford University Press, Oxford, 2011.

Useful websites

- The Declaration of Helsinki: http://www.who.int/bulletin/archives/79%284%29373.pdf
- The Medicines and Healthcare products regulatory authority: http://www.mhra.gov.uk
- Permissions and approvals: www.ethicsguidebook.ac.uk/Permission-and-approval-10
- The integrated research application system (IRAS): https://www.myresearchproject.org.uk

Questions

6.1 Explain the purpose and the relationship between the two types of research.

6.2 Why is the precise determination of the sample size so important?

6.3 The 95% confidence interval of a measurement is 22.5–27.5 units. What does this mean, and how can we improve this confidence?

6.4 What is the major difference between a case- control study and an intervention study?

6.5 A new drug has been synthesized in a chemistry laboratory. What are the first steps in determining whether it may one day become a useful therapeutic?

7

Research 2: The analysis of modest data sets

Learning objectives

After studying this chapter, you should be able to ...

- Understand the need to verify data
- Describe the analysis of categorical data
- Explain the analysis of data that is continuously variable
- Understand the difference in approach when analysing two groups, or three or more groups
- Appreciate the nature of the analysis of paired and serial measurements
- Recall the methods and purpose of correlation

It is likely that most laboratory scientist researchers will collect their own data (perhaps from their own experiments, and so testing relatively simple hypotheses), and that consequently the sample size will not be large, perhaps up to a hundred or so. Other researchers, possibly working on large spreadsheets with data derived elsewhere, may have thousands of items of data. There are specific (but not mutually exclusive) types of analysis for large and more modest sample sizes. This chapter will focus on the latter, leaving the analysis of large data sets for Chapter 8.

This chapter will begin by discussing aspects of research by considering the verification of the data and its subsequent analysis. Sections 7.2 and 7.3 outline the analysis of categorical data, whilst Section 7.4 refreshes on the importance of distribution, first broached in Chapter 1. These themes are developed in Sections 7.5 (analysis of two data sets) and 7.6 (three data sets), which rely heavily on the understanding of key words such as mean, standard deviation, median, and inter-quartile range, which are defined in Chapter 1, and are also part of Chapter 6. Section 7.7 looks at data linked by individuals or over several time periods, whilst the chapter concludes in Section 7.8 with an explanation of correlation.

7.1 Verification of data

In Chapter 6 we looked at the collection of data, and the placing of this data into a spreadsheet such as Excel. From this, data can be exported into a dedicated statistical analysis package such as Minitab or SPSS. However, because the most common source of error is human, the content of the spreadsheet must be verified. If the data set is small, with less than a few thousand data points (such as 10–15 measurements made in 100–200 subjects), then this can be done with relative ease by 'describing' the data.

Descriptive statistics refers to merely looking at certain features of data that has a continuous variation, and perhaps drawing a small number of conclusions based on the observations made. The first step when applying descriptive statistics is to summarize the data set with a number of mathematical attributes, or indices. The most common of these are the number of individual pieces of data (called 'n'), the mean, the standard deviation (SD), the minimum (smallest) value, the lower quartile (Q1), the median value, the upper quartile (Q3), and the maximum (largest) value.

Descriptive statistics
simple information regarding key aspects of a data set

As an example, we will look at the descriptive statistics of the ages of 45 parents collecting their children from primary school. The seven indices and the number of parents are shown in Table 7.1. The information has been arranged in two rows, with an initial and a subsequent description of the data.

Cross reference
Details of the seven attributes of a set of data (the mean, the standard deviation (SD), the minimum (smallest) value, the lower quartile (Q1), the median value, the upper quartile (Q3), and the maximum (largest) value) are illustrated in Chapter 1

The upper part of the printout (Initial description) tells us the mean age of the parents is 32.3 years (which is entirely likely), with a standard deviation (SD) of 39.3. The minimum age is 20, the lower quartile (Q1) is 24.5, the median is 26, and the higher quartile (Q3) is 29 years, all of which is likely. However, the maximum age is 289, which is impossible and is almost certain to be an error at data entry. It is likely that the real age of the subject is 28 years, and a slipped finger accidentally pressed 9 in addition to the 8. Note also that the SD is greater than the mean—we will return to this point in due course.

Identification and correction of the error will be followed by a repeat of the description of the data set, as shown in the lower part of Table 7.1 (Subsequent description). This shows that the corrected age and SD (26.5 and 3.1) are both lower, the SD markedly so. Note also that the minimum, Q1, median, and Q3 ages (20, 24, 26, and 29 respectively) are unchanged, but that the maximum age is now 34 years, which is far more believable.

The verification of categorical data, as discussed in Section 7.2, is generally more difficult. For example, in a study of heart disease we may find that there are 85 men and 23 women, and in a study of breast cancer, 583 women and two men. Both proportions seem appropriate. However, finding even one woman in a study of prostate disease betrays an error.

Partly because it is easier to control confounders, experiments using tissue culture, biochemistry, and animal models often have a much smaller sample size.

TABLE 7.1 **Descriptive statistics of parents' age**

Variable (age)	n	Mean	SD	Minimum	Q1	Median	Q3	Maximum
Initial description	45	32.3	39.3	20	24.5	26	29	289
Subsequent description	45	26.5	3.1	20	24	26	29	34

SD = standard deviation: Q1 = 25th (lower) quartile: Q3 = 75th (higher) quartile

Although simply describing data can be valuable, the true purpose of statistics is as a tool to confirm or refute an alternative hypothesis of a difference or within different sets of data. However, the analysis of data is fraught with difficulty: there are dozens of different statistical tests that could be applied. Selection of the most appropriate test relies on the nature of the data, which can be classified as being categorical or continuously variable. Other factors to be addressed are as follows:

Cross reference

Details of the nature of categorical and continuously variable data are presented in Chapter 1

- the number of groups of individuals or experiments being analysed
- whether the hypotheses address an intervention, the effect of time, or some other link between data sets
- the extent to which the effects of multiple factors are being examined
- the distribution of data as being normal or non-normal.

7.2 The analysis of categorical data: two groups

Categorical data can only exist in one of a set number of mutually exclusive groups. Examples include testing only two doses of a particular drug *in vitro*, another is sex and marital status. Regarding the latter, you are either male or female (leaving aside very rare cases) and are legally either single, married, divorced, or in widowhood (that is, being a widow or widower). Similarly, you are either alive or dead: there is no third category; and a woman has an obstetric history of an exact whole number of children—she can't have had 1.25 babies.

A great deal of medicine relies on attendance at certain clinics, such as for diabetes, hypertension, and cancer, where it is very likely that the particular disease is present in someone attending that clinic. A leading consequence of diabetes is cardiovascular disease such as a heart attack or a stroke, and again the clinical definition of these events is exact: you can't have had a bit of a heart attack—you either have or you haven't. These therefore are categories.

Let us consider an interventional case-control study where the hypothesis is that diabetics taking a certain drug (the cases) have a different risk of a cardiovascular event compared with diabetics not taking the drug (the controls). The null hypothesis is that there is no statistically significant difference in the frequency of a cardiovascular event; the alternative hypothesis is that the cases benefit from the drug compared with those not taking the drug. The degree of benefit we feel is important in that there is only a 5% chance that any difference found is coincidental; which is also to say that there is a 95% chance that any difference is genuine. This 5% translates to 0.05, which is our requirement for a probability (p) of $p<0.05$.

Patients with diabetes are enrolled, and half are placed on the drug in question, while the other half take an inactive placebo. Once the study (which may take several years) is complete, each patient will be in one of four groups:

- cases who have had a cardiovascular event
- cases who are free of a cardiovascular event
- controls who have had a cardiovascular event
- controls who are free of a cardiovascular event.

The nature of studies of this design is that there will have been roughly equal numbers of cases and controls (in this example, 400 of each, so that the total number of people is 800, as defined by the power calculation) at the beginning of the study.

At the end of the study we find that there have been 60 cardiovascular events. The null hypothesis states that there is no difference between cases and the controls. That is, if we have seen 60 cardiovascular events, we would expect 30 of these to be among the cases, and 30 among the controls. In reality, however, there have been 20 events in the cases and 40 in the controls. Is this difference significant? To determine this we apply the Chi-squared test.

The Chi-squared test

The most commonly used statistical test of a difference in the proportions of individuals in two groups is called the **Chi-squared** test (chi being a letter in the Greek alphabet, with the symbol χ). An alternative test is Fisher's exact test, which uses different mathematics, but which almost always arrives at a very similar result. A third, Barnard's test, is very rarely used. Tables 7.2 and 7.3 show the printout of a Chi-squared test of these data, laid out as a two-by-two table (the cases and controls arranged vertically, the presence of an event or no event arranged horizontally). When applying the Chi-squared test, the observed data (the 20 and 40 events in the cases and controls) is compared with the number expected if the null hypothesis were correct, which is 30 events per group (Table 7.2). In each part of the table—for cases and controls—a statistic is calculated as follows:

Chi-squared test
a statistical test that seeks difference in the categorical index in different groups

$$\frac{(\text{Observed} - \text{Expected})^2}{\text{Expected}}$$

So, for the observed and expected numbers of cases (that is, $(O-E)^2/E$) who have had an event the value is $(20-30)^2/30 = 3.333$. For the observed and expected number of controls the value is $(40-30)^2/30 = 3.333$. (Notice how both values are the same.)

In the same way a two-by-two table can be constructed for those 740 people from our overall study group who have not had an event, as shown in Table 7.3. In both cases, $(O-E)^2/E = 0.27$. Bringing together these two sets of mathematics from Tables 7.2 and 7.3 gives us a formal Chi-squared analysis (Table 7.4).

TABLE 7.2 **The Chi-squared mathematics for those who have had an event**

	Cases	Controls	Totals
Event observed (O)	20	40	60
Event expected (E)	30	30	60
$\frac{(\text{Observed} - \text{Expected})^2}{\text{Expected}}$	3.333	3.333	

TABLE 7.3 **The Chi-squared mathematics for those who have not had an event**

	Cases	Controls	Totals
No event observed (O)	380	360	740
No event expected (E)	370	370	740
$\frac{(\text{Observed} - \text{Expected})^2}{\text{Expected}}$	0.27	0.27	

TABLE 7.4 **The Chi-squared test**

	Cases	Controls	Totals
Event observed	20	40	60
No event observed	380	360	740
Total	400	400	

By bringing together each of the four individual $(O-E)^2/E$ sums (3.333 + 3.333 + 0.27 + 0.27) we arrive at the Chi-squared statistic, which is 7.207. Tests of this nature are inevitably done on statistical software; such software would report a p value for this statistic of 0.007—that is, the probability of these data having occurred purely at random is just 0.007. Since p=0.007 is around seven times smaller than p=0.05, we can deduce that the difference in the rate of events between the case and controls is markedly significant. If the two groups of 400 cases and 400 controls were matched for all other confounders (such as age, sex, disease duration, other medications, body mass index etc.), then we can be confident that the difference is due to the drug administered.

As such, we would conclude that the drug very probably (that is, p=0.007) protects against the development of cardiovascular disease. Put in terms of percentages, we are 99.3% certain the result truly reflects the effects of the drug, and we accept the (remote) possibility of 0.7% that the difference is due to chance.

The practicalities of the Chi-squared test

The preceding text has explained the mathematics behind the Chi-squared test, and how the Chi-squared statistic (7.207) can be obtained from a pencil, paper, and a hand-held calculator. But naturally all good statistical software packages can perform this analysis. To obtain a p value directly from a Chi-squared statistic, on-line calculators are available, but statistical tables may also be consulted. The latter are also available on-line, and in some textbooks (but not this one). Researchers accessing these tables will come upon a new feature, the **degrees of freedom**. In this test it relates to the nature of the number of analyses being performed. In the example above, we are considering the frequency of cardiovascular disease between two groups (cases and controls). The degrees of freedom is defined as the number of groups minus one. Since there are two groups, there is one degree of freedom (that is, n–1). As the number of types of analyses rises, then so do the degrees of freedom. We will return to this point in the section that follows, and in other sections.

Degrees of freedom
a measure of the number of values that are free to vary

7.3 The analysis of categorical data: three or more groups

Almost exactly the same analyses that we perform for two groups are performed when we look at three or more categorical groups. The two-by-two design of the Chi-squared table in the example given in Section 7.2 is extended to a three-by-two table, a three-by-three table etc. as the number of groups demands. The null hypothesis is that there is no difference in the proportions of people (or observations) in each particular box (that is, as are expected by the null

TABLE 7.5 **Number of children born to women living in different geographical areas**

	No children	One child	Two or more children	Total number of women
Women in a rural setting	35	66	42	143
Women in an urban setting	26	93	31	150
Total	61	159	73	293

hypothesis). The alternative hypothesis is that there is a difference in the proportions of people (or observations) in each particular box (that is, as are expected by the alternative hypothesis). The Chi-squared statistic is again the sum of all the individual instances of (observed – expected)^2/expected; the larger the value of Chi-squared, the more likely it is that we are seeing a genuine difference between data sets.

An example of a three-by-two design is the number of children born to 293 women who live in different geographical areas (with 143 living in a rural setting, and 150 in an urban setting). This data is summarized in Table 7.5.

If we simply look at the percentages of the women in each of the three 'number of children' groups (no children, one child, or two or more children), we find 24% of women living in a rural setting have no children, 46% have one, and 29% have two or more. By contrast, 17% of women living in an urban setting have no children, 62% have one, and 21% have two or more.

A quick view of this data seems to suggest a difference—for example, 46% would generally be thought of as being considerably less than 62% (in respect of those women having one child). But statisticians do not make claims simply by suggestion: a formal statistical test must be used to confirm our suspicions.

If we apply the Chi-squared test to this data we get p=0.025. This tells us that there is a high probability (i.e. 97.5%) that there is a meaningful difference between women living in a rural or an urban setting, and only a 2.5% chance that the difference is spurious, and has arisen by chance alone.

We return briefly to the topic of degrees of freedom. This, often abbreviated to DF, may appear in the output of software packages. If we were comparing no children to one child, there would be one degree of freedom, but in adding a further variable—two or more children—there are now two degrees of freedom. This arises from the short equation n–1, where n is the number of groups (that is: no child, one child, two or more children, hence 3–1 = 2 DFs).

The problem with this analysis is that it does not tell exactly where this difference is. Of those women living in a rural setting, is the group of 66 women (i.e. 46% of the whole group) with one child really different from the 93 women (62% of the whole group) with one child living in an urban setting? In order to do this we have to break up the data set and then perform individual Chi-squared tests on each pair of data points, as shown in Tables 7.6, 7.7, and 7.8.

First, let us compare those women with no children with those having one child (Table 7.6). A Chi-squared analysis of this data gives a probability of p=0.034. This means that it is 96.6% likely that the difference in proportions of children between the women in a rural setting compared with the women in an urban setting is due to their location and not due to some other unknown factor (which would have a likelihood of 3.4%).

TABLE 7.6 **Comparing women with no children or with one child**

	No children	One child
Women in a rural setting	35	66
Women in an urban setting	26	93

Table 7.7 compares women with one child with those having two or more children. A Chi-squared analysis of this data gives a probability of p=0.023. This means that it is 97.7% likely that the difference in proportions between the women in a rural setting compared with the women in an urban setting is due to the location and not due to some other unknown factor (which would have a likelihood of 2.3%). Table 7.8 compares women with no children with those having two or more children. A Chi-squared analysis of this data gives a probability of p=0.985. This means that the difference in proportions is too small to be statistically meaningful—indeed, the likelihood of the difference being due to the difference in area is only 1.5%.

TABLE 7.7 **Comparing women with one child or with two or more children**

	One child	Two or more children
Women in a rural setting	66	42
Women in an urban setting	93	31

TABLE 7.8 **Comparing women with no children or with two or more children**

	No children	Two or more children
Women in a rural setting	35	42
Women in an urban setting	26	31

So, in summary, we can say that the difference in the number of children born to 293 women living in two different areas is statistically significant to the level of p=0.025 (that is, we are 97.5% confident that the difference is genuine, so that there is just a 2.5% probability of this finding being due to chance). Looking at this group in more detail, we can say that this overall difference is due to both a difference between the proportion of women with no children versus women with one child (p=0.023) in urban and rural locations, and also in the proportion of women with one child or with two or more children (p=0.034) in the two locations. However, the difference in the proportion of women with no children compared with the proportion of women with two or more children is not statistically significant (p=0.985).

SELF-CHECK 7.1

Of a population of 100 elderly people with rheumatoid arthritis, 35 have suffered a major cardiovascular event such as a heart attack or stroke. Of 100 age and sex-matched people with osteoarthritis, 20 have suffered a similar cardiovascular event. Calculate the Chi-squared statistic, and comment on what it tells us.

7.4 Distribution of data with a continuous variation

Before we look at the analysis of data with a continuous variation, we must review the concept of the distribution of data, which is introduced in Chapter 1, and reinforced in Chapter 4. The importance of the distribution of data is that it determines the method of analysis. Consequently, determining the nature of the distribution is the first step in analysis. As far as most simple research is concerned, there are two types of distribution: normal and non-normal. If we consider the field of haematology, an example of a normal distribution is that of haemoglobin in a healthy population; an example of a non-normal distribution is the sedimentation rate of erythrocytes. From biochemistry, an example of a normal distribution is serum potassium, while a non-normal distribution is in levels of serum triglycerides. This section is pertinent only to clinical research on groups of people—because experiments on animals, in tissue culture, and on laboratory biochemistry are easier to control, data inevitably has a normal distribution (unless there is a major failure of the procedure, such as the bacterial contamination of cells in one well of a tissue culture plate).

Consider two groups of data, A and B; descriptive statistics are presented in Table 7.9, and graphical representations of the data sets and some additional descriptive statistics are presented in Figures 7.1 and 7.2, both of which are from Minitab software. The dominant feature of the four boxes in both figures is that of the upper left, which shows a series of grey columns.

A normal distribution

Focus on the large box at the top left of Figure 7.1. It consists of a histogram of a number of grey columns, the tallest of which are in the centre of the figure. These columns represent the number of data points in certain categories, and this is perhaps the most clear indication that the data has a normal distribution. The data points with the tallest columns are between 20 and 22, with a smaller number of data points less than 16 and greater than 26. This layout is typical of a normal distribution.

Cross reference

Figures 1.1 and 4.1 also show normal distributions

Cross reference

Histograms are explained in Section 4.3, where Figure 4.9 is relevant

Immediately below the graph in Figure 7.1 is a long rectangular box: this summarizes the data in the form of a box and whisker plot; below this, another long rectangular box brings together the 95% confidence intervals of the mean and the median value. The fact that these two intervals almost completely overlap is a further indication that the data set has a normal distribution.

The visual representation of the data is not the only way that the nature of the distribution can be determined. On the right of Figure 7.1 is a box with a series of terms and numbers, the topmost being the **Anderson–Darling test**, which is a formal mathematical test of distribution.

Anderson–Darling test
a statistic that describes the likelihood that a data set fits a normal distribution

Other formal tests of distribution include those of Ryan and Joiner, of Shapiro and Wilk, and of Kolmogorov and Smirnov. The mathematics of the Anderson–Darling test is essentially that it tests the degree to which the data tracks the theoretic line of best fit if the data were distributed normally.

TABLE 7.9 **Descriptive statistics of Groups A and B**

Variable	n	Mean	SD	Minimum	Q1	Median	Q3	Maximum
Group A	48	20.1	3.1	14.0	18.0	20.0	22.0	28.0
Group B	48	21.3	19.5	8.0	11.25	13.0	20.0	100.00

SD = standard deviation; Q1 = 25th (lower) quartile; Q3 = 75th (higher) quartile

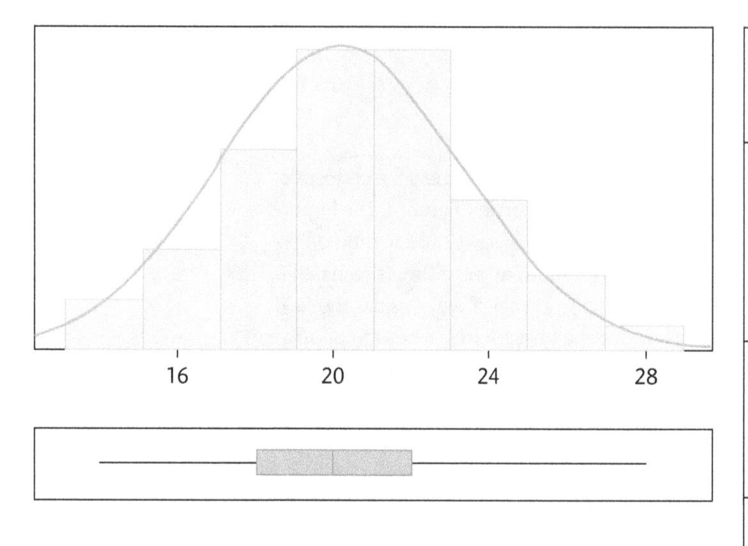

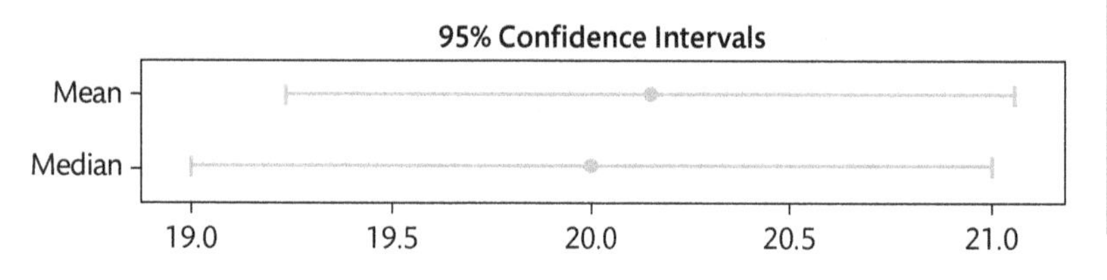

Anderson-Darling Normality Test	
A-Squared	0.27
P-Value	0.660
Mean	20.146
StDev	3.135
Variance	9.829
Skewness	0.066810
Kurtosis	-0.225363
N	48
Minimum	14.000
1st Quartile	18.000
Median	20.000
3rd Quartile	22.000
Maximum	28.000
95% Confidence Interval for Mean	
19.235	21.056
95% Confidence Interval for Median	
19.000	21.000
95% Confidence Interval for StDev	
2.610	3.927

FIGURE 7.1: GROUP A
Minitab printout of a normal distribution

The key value is the A^2 value, which if small, perhaps 0.25, tells us that the data is very likely to have a normal distribution. However, as the A^2 rises, then the likelihood of a normal distribution falls, so that an A^2 value greater than 0.75 implies that the data differs from a normal distribution, and so is non-normal. This line of best fit is present in Figures 7.1 and 7.2 as the blue line. In Figure 7.1, the Anderson-Darling statistic (A^2) is 0.27, which gives p=0.66. This tells us that the difference between the data set and the theoretical blue line of the bell-shaped curve is not significant, and so the data is very likely to have a normal distribution.

Other factors we can use to determine the nature of distribution are found in the box of data on the right of Figure 7.1. This box summarizes the numerical aspects of the data set. The mean (SD) of this data set is 20.146 (3.135); when expressed as a percentage, the SD is 15.6% of the mean. In determining the nature of distribution, there is no exact cut-off point for this crude percentage value, but if the SD is <30% of the mean, then the data set is likely to take a normal distribution. Below the 'StDev' in the data box is variance—this is simply the square of the SD.

Below 'variance' is '**skewness**'. This variable reflects the distribution of the data away from the bell-shaped curve of the normal distribution. Skewness can in theory vary from zero (implying a perfect normal distribution) to infinity (suggesting an impossibly non-normal distribution). In Figure 7.1, showing the group A data, the skewness is small, being 0.0668, which supports the view that the data set has a normal distribution. The final index is **kurtosis**, which we will come to shortly. 'N' is the sample size.

Skewness
the extent to which the major part of a data set is in the middle of the distribution, or is skewed over to the left or right

Kurtosis
the extent to which a data set has a 'sharp' distribution or a 'flat' distribution

The next block of data shows the minimum, 1st quartile, median, 3rd quartile, and maximum values. A quick check of the numbers shows that the median is right in the middle of both the 1st and 3rd quartiles, but also of the minimum and maximum values. Furthermore, the mean value (20.146) is only 0.73% higher than the median (20). Both of these features lend support to the likelihood that the data has a normal distribution.

The final aspects of the data box are the 95% confidence intervals (CIs) for the mean, median, and standard deviation. As is discussed in Chapter 6, these data tell us how confident we can be about the accuracy of the particular index. However, note that the 95% CIs for the mean and median are almost identical. This supports the view, as does the rectangular box at the bottom of the figure, that the data has a normal distribution.

A non-normal distribution

Figure 7.2 illustrates a typical non-normal distribution. The largest part of the figure is on the top left, which shows a histogram of a number of columns. The tallest columns (with results of 10 and 20) are over to the left side; the columns on the right are small (results of 80 and 100). Below this histogram plot, the grey box and whisker plot is also pushed over to the left, and in the final rectangular box at the bottom, the 95% confidence intervals of the mean and median almost fail to overlap at all. Compare this with the lower rectangle in Figure 7.1, where the confidence intervals almost completely overlap. All this is evidence that the data set illustrated by Figure 7.2 has a non-normal distribution.

Cross reference

Figures 4.2 and 4.3 show non-normal distributions

Returning to the histogram plot, the blue line denoting the theoretical best fit if the data were to have a normal distribution does not follow the heights of the columns of the histograms, and accordingly the Anderson–Darling statistic of 6.65 in the top of the data box on the right translates to $p<0.005$. This indicates that there is a 99.5% likelihood that the data has a non-normal distribution. Moving down the data box, the mean value is 21.271 and the standard deviation is 19.459. Thus, the standard deviation is 91.5% of the mean, very much larger than the 15.6% difference in data set A.

Below 'variance' is 'skewness'. In group A (Figure 7.1) the skewness is small, being 0.0668. However, in group B the skewness is much larger, being 2.558, which indicates that the tallest columns of

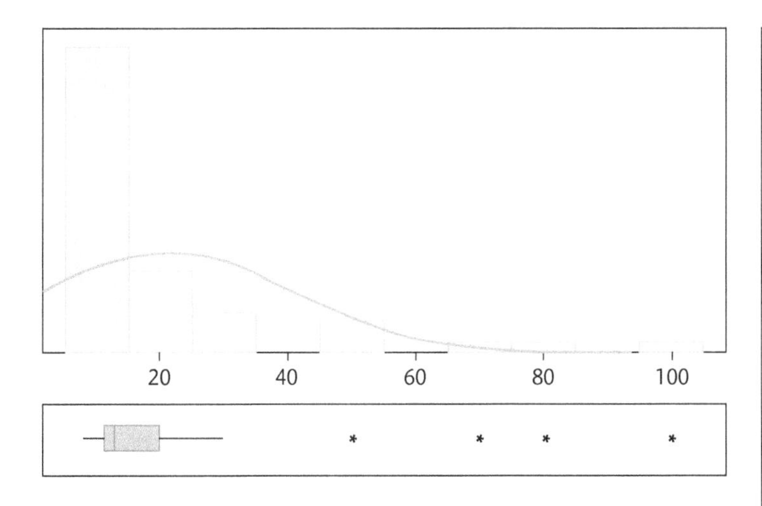

Anderson-Darling Normality Test	
A-Squared	6.65
P-Value <	0.005
Mean	21.271
StDev	19.459
Variance	378.670
Skewness	2.55824
Kurtosis	6.63672
N	48
Minimum	8.000
1st Quartile	11.250
Median	13.000
3rd Quartile	20.000
Maximum	100.000
95% Confidence Interval for Mean	
15.620	26.921
95% Confidence Interval for Median	
12.000	16.002
95% Confidence Interval for StDev	
16.199	24.375

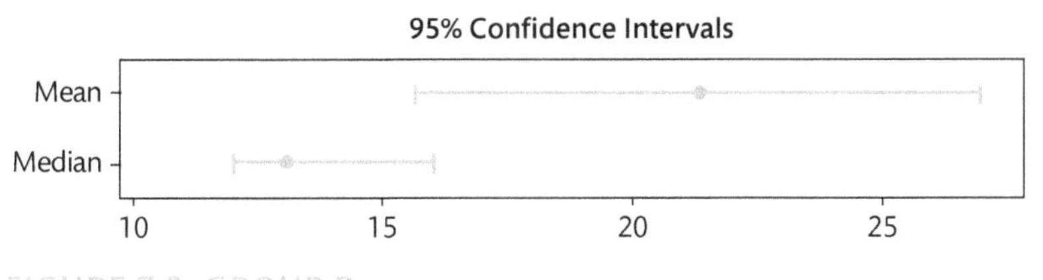

FIGURE 7.2: GROUP B

Minitab printout of a non-normal distribution

the histogram are not in the middle of the data set, but are shifted to the left or to the right. As is evident from the histogram plot, the data set is skewed strongly to the left. The kurtosis of data set B is 6.64, very much larger than the kurtosis figure of –0.22 in data set A. The sample size N is 48.

The next section of the numerical summary box shows the minimum, 1st quartile, median, 3rd quartile, and maximum values. Note that, unlike the data in Figure 7.1, the median is not in the middle of the 1st and 3rd quartiles: the median (13.00) is much closer to the 1st quartile (11.25) than to the 3rd quartile (20.00). Similarly, the median is far from the middle of the minimum and maximum values. Now compare the mean and the median. In Figure 7.1, the difference is only 0.73%, whereas in Figure 7.2, the difference is 63.6%.

The 95% CIs for the mean, median, and SD are provided at the bottom of the data box. They do have their uses, but these are rarely to help define distribution: the graphical representation of these data (on the immediate left) is far more helpful.

Kurtosis

As we have just noted, 'kurtosis' refers to the degree to which the distribution of a data set has a 'sharp' or a 'flat' shape, as is illustrated in Figure 7.3. It is important to note that each of these data sets has a completely symmetrical distribution, with an equal number of data points to the left and right of the mean point. Hence, for each, the skewness value is zero.

The distribution of the data in Figure 7.3a is over only five data points: 3, 4, 5, 6, and 7. Note that the height of column '5' clearly indicates there are more individual data points for this value than there are individual data points in '3', '4', '6', and '7'. The kurtosis value for this data is 2.41, suggesting a very 'sharp' distribution. Figure 7.3b shows a data set with a classical 'bell' shape. The heights of the columns are highest in the middle (the value 11) and fall gently to the left and right. The kurtosis value of this data set is zero. Figure 7.3c is almost completely 'flat': there are an equal number of data points for each of the individual values, 20 and 30, and one less for values 10 and 40. The kurtosis value is -1.34.

Kurtosis is an additional descriptor of the distribution of a data set. It is rarely used in relatively simple analyses, and is applicable mostly to data with a normal distribution. Almost by definition, kurtosis is markedly negative in data with a non-normal distribution. There is no definition of 'sharp' or 'flat' in the mathematical sense of kurtosis: these terms are used to help explain the types of distributions.

Comparing a normal and a non-normal distribution

Having now seen what characterizes a normal versus a non-normal distribution, how do we decide upon the nature of the distribution for a particular data set—whether it is normal or non-normal? There are several methods open to us, including an 'eyeball' test of the layout of the histogram (requiring a graphical representation); a formal test such as those of Anderson-Darling, Ryan-Joiner, Shapiro-Wilk, or Kolmogorov-Smirnov; the size of the SD compared with the mean; the extent to which the median is mathematically right between the 1st and 3rd quartiles; and the difference between the mean and the median, and skewness. Look at Table 7.10 which summarizes how each of these methods can be taken to suggest that your data is normally or non-normally distributed.

However, there are many instances where these different methods fail to agree. Indeed, all of these measures only give an *indication* of distribution. Accordingly, no single test is generally good enough for us to be confident about the nature of a distribution. Instead, each method should be seen as a useful tool for helping us to select the correct test to be used in analysing sets of data.

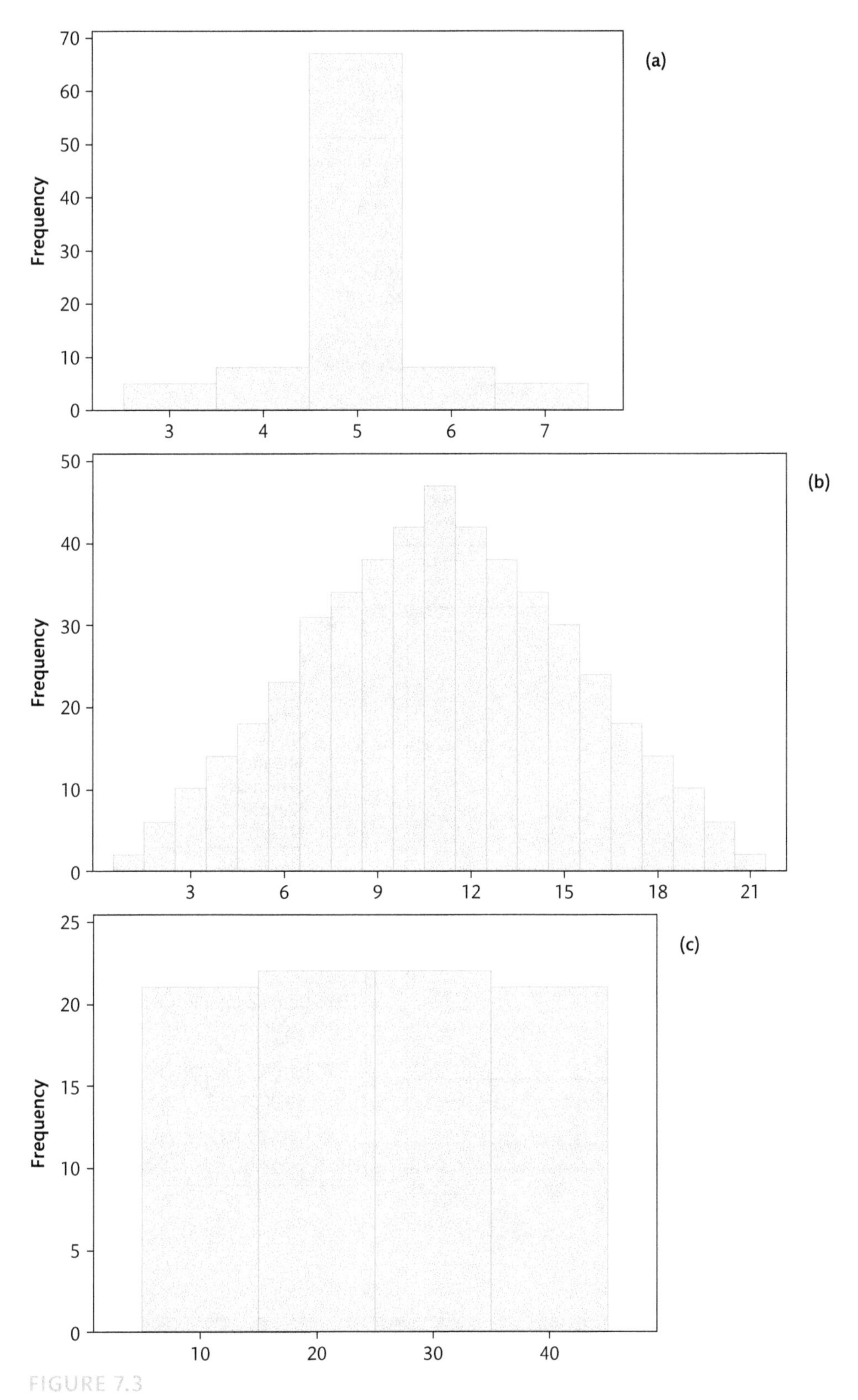

FIGURE 7.3

Kurtosis. (a) Positive kurtosis. (b) Neutral kurtosis. (c) Negative kurtosis.

TABLE 7.10 **Key features in the determination of distribution**

Method	Normal distribution (e.g. Figure 7.1)	Non-normal distribution (e.g. Figure 7.2)
'Eyeball' test of the layout of the histogram	Bell shaped: tallest columns in the centre	Tallest columns skewed over to the left or right
The Anderson–Darling test	Small A^2 $p>0.05$	Large A^2 $p<0.05$
The SD expressed in terms of the mean	The SD is small compared with the mean, e.g. <30%	The SD is large compared with the mean, e.g. >40%
The place of the median in terms of the 1st and 3rd quartile	The median is very close to being in the middle of the quartiles	The median is much closer to one quartile than the other
The mean and the median compared	The mean and the median are very close together	The mean and the median are far apart
Skewness	The skewness index is close to zero (often <0.1)	The skewness index is greater than zero (often >0.5)

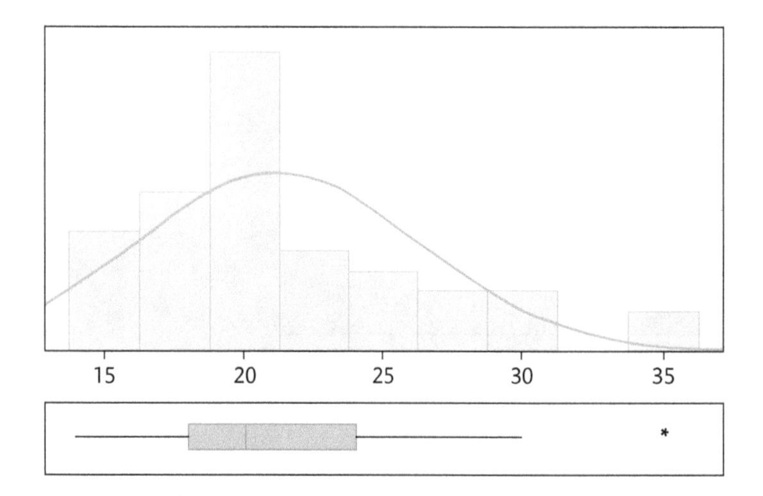

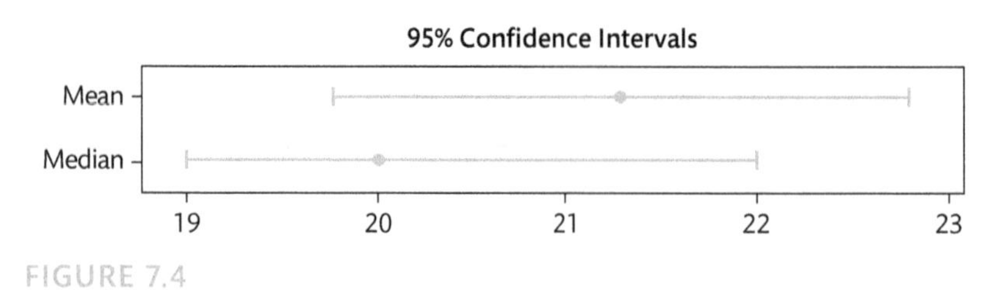

Anderson-Darling Normality Test	
A-Squared	1.44
P-Value <	0.005
Mean	21.283
StDev	5.093
Variance	25.941
Skewness	1.06249
Kurtosis	0.81745
N	46
Minimum	14.000
1st Quartile	18.000
Median	20.000
3rd Quartile	24.000
Maximum	35.000
95% Confidence Interval for Mean	
19.770	22.795
95% Confidence Interval for Median	
19.000	22.000
95% Confidence Interval for StDev	
4.224	6.415

FIGURE 7.4
Minitab graphical summary for the data in self-check 7.2.

SELF-CHECK 7.2

Look at the data set in Figure 7.4. In comparison with the data sets in Figures 7.1 and 7.2, comment on its distribution.

7.5 The analysis of continuously variable data: two groups

Let us now go on to consider how we can statistically analyse data in the light of whether its distribution is normal or non-normal. When comparing two sets of data, we use either Student's t test, or the Mann-Whitney U test. With three or more data sets, analysis is more complex, and will be considered in Section 7.6.

Student's t test

Student's t test is perhaps the best-known statistical test, and can only be used to compare two sets of data that are both normally distributed. The mathematics is relatively straightforward, and requires us to know the mean, the standard deviation (SD), and the sample size for each group. The latter two indices give the standard error (SE)—the SD divided by the square root of the sample size. These data are fed into an equation that gives a 't' value ...

Student's t test
a statistical test used to seek a difference between two sets of continuously variable data whose distributions are both normal

$$t = \frac{\text{mean of X} - \text{mean of Y}}{[(\text{SE of X})^2 + (\text{SE of Y})^2]^{1/2}}$$

By looking up the t value in statistical tables, the probability that any observed difference is significant can be determined. We also need to consider the degrees of freedom, derived from n–1, as in the Chi-squared test, but here n is related to the sample size and the number of analyses. Although all statistical programs can do this test, the t value can also be derived from the raw data with a hand-held calculator.

A worked example

Consider data on the white blood cell count from two different groups of patients as follows:

Group C: 6, 7, 8, 9, 7, 9, 10, 7, 9, 6, 10, 12, 7, 9

Group D: 9, 14, 17, 11, 8, 12, 9, 11, 10, 10, 13, 7, 9, 10

The unit for these data is 10^6 cells/mL. The first step is to define the distribution. This is done by finding the mean, standard deviation, median, and inter-quartile range. Although these can be obtained by using some statistical software, they can also be obtained using a standard hand-held calculator.

The mean value of Group C data is obtained by dividing the sum of the 14 data points (116) by 14, giving a value of 8.3 (to two significant figures).

Placing the 14 data points in rank order of 6, 6, 7, 7, 7, 7, 8, 9, 9, 9, 9, 10, 10, and 12 finds the median value to be 8.5 (that is, mid-way between the 7th and 8th point).

The lower quartile is 7; the upper quartile is 9. Hence the median and inter-quartile range is 8.5 (7–9). So we have a data set where the mean (8.3) is very close to the median (8.5), and the median is exactly equidistant between the IQR.

Similarly, the mean of Group D data is the sum (150) divided by 14, giving a value of 10.7. The median of the data is obtained from ranking: 7, 8, 9, 9, 9, 10, 10, 10, 11, 11, 12, 13, 14, and 17. The 7th and 8th values are both 10, so this is the median. The lower quartile is 9; the upper is 12. So the mean and median (10.7 and 10) are close, and although the median (10) is not perfectly in the middle of the inter-quartile range (9–12), it is acceptable.

TABLE 7.11 **Data on white blood cell count for Student's t test**

Data point of Group C	Difference from mean (8.3)	Square of difference from mean	Data point of Group D	Difference from mean (10.7)	Square of difference from mean
6	2.3	5.29	9	1.7	2.89
7	1.3	1.69	14	3.3	10.89
8	0.3	0.09	17	6.3	39.69
9	0.7	0.49	11	0.3	0.09
7	1.3	1.69	8	2.7	7.29
9	0.7	0.49	12	1.3	1.69
10	1.7	2.89	9	1.7	2.89
7	1.3	1.69	11	0.3	0.09
9	0.7	0.49	10	0.7	0.49
6	2.3	5.29	10	0.7	0.49
10	1.7	2.89	13	2.3	5.29
12	3.2	13.69	7	3.7	13.69
7	1.3	1.69	9	1.7	2.89
9	0.7	0.49	10	0.7	0.49
Sum of the squares		38.86			88.86
Mean sum of the squares		2.776			6.35
Standard deviation = square root of the mean sum of the squares		1.67			2.52

These values provide powerful evidence in favour of a normal distribution for both sets. But to be convinced we need to know the relationship between the mean and the SD.

To obtain the SD, we need to find the difference between the mean value and each of the individual data points (columns 2 and 5; Table 7.11). We then square each of these differences (columns 3 and 6, Table 7.11), and add them all up to obtain the sum of the squares. The sum of the squares for Group C (column 3) is 38.86. The mean sum of the squares is the sum of squares divided by the total number of data points = 38.86/14 = 2.776. The SD is the square root of this number, which is 1.67.

Similarly, the sum of the squares for Group D (column 6) is 88.86, and this divided by 14 gives the mean sum of the squares, which is 6.35. The SD is the square root of this, which is 2.52.

So the mean and SD for Group C is 8.3 (1.67), and for Group D these are 10.7 (2.52). Now we have this information we add it to the median and other data, and so summarize them with descriptive statistics (Table 7.12).

TABLE 7.12 **Descriptive statistics of the white cell count in Groups C and D**

	N	Mean	SD	Minimum	Q1	Median	Q3	Maximum
Group C	14	8.3	1.67	6	7	8.5	9.25	12
Group D	14	10.7	2.52	7	9	10	12.25	17

SD = standard deviation; Q1 = 25th (lower) quartile; Q3 = 75th (higher) quartile. Units: 10^6 cells/mL

In both groups, the SD is less than 25% of the mean and the mean and median are close to each other, and in each case the median is in the middle of the inter-quartile range, and the minimum and maximum. These observations lead us to be confident that both data sets have a normal distribution, and so a difference between the white blood cell counts can be determined using Student's t test.

The t value itself is obtained from the equation:

$$t = \frac{\text{mean of C} - \text{mean of D}}{[(\text{SE of C})^2 + (\text{SE of D})^2]^{1/2}}$$

Notice how the equation includes the SE (standard error)—the SD divided by the square root of the sample size (in the case of our example, the square root of 14, being 3.74). Hence, the SE for Group C is 1.67/3.74 = 0.446, and for Group D it is 2.52/3.74 = 0.674. We can now substitute these numbers into the equation, giving

$$t = \frac{10.7 - 8.3}{[(0.446)^2 + (0.674)^2]^{1/2}}$$

$$t = \frac{2.4}{[0.199 + 0.454]^{1/2}}$$

$$t = \frac{2.4}{0.81} = 2.96$$

The final step is to consult a statistical table of t values and their respective p values. These tables are readily available on-line and in major statistical textbooks. However, as discussed, all types of statistical software will automatically give the p value. If $p<0.05$, we can be over 95% confident that the difference is genuine.

To be able to look up our p value in a table of statistical values, we need to know the degrees of freedom, and as we have two variables then we have one degree of freedom. In the case of the analysis of the two groups, there are 14 data points in each group, so the total sample size is 28. However, we subtract one from this value for each group (C or D), and when consulting the tables, look for a difference with 26 DFs. From the table we obtain $p=0.008$, which means we can be 99.2% confident that the difference in the white blood count between the two groups is genuine.

An increased white blood cell count is commonly found in infections—the reference range is generally between 4 and 10×10^6 cells/mL. Accordingly, these data may be from patients with different infections, or perhaps taking different antibiotic drugs.

Figure 7.5 shows a Minitab printout of this analysis. The first part—Method—states that we define μ_1 as the mean of group C, and μ_2 as the mean of group D. The question is therefore whether there a difference between the two values of μ. Note that the software assumes that the variances (linked to the standard deviation) are equal. If the variances are unequal (perhaps

Method

μ_1: mean of Group C

μ_2: mean of Group D

Difference: $\mu_1 - \mu_2$

Equal variances are assumed for this analysis.

Descriptive Statistics

Sample	N	Mean	StDev	SE Mean
Group C	14	8.29	1.73	0.46
Group D	14	10.71	2.61	0.70

Estimation for Difference

Difference	Pooled StDev	95% CI for Difference
–2.429	2.216	(–4.150, –.707)

Test

Null hypothesis $H_0: \mu_1 - \mu_2 = 0$

Alternative hypothesis $H_1: \mu_1 - \mu_2 \neq 0$

T-Value	DF	P-Value
–2.90	26	0.008

FIGURE 7.5
Minitab printout of a Two-Sample T-Test and CI: Group C, Group D

with one SD being twice as large as the other) then an adjustment will be made. Minitab then summarizes the key features of the t test as descriptive statistics, which it uses in the next part to define an estimation of the difference. This is given as the difference between the means of C and D (that is, $\mu_1 - \mu_2$), and the 95% confidence interval (CI) for this difference. The pooled StDev is the mean of the two SDs. Note that two 95% CIs are both negative, as is the difference. If we placed D as the head of the analysis, we would get exactly the same numbers, but they would be positive, as D>C. Had one of the 95% CIs been negative, and the other positive, it would indicate no significant difference between C and D at the 0.05 level. The printout concludes with the test itself, which first states the null and alternative hypotheses, and then the t value, the degrees of freedom (that is 14–1 + 14–1) and the p value.

Other statistical packages (such as SPSS) will give different printouts, but the salient features will be the same as for Minitab printouts.

The Mann-Whitney U test

Mann-Whitney test
a statistical test used to seek a difference between two sets of continuously variable data where one or both distributions are non-normal

In contrast to Student's t test, the **Mann-Whitney test** can only be used when one or both sets of data have a non-normal distribution. The mathematics of this test are completely different from those of the Student's t test, and rely on the rank order: placing them in order, from smallest to largest. In Section 7.1 we were re-introduced to this system in the definition of the 1st and 3rd quartiles, and the median value of a data set.

Consider the following data sets, which may be the number of colonies of bacteria that were cultured from the urine of patients with a urinary tract infection:

Group E: 12, 34, 78, 56, 31, 59, 150, 42, 62, 46, 21, 87, 93, 43, 170

Group F: 86, 55, 42, 79, 156, 99, 70, 125, 225, 72, 130, 106, 68, 55, 87

Before we consider a possible difference between these sets we must determine their distribution. As in the other examples, we can obtain the mean, median, and inter-quartile ranges of the two sets of data. The median number of colonies of Group E is 56, but the mean is 66 (rounded up from 65.6), which is almost 17% higher. Furthermore, the median is not in the middle of the inter-quartile range of 34–87, as 56 is much closer to 34 than it is to 87.

Similarly, the mean of Group F (97) is 13% higher than the median (86), and the latter is similarly not in the middle of Group F's inter-quartile range of 68–125, being much closer to 68 than to 125. These data very convincingly support the view that the two data sets have a non-normal distribution. Accordingly, even though we don't yet have the SD, it is very likely that the t test is inappropriate.

If we now consider the SDs for Groups E and F, we find them to be 45 and 47 respectively; we can use these values to summarize the data with descriptive statistics (Table 7.13).

In both cases, the SD is around half of the mean. This further data unequivocally supports the view that both sets have a non-normal distribution (Table 7.10) so that the Mann–Whitney test must be applied. The mathematical basis of the Mann–Whitney U test is of ranking the data: placing them in order from smallest to largest. The individual data points in the two sets of data above are presented in random order.

We need to arrange the data sets so that the lowest number in either set takes rank 1, the second lowest = rank 2, etc. The largest number in both sets takes rank 30. The full set of rankings is shown in Table 7.14. Note that any duplicates (that is, the same number) are scored 0.5, and that the overall rank continues in order. These duplicates are called 'ties'.

The sum of the ranks when combined, which is column 3 (1 + 2 + 3 + 4 ... to 29 and 30), is 465. The null hypothesis states that there is no overall difference in the two data sets. If so, then there should be no difference in the sums of the 'Group E' ranks (the bottom of column 4) and the sum of the 'Group F' ranks (the bottom of column 5), which would be 232 for one set and 233 for the other (that is, roughly half of the overall total of 465). However, the alternative hypothesis states that there is a difference in the sum of the ranks between Groups E and F, that is, that either 182 or 283 is different from 232/233.

Clearly there is a difference, but is this difference statistically significant? The validity of the alternative hypothesis that there is a difference between Groups E and F can only be determined by looking up the relevant data in a statistical table. By convention, we take the smaller of the ranks as the test statistic (W), which is therefore 182. As with the t test, we can go to textbook or on-line tables to find out the probability that the W statistic is statistically significant.

TABLE 7.13 **Descriptive statistics for Groups E and F**

Variable	N	Mean	SD	Minimum	Q1	Median	Q3	Maximum
Group E	15	66	45	12	34	56	87	170
Group F	15	97	47	42	68	86	125	225

SD = standard deviation: Q1 = 25th (lower) quartile; Q3 = 75th (higher) quartile. Data are the number of bacterial colonies (two significant figures)

TABLE 7.14 **Data points for number of colonies in rank order**

Data point	Group	Rank	Rank by Group E	Rank by Group F
12	E	1	1	
21	E	2	2	
31	E	3	3	
34	E	4	4	
42	E	5 =	5.5	
42	F	5 =		5.5
43	E	7	7	
46	E	8	8	
55	F	9 =		9.5
55	F	9 =		9.5
56	E	11	11	
59	E	12	12	
62	E	13	13	
68	F	14		14
70	F	15		15
72	F	16		16
78	E	17	17	
79	F	18		18
86	F	19		19
87	E	20 =	20.5	
87	F	20 =		20.5
93	E	22	22	
99	F	23		23
106	F	24		24
125	F	25		25
130	F	26		26
150	E	27	27	
156	F	28		28
170	E	29	29	
225	F	30		30
Sum of the ranks		465	182	283

Method

η_1: median of Group E

η_2: median of Group F

Difference: $\eta_1 - \eta_2$

Descriptive Statistics

Sample	N	Median
Group E	15	56
Group F	15	86

Estimation for Difference

Difference	CI for Difference	Achieved Confidence
–30	(–58, –6)	95.35%

Test

Null hypothesis H_0: $\eta_1 - \eta_2 = 0$

Alternative hypothesis H_1: $\eta_1 - \eta_2 \neq 0$

Method	W-Value	P-Value
Not adjusted for ties	182.00	0.038
Adjusted for ties	182.00	0.038

FIGURE 7.6

A Minitab printout of the Mann-Whitney analysis of Groups E and F

This W value, alongside the sample size (15+15) gives us a probability, with a value p=0.038, that the difference between the two groups is genuine. In fact, we are 96.2% confident that this difference is genuine. We therefore reject the null hypothesis, but accept the alternative hypothesis that there is a difference between the sets.

Figure 7.6 shows a Minitab printout of the Mann-Whitney analysis of Groups E and F. As with the t printout (Figure 7.5), the method states that the test looks at a difference between medians, and then the descriptive statistics, which shows only the samples' size and median (represented by the Greek letter eta–η). An estimate of the difference shows the median of E minus the median of F (–30), and the 95% CI for that difference, which is –58 to –6. As with the t test analysis, as both these are negative, the probability is significant at the 0.05 level. Indeed, the printout states the confidence achieved is 95.35%. By subtracting this from 100, we get 4.65%, less than the crucial 5% (0.05). The final part is the test itself, with the null (that there is no difference between the medians) and alternative (that there is a difference) hypotheses, which gives the W value and the p value.

7.6 The analysis of continuously variable data: three or more groups

The Student's t test and Mann-Whitney U test can only be used to compare two groups. The situtation with three or more groups is more complicated, and depends on whether the three groups are independent of each other.

In clinical research, an example of three groups that are independent is patients with different diseases, such as cancer, heart disease, and skin disease. Another example is the ABO blood group system, with four groups, these being A, B, AB, and O. In both examples there is no natural or logical order or link (such as a progression) between the groups—you can't say that cancer is more important than heart disease, or that skin disease is a consequence of heart disease. Apart from frequency in a population, there is no natural order in the four ABO blood groups. Differences in data of this type (such as the age of the subjects concerned, or levels of particular substance in the blood) would be sought by **analysis of variance** (ANOVA, to be used where all the sets of data have a normal distribution), or by the **Kruskal-Wallis test** (where at least one set of data has a non-normal distribution).

Analysis of variance
a statistical test used to seek a difference between three or more sets of continuously variable data whose distributions are all normal

Kruskal-Wallis test
a statistical test used to seek a difference between three or more sets of continuously variable data where at least one set has a non-normal distribution

There are many situations where groups are not independent from each other, but are linked by an external factor. A good example of this is marital status—everybody starts life being single, then many get married. Eventally, married people will become single again because of divorce or the death of a spouse. This sequence is linked in time, and may be reversible (the divorced remarrying) or irreversible (you can't be divorced without first being married). Another example is of body mass index (BMI). Once more, everybody starts with a normal BMI, some then become overweight. This may progress to obesity and then morbid obesity. The link is therefore by increased grouping of BMI, but there may also be a link with time, although an individual may also revert back from a higher BMI group to a lower BMI group. A third example is of pain—that being absent, mild, moderate, severe and excruciating; a fourth is a laboratory experiment where the effects of increasing and specific doses of a drug are as being tested in tissue culture on a cell line. In each of these examples, the groups are not independent, but are arranged in an exact order, or in a linear trend or rank, generally thought of being from left to right. Accordingly, use of ANOVA or Kruskal-Wallis for this type of data would be inappropriate—the correct test is called **ordered groups**, and is applicable to data that has a normal distribution. Should the data have a non-normal distribution, it will need to be log transformed.

Ordered groups
a test where categorical (grouped) data can be ordered in a rank according to an external factor

Much basic laboratory research will be analysed by ordered groups as, for example, the effect of increasing doses of a drug on an animal or a cell line is clearly in an order (small dose—moderate dose—large dose). However, if the research is on the effect of three different drugs, then this is not ordered and an ANOVA would be appropriate.

Analysis of variance (ANOVA)

As the name implies, the ANOVA test relies on the variance of the data sets, where variance is related to the SD. Consider data of the levels of serum bilirubin in three small groups of subjects, G, H, and I, the descriptive statistics for which are shown in Table 7.15. Even without a histogram graphic and other information, we can be very confident that each of the three sets of data has a normal distribution.

SELF-CHECK 7.3

Look at the descriptive statistics in Table 7.15. How can you be sure each set has a normal distribution?

The mathematics of the ANOVA test is markedly more complicated than that of the Student's t test, and is best performed with appropriate statistical software. An alternative is a series of individual t tests, between Groups G and H, then between Groups G and I, and then between Groups H and I. However, there is a problem with this. So what form does this problem take?

TABLE 7.15 **Descriptive statistics for bilirubin**

Variable	N	Mean	SD	Minimum	Q1	Median	Q3	Maximum
Group G	19	17.2	1.5	14	16	17	18	20
Group H	19	19.8	2.7	17	19	20	23	30
Group I	19	18.7	1.6	16	17	19	20	25

SD = standard deviation; Q1 = 25th (lower) quartile; Q3 = 75th (higher) quartile. Units are nmol/L

The Bonferroni correction

The t test is for testing a null hypothesis by comparing two groups of data with a normal distribution, and we take a level of probability $p<0.05$ to minimize the risks of false positives. But with three groups there are three null hypotheses being tested: the null hypothesis between Groups G and H, that between Groups H and I, and a third hypothesis between Groups G and I. As a result, there is an increased likelihood that if there is a small difference between one of the three sets of data, it will be due to chance, and not because of a genuine difference. One way of getting around this is to demand a higher level of power, and so a lower level of probability.

A **Bonferroni correction** calls for the criterion of $p<0.05$ to be divided by the number of analyses. Since in our example we have three analyses (G with H, H with I, and G with I) then the required level of significance is corrected to $0.05/3 = 0.0167$. So if the difference between (say) Groups G and H according to a t test gives $p=0.032$, this is not significant because it needs to be <0.0167 to be significant. However, if the difference between (say) Groups H and I according to a t test is $p=0.012$, then this difference retains its significance because 0.012 is <0.0167.

Bonferroni correction
a statistical test used when there are multiple comparisons, which adjusts the level of probability

The whole point of ANOVA is that it automatically adjusts the overall data set (Groups G to I) for multiple comparisons so that a formal Bonferroni correction is not needed. Returning to our three groups, Figure 7.7 shows the Minitab printout of the ANOVA, which seeks to determine

One-way ANOVA: All data versus Group

```
Part (a) : Mathematics
Source    DF    SS      MS     F     P
Group      2    53.4    26.7   7.6   0.001
Error     54    189.9   3.5
Total     56    243.3

Part (b) : Graphic
Individual 95% CIs For Mean Based on pooled StDev
Group    N    Mean    StDev  -------+----------+----------+----------+------
G        19   17.2    1.6    (------*------)
H        19   19.5    2.4                                 (------*------)
I        19   18.4    1.5                      (-------*------)
                             -------+----------+----------+----------+------
                                  16.8       18.0       19.2       20.4

See text for details
(ss = sum of the squares, ms = mean squares)
```

FIGURE 7.7
Minitab printout of the analysis of variance of bilirubin data.

differences between the groups. Figure 7.7(a) (the upper part) shows the mathematics of the ANOVA, with the crucial probability (P) on the right, which is p=0.001. This indicates that there is a large difference between one or more of the sets of data.

The other information is linked to the mathematics of the Student's t test, where DF is the degrees of freedom, SS is the sum of the squares, and MS is the mean squares. Note that immediately below DF is the number 2, indicating 2 degrees of freedom (that is, the number of groups minus 1). Below that is the number 54, which is (19–1) × 3. The overall degrees of freedom (56) comes from adding together the individual degrees of freedom.

The value F is the equivalent of the t value in the Student's t test, and is obtained by dividing the larger mean squares value of 26.7 by the smaller mean squares value of 3.5. As with the Student's t value, the larger the F value, the greater is the probability that a significant difference is present.

Figure 7.7(b) is a graphical representation of the data set, which shows the sample size (n = 19), the mean, and the SD of each of the three groups. On the right of this data the mean value (*) and the 95% confidence intervals (CIs) (------) of that mean value are indicated.

The key point in interpreting this graphic is that the 'brackets' of the Group G data do not overlap with the brackets of the Group H data. In this setting, therefore, individuals in Group H have levels of bilirubin that are significantly higher than those in Group G. But notice also that the brackets of Group I do overlap with those of Groups G and H. Therefore, there is no difference in bilirubin between Groups G and I, nor between Groups H and I.

As for interpretation, Group G are reportedly healthy controls, and Groups H and I are from patients with a high daily intake of alcohol (where levels are highest) and with mild primary biliary cholangitis (where levels are intermediate between the other two groups) respectively.

The bilirubin data in Groups G to I represent a relatively simple analysis, as the difference lies between groups G and H, as the two groups of (----*----) brackets fail to overlap. However, in other data sets, any potential difference may not be as obvious. Consider the four data sets of total iron binding capacity from different groups of patients, J, K, L, and M, whose descriptive statistics are shown in Table 7.16.

These data are analysed by ANOVA in Figure 7.8. As we have p<0.001, we know that there is a very high probability that there is a difference between one or more of the groups. But exactly where are these differences? Looking at the graphical representation on the right-hand side of the figure, the 'brackets' between Groups K and L clearly fail to overlap, as do the brackets for Groups L and M. It is therefore unequivocal that these sets are significantly different.

TABLE 7.16 **Descriptive statistics for total iron-binding capacity**

Variable	N	Mean	SD	Minimum	Q1	Median	Q3	Maximum
Group J	20	56	3.5	51	54	55	57	66
Group K	20	52	4.7	42	48	51	54	66
Group L	20	60	7.1	49	55	59	64	80
Group M	20	53	3.9	47	50	53	56	63

SD = standard deviation; Q1 = 25th (lower) quartile; Q3 = 75th (higher) quartile. Units are μmol/L

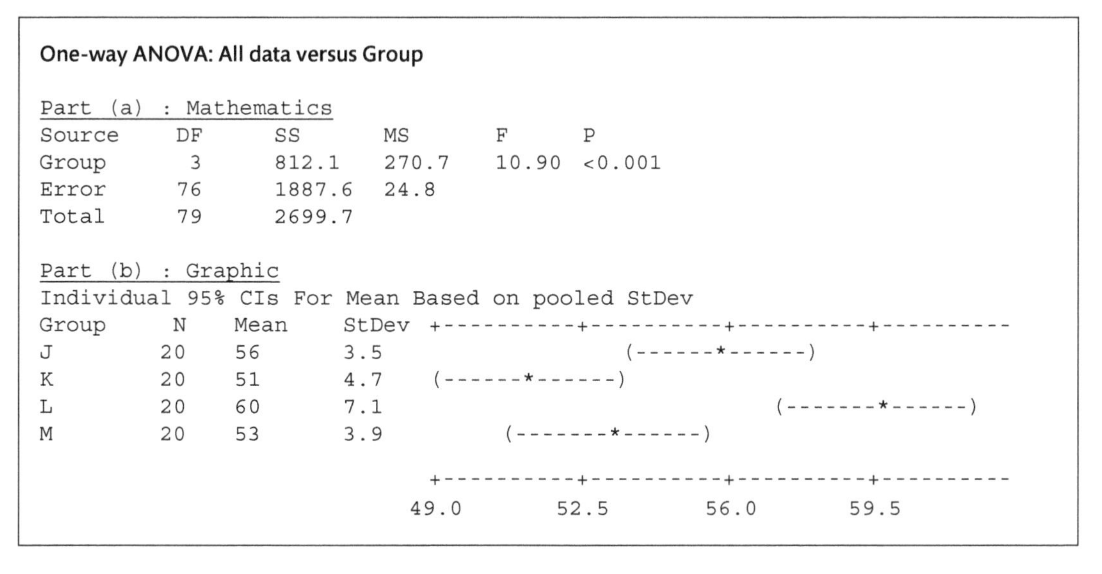

One-way ANOVA: All data versus Group

```
Part (a) : Mathematics
Source     DF      SS       MS       F      P
Group       3      812.1    270.7    10.90  <0.001
Error      76      1887.6   24.8
Total      79      2699.7

Part (b) : Graphic
Individual 95% CIs For Mean Based on pooled StDev
Group       N    Mean     StDev  +---------+---------+---------+---------
J          20    56       3.5                      (------*------)
K          20    51       4.7    (------*------)
L          20    60       7.1                                 (-------*------)
M          20    53       3.9         (-------*------)

                                 +---------+---------+---------+---------
                               49.0      52.5      56.0      59.5
```

FIGURE 7.8

Minitab printout of the analysis of variance of total iron-binding capacity data

However, close attention to the brackets for Groups J and K indicate that they may overlap, but the difference is too small for it to be determined by eye. For formal assurance of a difference, we need another test, one which follows the initial test (that is, ANOVA). These secondary tests are called 'post-hoc', and one of the most used is that developed by John Tukey.

Tukey's post-hoc test

Tukey's post-hoc test is a form of mathematics that seeks differences between multiple groups within an ANOVA. As with ANOVA, the test is available within all good statistical packages. Once we have $p<0.05$ from an ANOVA, the Tukey test can proceed. Table 7.17 shows a typical Tukey's test printout as applied to the data in the ANOVA of Groups J to M. Note the groups are not in alphabetic order (J-K-L-M) but as (L-J-M-K).

Tukey's post-hoc test a statistical test used to determine differences in multiple comparisons

The software has summarized the data by denoting the presence or absence of a difference with the sharing (or not) of a letter of the alphabet. Groups J and L both share the letter 'A',

TABLE 7.17 **Tukey's test for total iron-binding capacity data**

Grouping Information Using Tukey Method			
Group	**N**	**Mean**	**Grouping**
L	20	60	A
J	20	56	A B
M	20	53	B C
K	20	51	C

Groups that do not share a letter (such as L and M) are significantly different (see text for an explanation).

and so are not significantly different, as is clear from the graphic in Figure 7.8 (where the 95% confidence interval brackets overlap). Similarly, according to the graphic, Groups L and M are different, as their 95% confidence interval brackets are far apart, as is confirmed by the denotation that they do not share any of the letters 'A', 'B', or 'C'. However, in answering the question of a difference between Groups J and K, they do not share a letter in the Tukey test, so are indeed significantly different.

This data may be from patients with different forms of anaemia, perhaps some with liver disease, as this organ is important in red cell physiology since it stores iron and synthesizes molecules important in iron carriage and storage (transferrin and ferritin).

The Kruskal-Wallis test

Just as ANOVA is the test for three or more data sets, each of which has a normal distribution, the Kruskal-Wallis test is for three or more data sets, any one of which has a non-normal distribution. Inasmuch as an ANOVA is effectively a series of t tests with a Bonferroni correction, then the Kruskal-Wallis test is a series of Mann-Whitney U tests with a similar adjustment for multiple comparisons.

Consider the following data on the erythrocyte sedimentation rate (ESR) from three groups of patients, whose descriptive statistics are shown in Table 7.18. Applying the criteria regarding distribution in Table 7.10, Group N has a non-normal distribution because the SD is 69% of the mean; the mean and median differ by 17%; and although the median (27) is moderately within the middle of the inter-quartile range (16–43), it is far from the middle of the full range (5–84). Furthermore, the Anderson-Darling test of normality gives a probability of p=0.035 that the distribution is non-normal.

It is also very likely that Group O data has a non-normal distribution as the SD is 48% of the mean, and the mean and median differ by 14%. The median (43) is much closer to the lower quartile (Q1, 31) than it is to the third quartile (Q3, 62), and is also much closer to the minimum (23) than it is to the maximum (110). The Anderson-Darling test statistic is p=0.041.

However, examining the Group P data, the SD is 35% of the mean, and the mean and median differ by only 10%. The median is much closer to the middle of the IQR and the full range, and the Anderson-Darling test statistic is p=0.211. Therefore, we would describe Group P data as having a normal distribution. In spite of this, because data from Groups N and O both have a non-normal distribution, the entire data set must be analysed by non-normal methods: that is, the Kruskal-Wallis test (Table 7.19).

The mathematics of the determination of a p value in a Kruskal-Wallis analysis is similar to that of the Mann-Whitney test in that the entire data set is merged and then ranked. Note from Table 7.19 (modified from a Minitab printout) that the average rank of Group N is a little

TABLE 7.18 **Descriptive statistics for Groups N, O, and P**

Variable	N	Mean	SD	Minimum	Q1	Median	Q3	Maximum
Group N	15	32	22	5	16	27	43	84
Group O	16	49	23	23	31	43	62	110
Group P	17	40	14	22	30	36	48	72

SD = standard deviation; Q1 = 25th (lower) quartile; Q3 = 75th (higher) quartile

TABLE 7.19 **The Kruskal-Wallis test for the ESR data**

Kruskal-Wallis Test on All data				
Group	**N**	**Median**	**Average Rank**	**Z**
N	15	27	17.3	−2.40
O	16	43	30.1	1.95
P	17	36	25.6	0.41
Overall	48		24.5	
H = 6.60	P = 0.037			

over 17, while those of Groups O and P are 30.1 and 25.6 respectively. The average rank of the entire group of 48 items of data is 24.5 (that is, mid-way between 24 and 25).

The next step is to compare the average rank from each group (17.3, 30.1, and 25.6) with the average rank of the overall set (24.5). The statistical software (in this case, Minitab) performs the relevant calculation, and produces the 'Z' statistic, which is an indication of the degree to which the average rank of each group differs from the overall average rank. The Z values in Table 7.19 show that the average ranks of Groups N and O are much larger (−2.40 and 1.95 respectively) than that of Group P (Z = 0.41). The 'H' statistic (the Kruskal-Wallis equivalent of the F value of an ANOVA) is an adjusted sum of the Z values, and the p value (0.037) means that there is a 96.3% chance that a genuine statistical difference is present.

So, as with an ANOVA, we have a difference somewhere, but where is it? A clue is present as the average ranks of Groups N and O are furthest apart (17.3 and 30.1 respectively, with the most extreme Z values). One way of determining whether this difference is indeed significant is to carry out a series of individual Mann-Whitney analyses (N versus O, N versus P, and O versus P); but this is poor practice, for the same reason that a series of individual t tests is inappropriate. Fortunately, several software packages have the capacity to do this with a post-hoc test, and one package finds a significant difference between Groups N and P with a probability of p=0.011.

Transformation

An alternative to those statistical packages that do not perform a post-hoc test is to do an ANOVA to look for differences between the three groups. However, this test is for data whose distribution is normal, and we have already established that Groups N and O have a non-normal distribution. One way around this is to convert the data to a normal distribution, a process called transformation. One of the most common methods for transforming data is by taking the logarithm of each value, and then doing an ANOVA (Figure 7.9).

The overall p value in this ANOVA is 0.014, telling us that there is a 98.6% chance that a genuine difference exists. Notably, this p value is smaller than that from the Kruskal-Wallis analysis of p=0.037, demonstrating the power of the ANOVA. The 'proof' that taking the log of the raw data of Groups N to P converts it to a normal distribution is that the SD values are now between 8% and 18% of the mean values—much smaller than that of the raw data.

One-way ANOVA: Log of Data versus Group

```
Part (a) : Mathematics
Source     DF       SS        MS       F      P
Group       2     2.452    1.226    4.72   0.014
Error      45    11.696    0.260
Total      47    14.148

Part (b) : Graphic
Individual 95% CIs For Mean Based on pooled StDev
Group      N     Mean      StDev  -------+---------+---------+---------+------
N         15     3.9       0.72   (--------*--------)
O         16     4.5       0.42                     (--------*--------)
P         17     4.3       0.33               (--------*--------)
                                  -------+---------+---------+---------+------
                                       3.90      4.20      4.50      4.80
```

FIGURE 7.9

Minitab printout of the analysis of variance of the log of the ESR data.

Further examination of the graphical representation reveals that the 95% confidence intervals (as brackets) of Groups N and O fail to overlap, and so are statistically different. Some may prefer to confirm this with Tukey's post-hoc test, but the difference in this case is clear and unequivocal.

This data may be from patients with different forms of inflammatory disease, cancer, or anaemia, all conditions known to be associated with an abnormal ESR. Alternatively, the data may represent ESR values from patients with the same disease, but treated with different drugs, or perhaps at different stages of the disease.

Ordered groups

This analysis is to be used where groups of data have natural linear order, perhaps a progression, that are linked by a factor such as time, the age of the subjects, disease severity or in the effects of an increasing dose of a drug. The process of analysing data of this nature is best served with examples.

Example 1: clinical research

Suppose a researcher wishes to test the hypothesis that a certain molecule in the blood (Substance XYZ) rises during pregnancy. Ideally, the researcher would test this in a group of women, each of whom would be followed over a long period of time before and as her pregnancy develops. Indeed, if data of this nature were obtained, it would be analysed in a completely different manner–that of serial repeated measures, as explained in Section 7.7 that follows. However, in this example the researcher has only a limited period of time, so that he/she can only sample women on a single occasion.

The key mathematical point with this analysis is that the hypothesis predicts that levels of XYZ rise from left to right across these stages of pregnancy with a linear trend. The increase may not be perfectly symmetrical, increasing in equal increments at each stage, but it is the overall picture that we consider. This is illustrated in Figures 7.10 (a) and (b), showing the mean and standard deviation (SD) of two sets of XYZ data. The horizontal line is the mean of the entire data set. For each stage, the mean values indeed rise from left to right, and a line of best fit has been drawn linking each set.

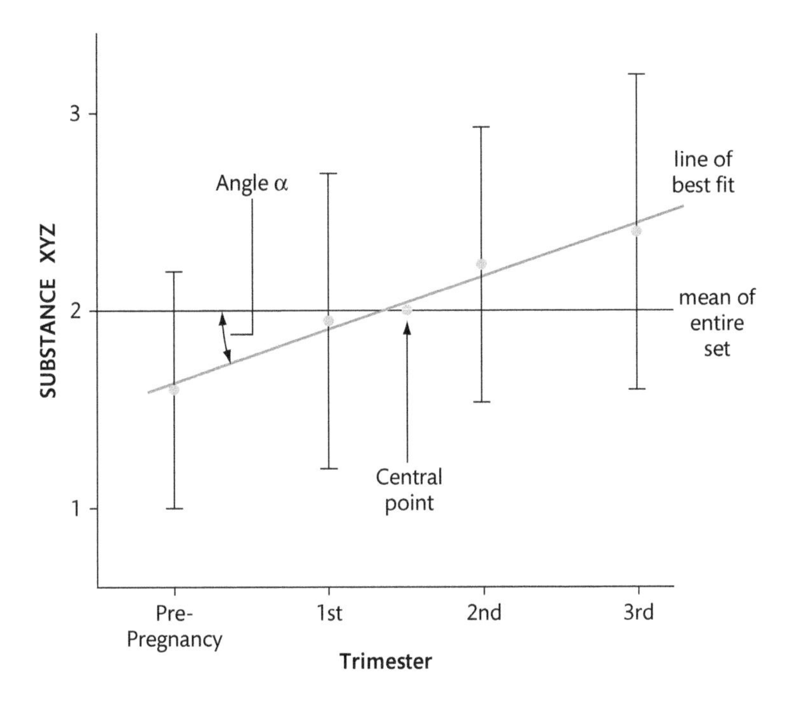

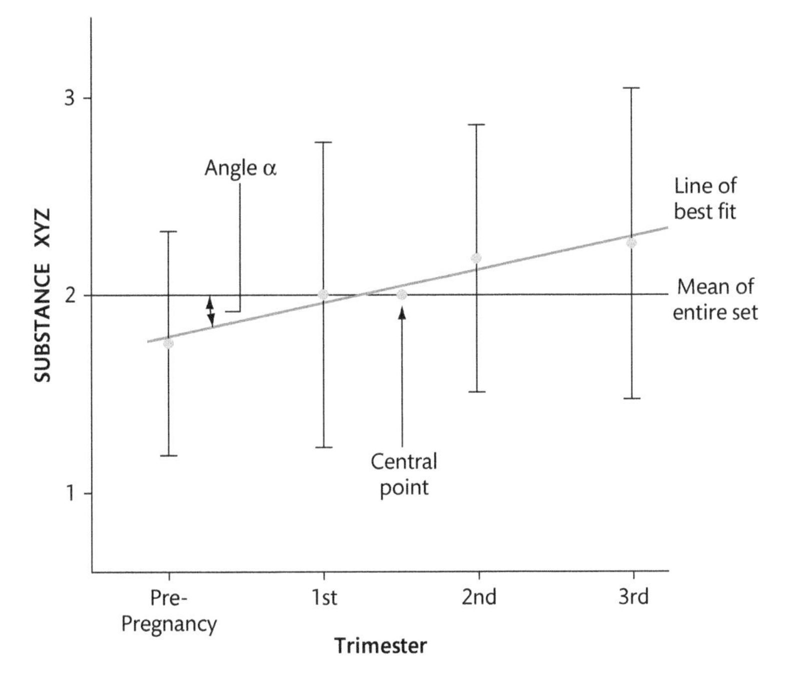

FIGURE 7.10

Mean and standard deviation data on Substance XYZ from women at four stages of pregnancy.

The mathematics of whether this increase is statistically significant makes a number of assumptions, such as that the 'distance' between each of the four groupings is equal, and that as XYZ rises there is a point in the data set that the line of best fit linking the groups crosses the line of the mean of the entire data set, which may be seen as the central point of the whole group. The mathematics then places weight on the mean/SD of each group: that to the far left is weighted −3, mid left −1, mid right +1 and far right +3, so that each point 'rotates' around the central point, and the sum is (in theory) zero. Should there be six groups, then the weights −5, −3, −1, +1, +3 and +5 are applied. Should there be an odd number of groups, the middle group is counted as having a zero weight, and so is ignored, so that only those points above and below it are considered as if there is an even number of groups. Of interest is the angle (α) made between the line of best fit and the line of the mean XYZ of the entire group, which we will return to in due course.

Let us now look at a worked example, using the two sets of data from Figure 7.10, as presented in Table 7.20. In order to determine if the linear trend is significant, we need to generate a statistic called *L*, and its standard error, that is, the se(*L*).

TABLE 7.20 **Data in an ordered groups analysis**

		Not pregnant	1st trimester	2nd trimester	3rd trimester
Set A	Mean	1.60	1.94	2.22	2.40
	Standard deviation	0.56	0.75	0.66	0.79
Set B	Mean	1.74	2.00	2.17	2.28
	Standard deviation	0.57	0.77	0.68	0.79
Both sets	Sample size	10	9	8	9

The *L* value is obtained from the sum of the means on the left-hand side (which will be weighted with −3 or −1) of the central point and those on the right-hand side (which will be weighted with +1 or +3). Thus ...

$$L = (-3 \times 1.60) + (-1 \times 1.94) + (1 \times 2.22) + (+3 \times 2.41)$$
$$= -4.8 + (-1.94) + 2.22 + 7.23 = +2.68$$

So far so good. Next we need se(*L*), which is a little more complicated. First, we need a statistic from an ANOVA, which tells us of the overall variance of the entire data set, exactly as in a standard ANOVA. The data we need is the 'within groups mean squares'. In this situation this number is 0.4811. We put this into an equation that includes the sample size of each group, that is, *n*, and then take square roots. Thus ...

$$se(L) = (\text{within groups mean squares})^{1/2} \times (-3^2/n_1 + -1^2/n_2 + 1^2/n_3 + 3^2/n_4)^{1/2}$$

where n_1, n_2, n_3, and n_4 are the sample sizes for the four groups from left to right.

So putting these data together we have ...

$$se(L) = (0.4811)^{1/2} \times (9/10 + 1/9 + 1/8 + 9/9)^{1/2}$$
$$= (0.694) \times (2.136)^{1/2} = 0.694 \times 1.462 = 1.014$$

We then divide *L* by the se(*L*), that being 2.68/1.014 = 2.64, which is a t value (as in Student's t test). Since we have a sample size of 36, then we have 35 degrees of freedom. So by consulting statistical tables, or by going on-line to one of the free calculators, we find p=0.012. So there is a strong probability of 98.8% that the linear trend of these four ordered groups is genuine.

A further proof of the power of this method is that if an incorrect standard ANOVA were to be performed, p=0.091 would be obtained. This underlines the importance of knowing the correct test to apply to a particular set of data.

We can also consider a second set of data from four ordered groups: Figure 7.10(b) and the lowest set of data in Table 7.20. By applying the same arithmetic as above ...

$$L = (-3 \times 1.74) + (-1 \times 2.00) + (1 \times 2.17) + (+3 \times 2.28) = +1.79$$

The within groups mean squares from a standard ANOVA is 0.5. Since the sample size of the number of women per group is the same as the data set (A), we have ...

$$\begin{aligned} se(L) &= (0.5)^{1/2} \times (9/10 + 1/9 + 1/8 + 9/9)^{1/2} \\ &= (0.707) \times (2.136)^{1/2} = 0.707 \times 1.461 = 1.03 \end{aligned}$$

And so $L/se(L)$ becomes 1.79/1.03 and so a t value of 1.738. Once more, consulting tables or an on-line calculator, then with 35 degree of freedom we obtain p=0.091. Thus the likelihood that this linear trend is genuine is only 90.8%. A standard (but incorrectly applied) ANOVA for this data gives p=0.380, again underlining the need to use the correct test.

Example 2: basic laboratory research

A researcher wishes to test the hypothesis that increasing doses of a drug have an effect on a cell line that she is growing in tissue culture. The cells are being grown in a 12-well plastic tray, and all cells in each well are growing well and are free of bacterial infection. She adds a drug at increasing concentrations (1 mg/mL, 2mg/mL, 5 mg/mL and 10 mg/mL) to each of three wells. After a set period of time, the cells, or their tissue culture supernatant, are harvested, and an analyte is measured. The nature of this analyte is not important–it may be a cytokine or other cell product shed into the supernatant, a nuclear factor within the cell, or the amount of mRNA in the cell cytoplasm. The important aspect is that the analyte can be accurately measured. The results are shown in Table 7.21, descriptive statistics in Table 7.22.

The raw data and the descriptive statistics all indicate a positive effect of the dose of the drug, but 'indicate' is not enough, we must test the hypothesis formally. A formal test of distribution would not be necessary as the data clearly (ratio of SD to mean is very small) has a normal distribution. As the doses of the drug clearly increase in discrete intervals, then we use ordered groups, as in the pregnancy example, not an ANOVA (and certainly not a series of t tests).

TABLE 7.21 **Tissue culture raw data**

Dose	Well 1	Well 2	Well 3
1 mg/mL	52.3	50.1	47.6
2 mg/mL	69.5	55.8	72.3
5 mg/mL	71.2	66.8	75.0
10 mg/mL	74.5	78.0	77.3

Units: arbitrary

TABLE 7.22 **Descriptive statistics for the tissue culture experiment**

Variable	N	Mean	SD	Minimum	Q1	Median	Q3	Maximum
1 mg/mL	3	50.0	2.3	47.6	47.6	50.1	52.3	52.3
2 mg/mL	3	65.9	8.8	55.8	55.8	69.5	72.3	73.4
5 mg/mL	3	71.0	4.1	66.8	66.8	71.2	75.0	75.0
10 mg/mL	3	76.6	1.85	74.5	74.5	77.3	78.0	78.0

SD = standard deviation; Q1 = 25th (lower) quartile; Q3 = 75th (higher) quartile. Units: arbitrary.

We first need the *L* statistic, which is the weighted sum of (–3 times) the 1mg/mL mean plus (–1 times) the 2 mg/mL mean compared to the weighted sum of (+1 times) the 5 mg/mL mean plus (+3 times) the 10 mg/mL mean, so that ...

$$L = [(-3 \times 50) + (-1 \times 65.9)] - [(1 \times 71) + (3 \times 76.6)] = (-215.9) + (300.8) = 84.9.$$

Next, we take the standard error of *L*, that being (se)*L*, for which we need the within groups mean squares from a standard ANOVA, the Minitab printout being shown in Figure 7.11. This translates as the AdjMS at the error level, that being 25.94. This number goes into an equation with the sample size of each group:

$$se(L) = (\text{within groups mean squares})^{1/2} \times (-3^2/n_1 + -1^2/n_2 + 1^2/n_3 + 3^2/n_4)^{1/2}$$
$$se(L) = (25.94)^{1/2} \times (-3^2/3 + -1^2/3 + 1^2/3 + 3^2/3)^{1/2} = 5.093 \times 2.582 = 13.15$$

Dividing *L* (84.9) by the (se)*L* (13.15) gives a t value of 6.46. Consulting tables give us p=0.00005 from 11 degrees of freedom. Therefore, there is considerable evidence of a linear trend in the effect of the drug. We can also determine differences between the doses with Tukey's test, as shown in Figure 7.12. This tells us that doses 2, 5, and 10 mg/mL are all more effective than 1 mg/mL, but there is no statistical difference between the three high doses. These data are shown graphically in Figure 7.13. Incidentally, note also from Figure 7.12 that those analyses where the 95% CI crosses the positive/negative boundary are not significantly different, whereas those where the 95% CI is entirely positive are significant.

Clearly, the variation in the 2 mg/mL result (that is, the standard deviation) is markedly larger than those of the other doses (8.8, compared to 1.85, 2.3, 4.1). It follows that there may well

Analysis of Variance

Source	DF	Adj SS	Adj MS	F-Value	P-Value
Dose	3	1179.9	393.31	15.16	0.001
Error	8	207.5	25.94		
Total	11	1387.4			

Printout from Minitab software

FIGURE 7.11
ANOVA on tissue culture data.

Tukey Simultaneous Tests for Differences of Means

Difference of Levels	Difference of Means	SE of Difference	95% CI	T-Value	Adjusted P-Value
1-2 mg/mL	15.87	4.16	(2.55, 29.19)	3.82	0.021
1-5 mg/mL	21.00	4.16	(7.68, 34.32)	5.05	0.004
1-10 mg/mL	26.60	4.16	(13.28, 39.92)	6.40	0.001
2-5 mg/mL	5.13	4.16	(−8.19, 18.45)	1.23	0.624
2-10 mg/mL	10.73	4.16	(−2.59, 24.05)	2.58	0.120
5-10 mg/mL	5.60	4.16	(−7.72, 18.92)	1.35	0.562

Individual confidence level = 98.74%. Modified printout from Minitab software

FIGURE 7.12
Tukey analysis of tissue culture data.

have been a problem with this particular part of the experiment. If this 2 mg/mL data set did indeed have errors (we speculate), and so *should* have a standard deviation in line with the other doses, there may well have been a statistical difference between the 2 and 10 mg/mL doses. It may be that well 2 of the 2 mg/mL part of the experiment was somehow atypical, as the level of the analyte (55.8 units) is markedly lower (by 21%) than the mean (70.9 units/mL) of the other two wells (69.5 units and 72.3 units).

In further discussion of this experiment, the data would have been stronger had four wells per dose and/or had only three different doses been used. This would have increased the overall power of the experiment. Furthermore, the experiment may be criticized in its lack of a control group of cells that were not exposed to the drug (i.e. 0 mg/mL).

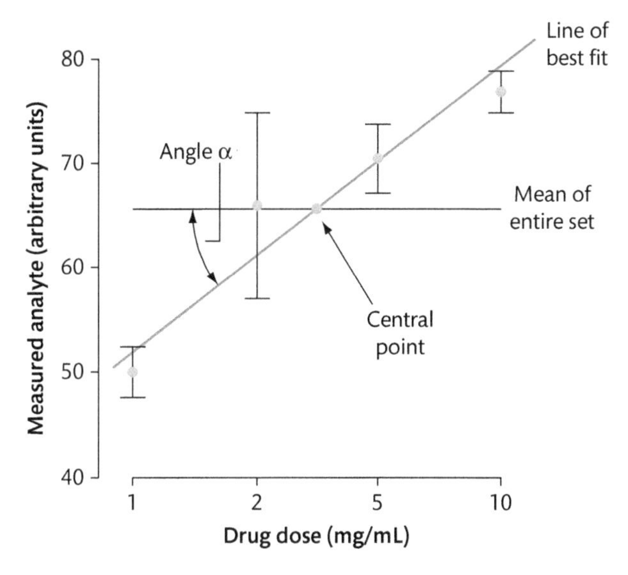

FIGURE 7.13
Graphical representation of the tissue culture data.

The same design can be used in laboratory biochemistry, where the effect of increasing doses of a drug on various models can be determined. Examples of these include kinetics of an enzyme/substrate complex (such as trypsin and a protein solution), the modification of a molecule (the glycation of haemoglobin) or the generation/consumption of metabolites ($NADH/NAD^+$).

Discussion

A number of points are worthy of discussion with respect to these two data sets and their analyses.

1. The first key data point is the weighted sums of all the right-hand means minus the weighted sums of all the left-hand means. The bigger this difference is, then the larger the *L* value, so the larger the t value, and greater significance.
2. This feeds directly into the angle α, formed between the line of best fit between the individual means and the mean of the entire set. In the pregnancy example, the data set (a) has an α angle around 20°, but in set (b), it is around 11.5°. Therefore (and intuitively) the greater the α angle, the greater the likelihood of significance. In the tissue culture experiment, the angle α is around 40°, and so the effect seems greater.
3. The standard error of *L* relies on the within groups mean squares (from an ANOVA). This is effectively a measure of the precison of the data, and ideally a smaller within groups mean squares (implying low variance) will give a low standard error.
4. The second part of the mathematics of se(*L*) is the sample size (*n*), that being $(-3^2/n_1 + (-1^2/n_2) + 1^2/n_3 + 3^2/n_4)^{1/2}$. If *n* is larger, this will have a direct impact on the standard error, and this too is intuitive, as a large sample size automatically gives more confidence.
5. Incidentally, how do we know the sample size is large enough? In both examples, there is no power calculation to justify what appears to be a small sample size. In the pregnancy example, yes, a positive result with set (A) was obtained, but what of set (B)? By simply quadrupling the sample size per group to 144 women altogether, with exactly the same mean and standard deviation for Substance XYZ, the linear trend for B would be extremely significant at $p=0.0001$. This is because, although the *L* value would be exactly the same, the standard error of *L* would be much smaller, leading to a t value of 4.72. This again reinforces a powerful point–that of the likelihood of a false negative if the sample size is small. In the tissue culture experiment, increasing the number of wells for each drug would drive down the standard error, and perhaps there would be a difference between the 5 and 10 mg/mL doses. With greater resources available in an industrial setting, this crucial result would be pursued.
6. The length of a pregnancy is of course a continuously variable index, generally measured in weeks. If we have the exact time from conception to venepuncture, we could perform a Pearson correlation (assuming both indices have a normal distribution) between Substance XYZ and weeks of pregnancy, expecting a positive and significant *r* value. But this correlation would exclude the pre-pregnancy data of zero weeks.
7. What we can't do in either example is to try to correlate the measured analyte on the vertical axis with the four stages on the horizontal, as the latter is a categorical and not a continuous index (as discussed in Section 7.8).

A number of commercial statistical packages are available for the computation of dose-response curves. However, these packages (as all packages) will do what they are told, and if incorrectly programmed with inappropriate data, will give incorrect results. The researcher must (as always) know exactly what he or she is doing, or get expert advice.

7.7 The analysis of paired data

Perhaps the most common analysis is two sets of data from independent, unlinked groups, such as the heights of randomly selected men and women, or levels of a molecule in the blood of patients with heart disease compared with patients free of heart disease. In another setting, two sets of data may be linked by a certain factor, and so exist in pairs. An example of this may be height of an adult man and his adult sister, where the cases are linked by genetics. In a laboratory setting, there are several ways in which this linking factor can present itself, such as: blood pressure before and after a particular intervention (a drug, for example); the same items of data analysed independently by two different scientists or on two different analysers; or the same molecule or ion measured in both serum and plasma.

If there is a clear link between two sets of data we say that the data are paired. Where there is no clear link (as in the height of unrelated men and women) the data is described as unpaired. The exact method of analysis of paired data depends on the distribution of the data. Where the difference between the two sets of data has a normal distribution, a paired t test is appropriate. However, if the difference between the two sets of data has a non-normal distribution, we use Wilcoxon's test. The choice of tests again emphasizes the importance of understanding the distribution of a set of data.

Cross reference

Section 4.6 (Figure 4.14) of Chapter 4 shows a graphical representation of paired, linked data, that of systolic blood pressure before and after the use of a drug to treat hypertension

Paired t test

The **paired t test** is used when the difference between two data sets has a normal distribution. As an example, consider the case of a potential difference between levels of cholesterol before and six weeks after the daily use of a drug, such as a statin, to reduce levels of cholesterol in patients with hypercholesterolaemia (Table 7.23). The null hypothesis states that there will be no change in levels of the lipid. The alternative hypothesis states that there will be a difference in the serum cholesterol in the subsequent sample compared with the initial sample.

Paired t test

a statistical test used to determine differences between two sets of paired data where the distribution of the difference is normal

TABLE 7.23 **Serum cholesterol**

Initial sample	Subsequent sample	Difference
7.6	6.7	0.9
7.1	6.1	1.0
8.4	7.2	1.2
7.0	5.4	1.6
7.2	7.4	–0.2
6.5	6.4	0.1
7.3	6.5	0.8
6.2	6.3	–0.1
7.0	6.5	0.5

Units are mmol/L

In summarizing the cholesterol levels, the initial sample is mean (standard deviation) 7.1 (0.63) mmol/L, whilst in the subsequent sample, levels are 6.5 (0.59) mmol/L. Is this reduction statistically significant? To find out we must first determine the distribution of the difference between the initial and subsequent samples.

The descriptive statistics for the difference between the first and second sample are: mean 0.64, standard deviation 0.65, median 0.8, inter-quartile range 0.0 to 1.1. Although the standard deviation is greater than the mean, this does not mean that the data is non-normally distributed. Rather, it is because there are some negative values in the data set. Applying this data to the Anderson–Darling test to determine the distribution gives an A^2 value of 0.23, and so $p=0.74$, which we translate as a normal distribution. Therefore, a paired t test is appropriate.

The mathematics of the paired t test are reasonably straightforward. In considering the changes in the data between the two cholesterol results, if there are as many instances of cholesterol levels increasing as there are of cholesterol levels decreasing, all the differences will cancel each other out. This is the null hypothesis—that the sum of the differences where cholesterol goes up equals the sum of the differences where cholesterol goes down, such that the net difference is zero.

The alternative hypothesis is that one of the two differences is statistically larger than the other, so that those cases where cholesterol goes up do not cancel out those cases where cholesterol goes down (or vice versa). Whether such a difference is present can be determined from the 95% confidence interval (CI) of this difference. The mathematics of deriving a 95% CI has been explained in Chapter 6, the equation being...

$$95\%\text{ confidence interval} = \text{mean} \pm 1.96 \times (\text{standard deviation} / \text{square root of n})$$

where n is the sample size. So in our case the equation takes the form...

$$\begin{aligned}95\%\text{ confidence interval} &= 0.64 \pm 1.96 \times (0.64 / \text{square root of } 9)\\ &= 0.64 \pm 1.96 \times (0.20)\\ &= 0.64 \pm 0.40\end{aligned}$$

This tells us that the 95% CI interval of the difference is 0.20 to 1.04. As this confidence interval does not include zero, the difference is significant with a probability of $p<0.05$. In fact, feeding this data into some statistical software gives this difference to be significant with $p=0.02$. This means that we are 98% confident that the overall difference in the two values for cholesterol is genuine, and only 2% confident that this difference is due to chance. Put another way, levels of cholesterol by this analysis have decreased significantly.

Wilcoxon's test

Wilcoxon's test
a statistical test used to determine differences between two sets of paired data where the distribution of the difference is non-normal

Wilcoxon's test is the non-normal version of the paired t test. Consider the following data on levels of the inflammatory marker CRP in the serum of patients with an autoimmune disease such as rheumatoid vasculitis (Table 7.24). The initial sample is taken before the daily use of an anti-inflammatory drug such as a steroid. The second sample is taken six weeks later. As with the cholesterol example, the null hypothesis states that there will be no change in CRP levels. The alternative hypothesis states that there will be a change in CRP levels.

As with almost all analyses, we first summarize the data with descriptive statistics. Levels of serum CRP in the initial sample are mean 128, standard deviation 181, median 50, inter-quartile range 37.5 to 137. These descriptive statistics strongly suggest a non-normal distribution. Values in the subsequent sample are mean 34, standard deviation 20, median 26, inter-quartile

TABLE 7.24 **Serum CRP**

Initial sample	Subsequent sample	Difference
39	14	25
163	43	120
36	21	15
50	12	38
40	70	–30
27	20	7
111	26	85
87	58	29
596	38	558

Units are mg/L

range 17 to 50.5; these values are also strongly suggestive of a non-normal distribution. We again ask if this fall in levels of CRP is significant, and to do so we would apply a paired test—but which one?

This question is answered, as in the cholesterol example, not by looking at the distribution of the initial and subsequent samples, but by looking at the distribution of the difference between them. The descriptive statistics of the difference in CRP levels are mean 94, standard deviation 179, median 29, inter-quartile range 11 to 102. Once more, these data are strongly suggestive of a non-normal distribution. In support of this is the result of the Anderson-Darling test of $A^2 = 1.54$, which gives a strong probability ($p<0.005$) that the data has a non-normal distribution.

SELF-CHECK 7.4

Without the benefit of the Anderson-Darling test, how would you determine the type of distribution of the cholesterol and CRP data in Tables 7.23 and 7.24?

We begin by revisiting the data from Table 7.24, and determine not simply the difference between the two samples, but the rank order of the difference between the two, as in the mathematics of the Mann-Whitney U test (Table 7.14). Columns 1 to 3 of Table 7.25 carry the same information as the columns of Table 7.24, but the data are laid out in rank order, with the lowest initial sample at the top.

Column 4 of Table 7.25 ranks the differences between the initial and subsequent samples regardless of the sign (positive or negative), whilst column 5 returns the sign (positive or negative) to the rank.

If we sum all the possible rank positions (1 + 2 + 3 + 4 + ... to 9) we get 45. If there were as many increases in CRP in the initial/subsequent pairs as decreases, then there would be the same number of positive ranks as negative ranks, which would theoretically be +22.5 and −22.5. The degree of significance of the difference between the two samples therefore depends on

TABLE 7.25 **Serum CRP**

Initial sample	Subsequent sample	Difference	Rank of difference	Sign of rank of difference
27	20	7	1	+1
36	21	15	2	+2
39	14	25	3	+3
40	70	–30	5	–5
50	12	38	6	+6
87	58	29	4	+4
111	26	85	7	+7
163	43	120	8	+8
596	38	558	9	+9

Units are mg/L

the extent to which the sum of the positive and/or negative ranks differs from 22.5. This is very close to the mathematics of the Mann-Whitney U test. Put another way, the null hypothesis states that (in this case) the sum of either rank will be 22.5. The alternative hypothesis is that the sum of either rank is not 22.5. By summing the positive values and negative values in column 5 of Table 7.25, we find the rank sum of 5 for the negative values and 40 for the positive values.

As with the Mann-Whitney test, by convention, statisticians focus on the smaller of the sums of the ranks. The final step is (once more) to consult statistical tables or use statistical software to determine the extent to which the rank sum of 5 is statistically different from 22.5 given 9 pairs of data. In this case, we apply Wilcoxon's test on Minitab software, and obtain $p=0.044$, so we are 95.6% confident that the difference is genuine, but only 4.4% confident that the difference is due to chance. As $p<0.05$, we conclude that the apparent decrease from the initial to the subsequent sample is significant.

Incidentally, what we have not done in these two examples is a power calculation to determine whether the sample sizes are large enough to test the hypotheses. In looking again at the cholesterol and CRP comparisons, the p values are quite close to 0.05, so it is possible that we may be at risk of a false positive error brought about by a small sample size.

Proof of concept

A further aspect of analysis by paired t test and Wilcoxon's test is to perform both tests and compare the results. In the case of the cholesterol comparison, we obtain $p=0.02$ by the correct paired t test, whilst the inappropriate Wilcoxon's test gives $p=0.044$. Thus, the difference is significant regardless of the test used, although more significant using a paired t test.

Similarly, an inappropriate paired t test on the CRP data gives a non-significant $p=0.154$, whilst the appropriate Wilcoxon's test gives a significant difference at $p=0.044$. This is reinforced by the 95% confidence interval of the mean difference in the CRP data, which is –44 to 232 mg/L. Data with a normal distribution inevitably has a confidence interval that is either all negative (such as −44 to −232) or all positive (such as 44 to 232). These results underline the need to fully understand which test should be used.

SELF-CHECK 7.5

Under what set of conditions would data be analysed by a paired t test or by Wilcoxon's test?

Serial measurement of data

The principle of paired data can be extended to the linked analysis of a series of data obtained on three, four, or many time points or situations. An example of this may be changes in a particular index before, and at two additional time points after, an intervention. In Section 7.6 we looked at changes in a molecule in women in three stages of a pregnancy, and used the method of ordered groups. In an ideal world, a blood sample would have been taken from every woman at each time point. One such data set may be changes in the body mass index (BMI) of a small group of obese people as they attempt to lose weight. In this example, obesity is defined as a BMI >30 kg/m^2. The descriptive statistics of such a group are shown in Table 7.26, the raw data in Table 7.27.

As regards the former, following the concepts of the determination of distribution in Table 7.10, each data set seems likely to have a normal distribution, and BMI seems to fall from baseline to time point 1 and again to time point 2. However, there is a problem, in that we have two variables—the subjects and the time points—and so need some ANOVA software that can accommodate these variables. Several programs are available, with names such as repeated measures ANOVA and two-way ANOVA, whilst Minitab has a program called **General Linear Model** (GLM) that is powerful, yet simple.

General Linear Model
a procedure to describe the statistical relationship between one or more predictors and a continuous response variable

The mathematics of the GLM considers whether the variation in data from the subjects and the variation in data from the three time points (described as predictors) have an influence on the variation in BMI (described as a response variable).

The abbreviated results of a Minitab GLM on the data in Table 7.27 is presented in Figure 7.14. The uppermost section (the ANOVA) shows that both 'Subject' and 'Time' are independent influences (at $p<0.001$ and $p=0.002$ respectively) on BMI. This could have been predicted by close inspection of the data in Table 7.27, which shows that all three BMI data points of subject five are higher than those of subjects two and three. Similarly, Table 7.26 shows that the mean BMI of timepoint 1 is lower than that of the baseline BMI by one SD (1.4 kg/m^2), a strong indicator of a statistically significant difference. Recall that DF represents the degrees of freedom (generally, number of variables−1): the DF for the 'Subject' line is 6 as there are 7 obese people, the DF for the 'Time' line is 2 because there are 3 time points. The 'Model Summary' suggests

TABLE 7.26 **Descriptive statistics of BMI data**

Variable	N	Mean	SD	Minimum	Q1	Median	Q2	Maximum
Baseline	7	32.3	1.4	30.5	30.8	32.6	33.6	33.7
Time point 1	7	30.8	2.2	27.8	28.4	30.9	33.1	33.9
Time point 2	7	30.2	1.7	28.2	28.7	29.5	32.2	32.2

Units: kg/m^2.

TABLE 7.27 **Serial measurement of data (units: kg/m^2)**

Subject	Baseline	Time point 1	Time point 2
1	33.7	30.3	32.2
2	30.8	28.4	29.0
3	30.5	27.8	28.2
4	32.6	31.0	29.5
5	33.6	33.9	31.5
6	31.5	30.9	28.7
7	33.5	33.1	32.2

that almost 61% of the variation of BMI can be accounted for by Subject and Time Point. It follows that some 39% of the variation in BMI must be due to other factors, such as the sex of the subject and whether they have diabetes (an established risk factor for obesity).

Although the GLM tells us there are strong links between the subject ($p<0.001$), the time points ($p=0.002$), and BMI, it does not tell us exactly where these differences are present. To do so, formal post-hoc testing, such as those of Tukey, Fisher, Bonferroni, and Sidak, are required. The Minitab printouts of Tukey analyses are shown in Figures 7.15 and 7.16, the former showing that the difference between baseline and time point 1 is significant with $p=0.019$, whilst the difference between baseline and time point 2 is even more significant at $p=0.002$. However, the difference between time points 1 and 2 is not significant ($p=0.465$). Figure 7.16 shows a graphic of this result, expressed as 95% confidence intervals (95% CI).

General Linear Model: All data versus Subject, Time

Analysis of Variance

Source	DF	Adj SS	Adj MS	F-Value	P-Value
Subject	6	49.425	8.2375	10.18	<0.001
Time	2	16.927	8.4633	10.46	0.002
Error	12	9.707	0.8089		
Total	20	76.058			

Model Summary

S	R-sq	R-sq(adj)	R-sq(pred)
.899383	87.24%	78.73%	60.92%

FIGURE 7.14
GLM of the BMI data.

Tukey Simultaneous Tests for Differences of Means

Difference of Time Levels	Difference of Means	SE of Difference	Simultaneous 95% CI	T-Value	Adjusted P-Value
1-2	−1.543	0.481	(−2.824, −0.261)	−3.21	0.019
1-3	−2.129	0.481	(−3.410, −0.847)	−4.43	0.002
2-3	−0.586	0.481	(−1.867, 0.696)	−1.22	0.465

Time level 1=baseline, 2= time point 1, 3=time point 2

FIGURE 7.15

Tukey analysis of differences of means.

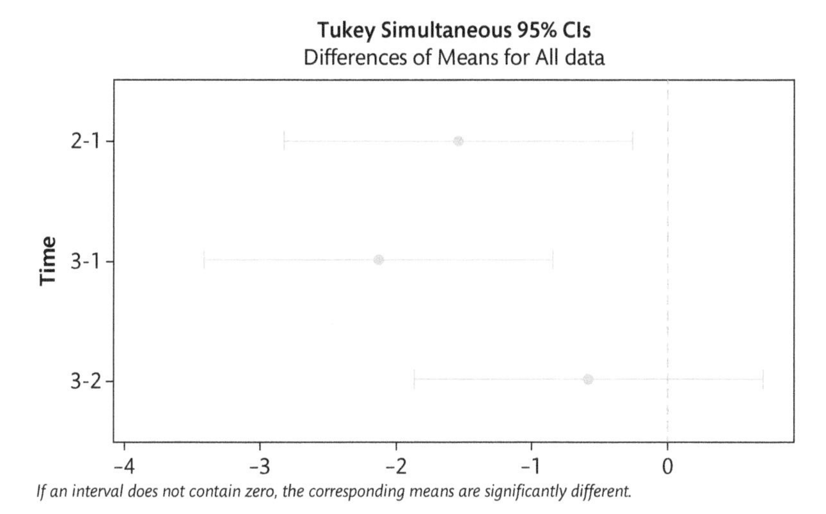

FIGURE 7.16

Tukey analysis with confidence intervals.

Time level 1=baseline, 2= time point 1, 3=time point 2. Figure by Minitab software.

Because the upper two interval bars (baseline-time point 1 and baseline-time point 2) are both to the left of the dashed green line (indicating a difference), the differences they represent are significant (p=0.019 and p=0.002 respectively). However, the lower interval bar straddles zero, so 95% CI will contain a positive and a negative number, so the difference is not significant (p=0.465). Despite the effective mathematics of this analysis, we still may harbour some doubts about the overall results as there is no power calculation, and 7 is a very small number on which to base our conclusions—we may be looking at a false positive.

The simplest and most popular procedure for data where at least one set has a non-normal distribution is the Friedman test, and as with the Kruskal-Wallis test, it is a test of medians. However, should this test not be available, an alternative is to log or otherwise transform the data so it has a normal distribution, and so be amenable to GLM analysis.

Other forms of data, where there is no time aspect, can also be analysed using these methods. Such an example may be the measurement of a molecule in the serum and in plasma with different anticoagulants (EDTA, citrate, heparin) from, shall we say, twenty or thirty people.

The power of the GLM is such that it can also be used in basic laboratory research, where the effects of two independent indices (drug dose, time) can be determined simultaneously.

7.8 Correlation

In Section 7.5 we looked at the comparison of two sets of data from independent groups, such as serum cholesterol from two different populations (for example, factory workers and office workers). Section 7.7 considered sets of data that were in pairs, such as serum cholesterol before and after several months' prescription of an experimental cholesterol-lowering drug. We can also look at two different sets of data (such as serum cholesterol versus blood glucose) *within* a single population (perhaps just the office workers). For example, it may be that someone with a high level of serum cholesterol also has raised blood glucose. We can determine whether this is the case by the process of **correlation**.

Correlation

a statistical test used to determine whether there is a relationship between two different indices taken from a single group

Correlation addresses the need to be able to define the extent to which two indices correlate. What this means is that when one index is high, then the second, paired index is also high, and similarly, when one index is low, then the second index is also low. A suitable example of this is the relationship between height and weight, as described in Section 4.2, where Figures 4.5 and 4.6 apply. It is well established that, in general, tall people are heavier than short people, but it is just as well known that two people of the same height can have very different weights (and vice versa). It is also very evident to most people that, in general, there is a clear relationship between height and weight.

Accordingly, statisticians say that height and weight correlate with each other.

But is this relationship present because the taller someone is, the heavier they become, or is it that the heavier one is, the taller they become? The former seems more likely. We must be cautious about interpreting correlations. If two indices correlate strongly, it does not mean that one causes the other, or that they are related in some pathological or physiological manner. A crucial comment is always to recall that ...

correlation does not imply causation

A better example is from human epidemiology and clinical studies where it is known that, in a large population, systolic blood pressure generally correlates very strongly with diastolic blood pressure. But the increased systolic value does not *cause* the diastolic value to rise or vice versa. In fact, those factors which act to increase one of these blood pressure indices also act to increase the other index. As a result, both systolic and diastolic blood pressure rise in parallel—but do so independently of one another.

Cross reference

Section 3.5 of Chapter 3 and Section 4.2 of Chapter 4 also describe correlation

Perhaps the most amusing example of a relationship between correlation and causation is that, during the 1950s, the rise in the import of bananas in the United Kingdom correlated very strongly with the rise in the divorce rate. It could be argued that the increased consumption of bananas *caused* marital problems and hence a rise in the divorce rate or, equally, that newly divorced people eat more bananas. Both hypotheses are interesting.

Correlation coefficient

a number used to denote the strength of the correlation between two sets of data

To gauge the strength of the correlation between two indices, we use a **correlation coefficient**, represented by the Greek letter *rho* (r), which varies between 0 and 1. A very strong association between two sets of indices would be represented by a correlation coefficient where r is close to 1 (say, 0.92). By contrast, one would consider the relationship to be weak if a small r value of perhaps 0.15 was obtained. In practice, most correlation coefficients between indices in biological and biomedical sciences are in the range 0.3 to 0.6, although in some cases (such as the correlation between systolic and diastolic blood pressure) this may be as strong as 0.8 or 0.9.

The importance of distribution

As with comparing two groups of data from different populations—where we use the Student's t test if the data is normally distributed, and the Mann-Whitney U test if it is non-normally distributed—we have a choice of different tests of correlation. If the two sets of data have a normal distribution (such as height and weight), we use **Pearson's correlation method**. However, if one or both sets of data have a non-normal distribution (such as total cholesterol and triglycerides), we have to use **Spearman's correlation method**.

Pearson's correlation method
a statistical test used to determine whether there is a relationship between two different indices when both indices have a normal distribution

Spearman's correlation method
a statistical test used to determine whether there is a relationship between two different indices when one or both of the indices have a non-normal distribution

Consider data from a group of patients with liver disease, and so abnormal liver function tests (bilirubin, gamma-glutamyl transferase (GGT), and aspartate aminotransferase (AAT)) and whose raw data are set out in Table 7.28, and their descriptive statistics in Table 7.29. A statistical software package can easily produce correlation coefficients of these data, as shown in Table 7.30. The two sets of figures are the correlation coefficient (between 0 and 1) and the related p value.

The correlation matrix of Table 7.30 tells us that the relationship between bilirubin and GGT is very strong (probability 99.8%), that the relationship between bilirubin and AAT is weak (probability 96.2%), but that the relationship between GGT and AAT is not significant, with a probability of 91.3% that the relationship is genuine.

TABLE 7.28 **Raw data of liver function tests**

Bilirubin	GGT	AAT
28	19	39
18	20	27
20	19	27
17	27	25
28	26	33
30	36	43
32	32	40
35	33	100
40	45	70
45	43	50

GGT = gamma-glutamyl transferase, AAT = aspartate aminotransferase

TABLE 7.29 **Descriptive statistics for liver function tests**

Variable	N	Mean	SD	Minimum	Q1	Median	Q3	Maximum
Bilirubin	10	29.3	9.25	17	19.5	29	36.25	45
GGT	10	30.0	9.5	19	19.75	29.5	37.75	45
AAT	10	45.4	23.4	25	27	39.5	55	100

SD = standard deviation; Q1 = 25th (lower) quartile; Q3 = 75th (higher) quartile. Units: bilirubin µmol/L, GGT IU/L, AAT IU/L

TABLE 7.30 **Pearson correlation of liver function tests**

	GGT	AAT
Bilirubin	*r*=0.84: p=0.002	*r* = 0.66: p=0.038
AAT	*r*=0.57: p=0.087	

r = correlation coefficient

However, close attention to the descriptive statistics in Table 7.29 reveals that AAT is very likely to have a non-normal distribution (the standard deviation is over half of the mean and the mean is not in the middle of the full range). As the correlation matrix in Table 7.30 is according to Pearson's method, this analysis is therefore inappropriate in seeking correlations using a non-normal data set. The correct method for any analysis using AAT is Spearman's method. The mathematics of the two methods are markedly different, as Spearman's method does not consider the numerical values of each data set, but rather considers them in rank order. If we return to the software and instruct it to perform a Spearman correlation, we get the following results (Table 7.31).

TABLE 7.31 **Spearman correlation of the ranks of the liver function tests**

	GGT	AAT
Bilirubin	*r*=0.78: p=0.008	*r*=0.93: p<0.001
AAT	*r*=0.75: p=0.013	

Using Spearman's method, the relationship between bilirubin and GGT is still very strong (p=0.008), but it is slightly weaker (*r*=0.78) than using Pearson's method (*r*=0.84). Notably, the relationship between bilirubin and AAT has become stronger, with the correlation coefficient increasing from *r*=0.66 to *r*=0.93, and in parallel the probability that the relationship is genuine has increased from 96.2% (that is, p=0.038) to over 99.9% (being p<0.001). Therefore, Spearman's method has reversed the relationships between bilirubin and the two other liver function tests.

However, of greater interest is the relationship between GGT and AAT. Although this relationship was not significant using Pearson's method (*r*=0.57, p=0.087), using Spearman's method the relationship has become significant (*r*=0.75, p=0.013). This brings a 98.7% confidence that the relationship between AAT and GGT is genuine, and illustrates the importance of choosing the correct method of analysis.

A worked example: athletics

Consider the data in Table 7.32 from 12 people taking part in an athletics event. These data are their height, weight, age, and distance that they can run in a certain fixed time period such as 30 minutes. The descriptive statistics are shown in Table 7.33. We have briefly used this model in Figure 4.7.

In each case the SD is less than 25% of the mean, and alongside other factors outlined in Table 7.10, we are confident that each set has a normal distribution. Accordingly, a Pearson correlation is appropriate. Table 7.34 is a correlation matrix of the four sets of data. The largest correlation coefficients are between height and weight (*r*=0.82) and between age and distance run (*r*=–0.89), both of which are highly significant, indicating a very powerful relationship.

TABLE 7.32 **Data on height, weight, age, and distance run**

Subject	Height (metres)	Weight (kilograms)	Age (years)	Distance run (metres)
1	1.56	70	43	6100
2	1.23	65	56	5200
3	1.70	72	29	6900
4	1.81	84	59	4800
5	1.46	72	34	7100
6	1.50	70	56	4800
7	1.59	66	45	4500
8	1.66	79	72	3700
9	1.70	74	56	5100
10	1.48	69	45	4800
11	1.85	88	67	4200
12	1.66	75	55	5250

These relationships can be shown graphically with a scatterplot: Figure 7.17 shows the relationship between height and weight, and Figure 7.18 shows the relationship between age and distance run.

Height and weight

Close attention to Figure 7.17 shows that the tallest person (1.85 m) is also the heaviest (88 kg), whilst the shortest (1.23 m) is also the lightest (65 kg). However, in between these two extremes, the relationship is not clear-cut, despite an overall trend being present. Indeed, several people who are around 1.5 m in height weigh more than a person almost 1.6 m tall. But, overall, for each of the pairs of data (an individual's height and their weight), as height increases, so does weight. Further, the line of dots rises (broadly speaking) from the lower left to the upper right, so the relationship is positive (that is, as one increases, so does the other).

TABLE 7.33 **Descriptive statistics for the data in Table 7.32**

Variable	N	Mean	SD	Minimum	Q1	Median	Q3	Maximum
Height	12	1.6	0.17	1.23	1.485	1.625	1.70	1.85
Weight	12	73.7	6.95	65.0	69.25	72.0	78.0	88.0
Age	12	51.4	12.65	29.0	43.5	55.5	58.2	72.0
Distance run	12	5204	1024	3700	4575	4950	5888	7100

SD = standard deviation; Q1 = 25th (lower) quartile; Q3 = 75th (higher) quartile.

TABLE 7.34 **Correlation matrix of height, weight, age, and distance run**

	Weight	Distance run	Age
Height	r=0.82: p=0.001	r=–0.24: p=0.443	r=0.28: p=0.378
Age	r=0.56: p=0.06	r=–0.89 p<0.001	
Distance run	r=–0.34: p=0.271		

Age and distance run

A major difference between Figures 7.17 and 7.18 is that in the latter the slope of the line of dots goes in the opposite direction: it slopes downwards from upper left to lower right, telling us that—in general terms—as people get older, they run shorter distances. We call this an inverse relationship, which is why the *r* value has a minus sign, that is, $r = -.89$. If we look again at Figure 7.18, however, we see that, as with the height and weight example, there are several cases where some older people run further than those who are younger.

Interpretation

These examples of the relationships between height, weight, age, and athleticism illustrate a number of key points in using correlations to seek significant relationships between two indices. These are:

- The distribution of the data sets drives the method of correlation (Pearson or Spearman's).
- The correlation coefficient (*r*) marks the strength of the relationship between the indices.
- Relationships can be positive (where both indices increase together) or negative (where one index increases as the other index decreases).
- Merely because two indices correlate strongly does not imply a causal relationship.

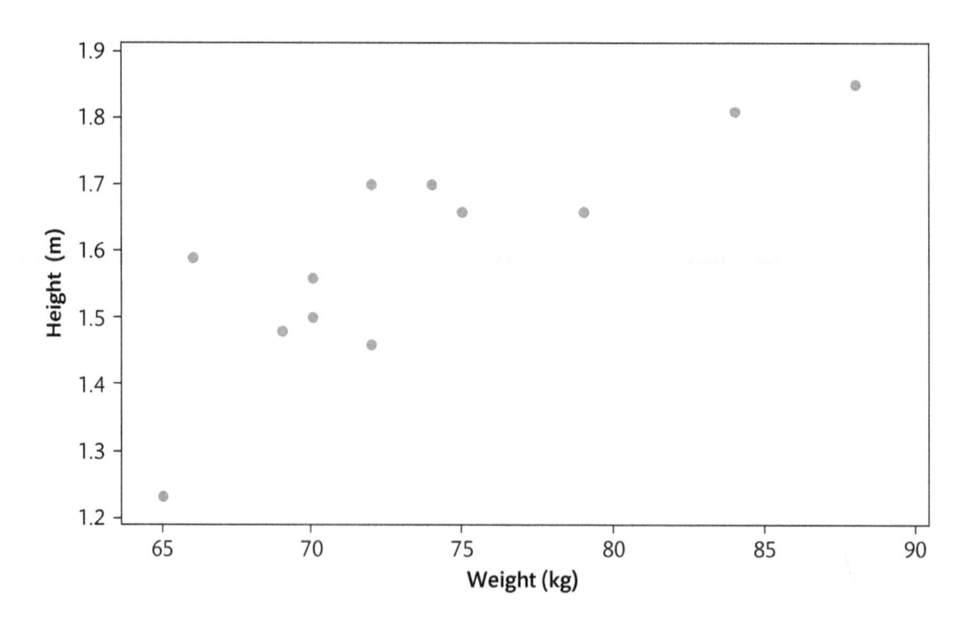

FIGURE 7.17
Minitab scatterplot of height versus weight.

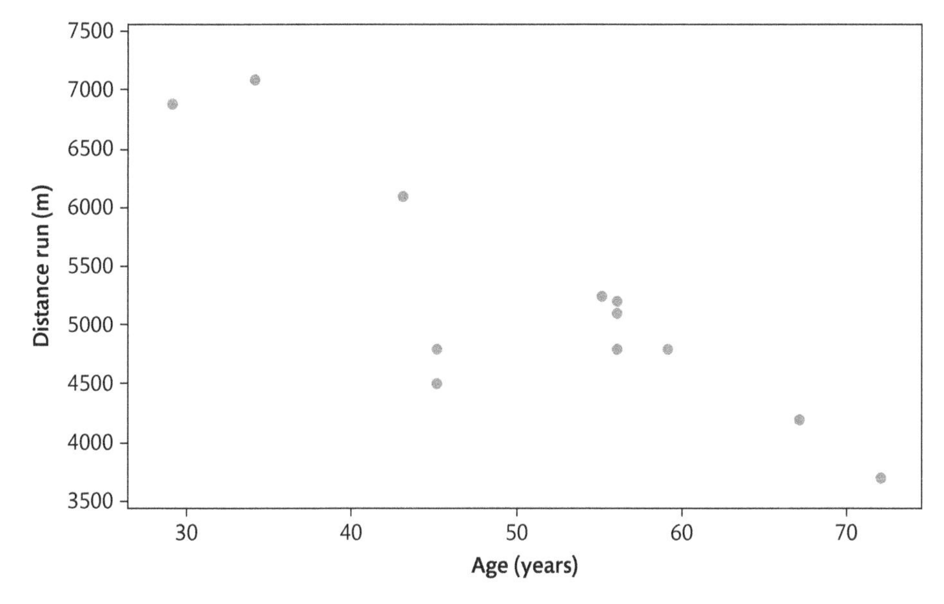

FIGURE 7.18
Minitab scatterplot of age versus distance run.

We also need to consider the statistical power of the study. It is clear that we can confirm the obvious statements that height and weight are strongly related, and that one's athletic prowess seems to decrease with age. But consider the relationship between age and weight (Table 7.34), which, although not significant, has a correlation coefficient (r=0.56) that in another setting would be readily acceptable. Indeed, the probability is almost significant, with p=0.06.

The reason that we cannot confirm a relationship between increasing age and increasing weight is that the sample size is too small. Adding only one other person (with average age and weight) to this study reduces the probability to p=0.048, with exactly the same correlation coefficient of r=0.56. Therefore, our failure to find a significant relationship in a study of n = 12 is a false negative. What we should have done, of course, is to first define our original quantified hypothesis, and then undertake a formal power calculation to find the correct sample size.

Chapter summary

- The most frequent source of error in statistics is human, inevitably the incorrect input of raw data into a spreadsheet. The first step in analysis is verification of the integrity of the spreadsheet.
- Descriptive statistics summarizes a data set in terms of the number of observations, the mean, the median, the standard deviation, and the inter-quartile range, etc.
- Categorical data is analysed by tests such as the Chi-squared test.
- The distribution of continuously variable data must be determined: this may be normal or non-normal.
- Continuously variable independent data with a normal distribution is summarized with the mean and standard deviation. Two such groups are compared by the Student's t test, three or more groups by analysis of variance (ANOVA).

- Continuously variable independent data with a non-normal distribution is summarized with the median and inter-quartile range. Two groups where one or more has a non-normal distribution are compared by the Mann–Whitney U test; three or more groups by the Kruskal–Wallis test.
- Continuously variable data that is linked in a trend may be analysed by the method of ordered groups. Differences between three or more groups may be sought with a post-hoc test such as that of Tukey.
- Data of a single nature that are linked together in pairs are analysed by the Student's t test (when normally distributed) or by Wilcoxon's test (when non-normally distributed). Data from several time points, where there are repeated measurements, can be analysed by a General Linear Model.
- Relationships between different sets of data may be sought by Pearson's correlation method (if both sets have a normal distribution) or by Spearman's rank correlation method (if one or both sets have a non-normal distribution).
- A flowchart that summarizes the thought process behind the choice of statistical test for simple data sets is shown in Figure 7.19.

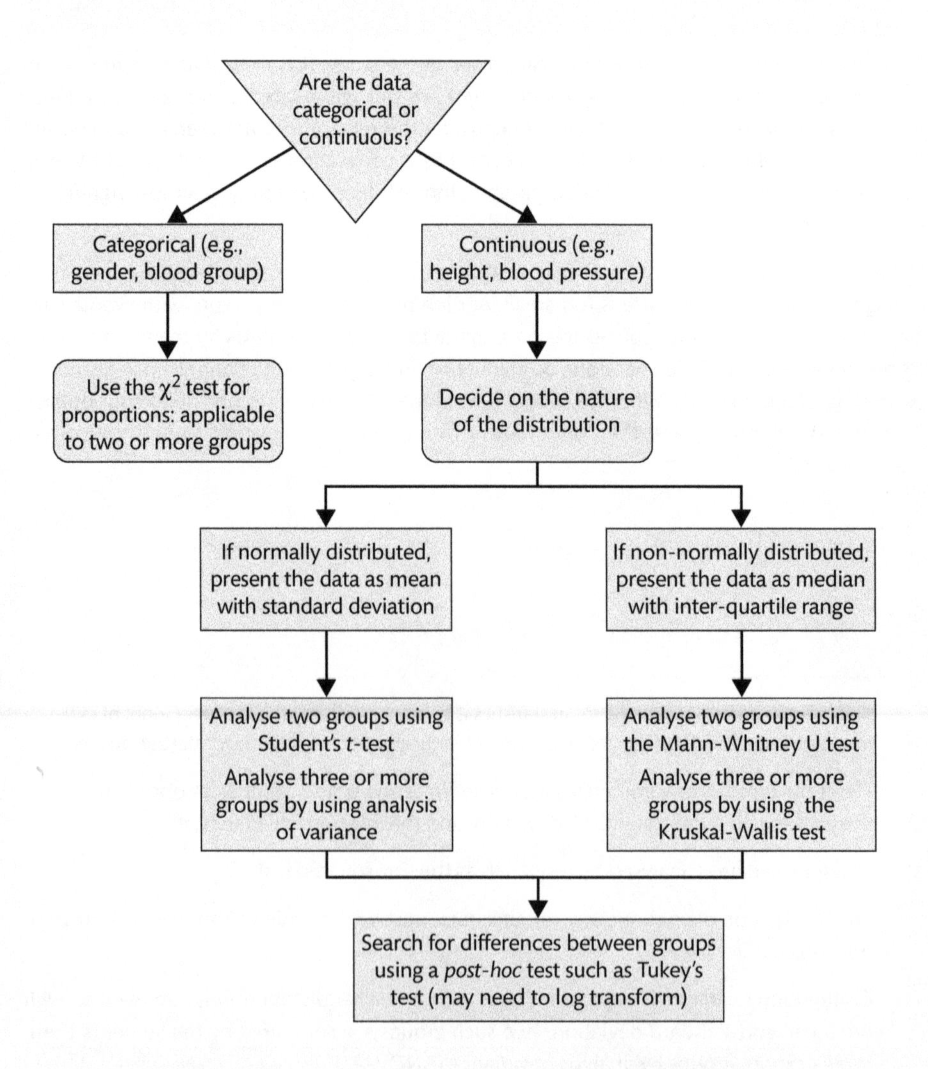

FIGURE 7.19
Flowchart illustrating the mechanism of choice of a statistical test.

Suggested reading

- Altman, D.G. *Practical Statistics for Medical Research*. Chapman & Hall, London, 1991.
- Blann, A.D., and Nation, B.R. (2008). Good analytical practice: statistics and handling data in biomedical science. A primer and directions for authors. Part 1: Introduction. Data within and between one or two sets of individuals. *Br J Biomed Sci* 65, 209–217.
- Blann, A.D., and Nation, B.R. (2009). Good analytical practice: statistics and handling data in biomedical science. A primer and directions for authors. Part 2: Analysis of data from three or more groups, and instructions for authors. *Br J Biomed Sci 66*, 1–5.
- Holmes, D., Moody, P., and Dine, D. *Research Methods for the Biosciences*. Oxford University Press, Oxford, 2006.
- Peacock, J.L., and Peacock, P.J. *Oxford Handbook of Medical Statistics*. Oxford University Press, Oxford, 2011.

Useful websites

- www.minitab.com
- https://www.ibm.com/analytics/data-science/predictive-analytics/spss-statistical-software
- www.stata.com/
- https://www.danielsoper.com/statcalc/calculator.aspx?id=8

Questions

7.1 Calculate the t value for the difference between one data set whose mean (SD) is 105 (25), and another with a mean (SD) of 119 (17). The sample sizes are 22 and 27 respectively.

7.2 Calculate the W value for one data set comprising 23, 86, 45, 79, 123, and 44 and another of 67, 134, 87, 42, 77, and 92.

7.3 Comment on the following correlation matrix on this hypothetical set of data.

	Albumin	Basophils	Cholesterol
D-dimers	r=–0.196 p=0.395	r=–0.486 p=0.026	r=–0.115 p=0.620
Cholesterol	r=0.083 p=0.719	r=0.472 p=0.031	
Basophil	r=–0.123 p=0.597		

7.4 Which of these tests should be used to determine distribution? Anderson-Darling, Bonferroni, Chi-squared, General Linear Model, Mann-Whitney.

7.5 Choose which of the following tests are best suited to the type of data: Wilcoxon's test, ordered groups, Tukey's test, Spearman's correlation, Student's t test, Pearson's correlation, Kruskal-Wallis, paired t test, ANOVA.

Type of data	Test to be used
Differences between independent groups after the use of an ANOVA	
Seeking a relationship between linked pairs of data (e.g. height and weight) where both have a normal distribution	
Difference between two independent groups that both have a normal distribution	
Difference between three or more groups that are linked by a common factor (i.e. are not independent)	
Difference between three or more independent groups where at least one has a non-normal distribution	
Difference between two sets of paired data (e.g. before/after an intervention) where the difference has a normal distribution	
Seeking a relationship between linked pairs of data (e.g. height and weight) where at least one has a non-normal distribution	
Difference between three or more independent groups that all have a normal distribution	
Difference between two sets of paired data (e.g. before/after an intervention) where the difference has a non-normal distribution	

8

Research 3: Large data sets

Learning objectives

After studying this chapter, you should be able to ...

- Understand the need for complex analyses in large data sets
- Explain the basis of multivariate regression analysis
- Understand the concepts of a dependent variable and independent variables
- Recall differences in the approach to linear and logistic analyses
- Outline the value of the odds ratio in assessing risk
- Describe the important aspects of survival analysis and hazard ratio
- Recognize essential methods in epidemiology

In Chapters 6 and 7 we looked at small data sets—those with only a small number of different groups and up to perhaps a hundred data points. These sample sizes have the power to test only a small number of hypotheses. By increasing the sample size to perhaps hundreds of data points, however, more complex hypotheses can be tested. Such testing calls for more complicated methods, such as **multivariate regression analysis**, of which there are two major types. In Section 8.1 we look at situations where the index we are most interested in has a continuous variation—and so where linear analysis is used. However, if our index of interest has a categorical distribution, logistic analysis is the method of choice, as will be explained in Section 8.2.

Multivariate regression analysis
a complex method for determining which of several indices are most closely linked to each other

Some studies are designed to answer important questions regarding health care, such as the factors influencing the progression of cancer, or the efficacy of new drugs. These often call for analyses over a long period of time—for example, the monitoring of the outcome of a group of patients randomized to different treatments. Some of these may be described as **survival analysis**, and will be discussed in Section 8.3.

Survival analysis
a method for determining factors that influence the outcome of a group of subjects over a period of time

As the number of subjects in a study gets larger, with perhaps thousands or tens of thousands of people, we enter the realm of **epidemiology**. Related to the word 'epidemic' (itself derived from 'epi', meaning upon, and 'demos' meaning the people), epidemiology is the study of patterns of health and disease within large populations, often comprising

Epidemiology
the study of the health and disease of large populations

hundreds of thousands of people. Epidemiologists can form international collaborations and compare populations across the world, and so draw conclusions regarding health and disease on different continents and between different peoples. All such studies call for analysis with powerful statistical software, and essential topics will be described in Section 8.4.

8.1 Linear regression analysis

In this section we consider data that has a continuous variation, such as blood pressure, or levels of a molecule in the blood. In Section 7.8 we looked at four such indices: the age, height, weight, and distance run by a group of athletes. There were powerful relationships between height and weight ($r = 0.82$, $p = 0.001$), and between the age of the athlete and the distance they were able to run ($r = -0.89$, $p < 0.001$). Since the relationships between distance run and both height and weight were not significant, we can assume that height and weight do not have a great bearing on the distance the athlete could run. But suppose that the weight of the athlete had a modest relationship with distance run (perhaps $r = 0.7$, $p = 0.02$). Because this relationship, although significant, is less strong than that between age and distance run, we can rank those factors that are most important: age is strongest, followed by weight, followed by height. To formally determine the impact of each of these factors (age, height, weight) on another factor (distance run), we use linear regression analysis.

Linear regression analysis
a method for determining which of a series of factors independently influence the variability of a single factor, which must have a continuously variable distribution

The purpose of **linear regression analysis** is to determine which of a number of factors influence the levels of another factor. Let us take as an example those factors that influence renal function in diabetics, a convenient marker of which is the estimated glomerular filtration rate (eGFR). Let us hypothesize that serum albumin, serum cholesterol, systolic blood pressure (SBP), and glycated haemoglobin (HbA1c) all contribute significantly to the variation in eGFR.

We test this hypothesis by seeking relationships between these indices in patients with diabetes. But how many patients should be recruited? Although we have a hypothesis, it is not quantified. But a rule of thumb is that there should be at least 15 subjects for each index, although up to 20 subjects per index would generally be advocated. Therefore, with five indices in an analysis we need good data from a minimum of 75 patients, a process that must begin by obtaining permissions, such as the approval of a Research Ethics Committee. With approval secured, the patients can then be recruited, and data obtained, entered onto an electronic spreadsheet, and then analysed.

Cross reference
The need for permissions is explained in Section 6.3

In order to understand the theory of the mathematics of linear regression analysis, which is moderately complicated, we must begin with the equation of the straight line.

The equation of the straight line

A fundamental mathematical concept is the equation of a straight line, $y = mx + c$. This equation gives us a way of evaluating the relationship between an index on the y (vertical) axis and that on the x (horizontal) axis. We use 'dependent' to describe the y value, and 'independent' to describe the x value. The importance of these terms will be discussed shortly. The m value is the gradient, or slope, of the straight line, and c is the point where the line crosses the y axis. This can be illustrated by the simple relationship between two indices, A and B, whose descriptive statistics are shown in Table 8.1.

The graphic of this relationship is shown in Figure 8.1, and includes the line of best fit, which adheres to the equation $y = mx + c$ (or in this case, $A = mB + c$). It is possible to interrogate the

TABLE 8.1 **Descriptive statistics of data sets A and B**

Variable	N	Mean	SD	Minimum	Q1	Median	Q3	Maximum
A	25	40.1	10.7	20	31	40	51	60
B	25	79.9	21.3	47	61	78	99	125

SD = standard deviation; Q1 = 25th (lower) quartile; Q3 = 75th (higher) quartile

data and determine the exact values of m and c, which are 0.44 and 5 respectively, giving the ultimate relationship equation as $A = 0.44B + 5$. This equation also allows us to:

- estimate any value of A or B given the other data point. For example, if B is 100 units then A may be estimated as $(0.44 \times 100) + 5$, which gives 49 units;
- estimate the degree to which an increase in one index leads to an increase in the other. For an increase in one unit of B, the value of A increases by 0.44 units. Likewise, if A increases by 1 unit, then B increases by 1/0.44, which is 2.27 units.

But how closely do our two sets of data—the individual values of A and B plotted in Figure 8.1—actually adhere to the relationship $A = mB + c$? That is, how strongly are the two data sets correlated? Close attention to these data indicate that they both have a normal distribution, so that the existence of a correlation between the two should be sought using the **Pearson correlation method**, as explained in Chapter 7. This method gives a correlation coefficient (*r*) of 0.85, indicating a powerful relationship. Indeed, we find a probability of $p < 0.001$ (and so a confidence of over 99.9%) that the relationship between A and B is genuine and is not spurious.

Pearson correlation method
a technique for determining the relationship between two indices, both of which have a normal distribution

There are additional mathematics we can employ to further dissect the relationship between A and B. There are three points that lie exactly on the line of best fit; many lie close by; but some lie a long way from this line. By measuring the distances between all the data points and the line of best fit, we can derive a further index of the 'closeness' of the relationship between A and B. Although the relationship between A and B is very strong, with a correlation coefficient of 0.85, it is far from perfect: it does not have a correlation coefficient of 1. In addition,

Cross reference
Chapters 3, 4, and 7 have details of the process of correlation and the determination of the correlation coefficient, and see especially Section 7.8

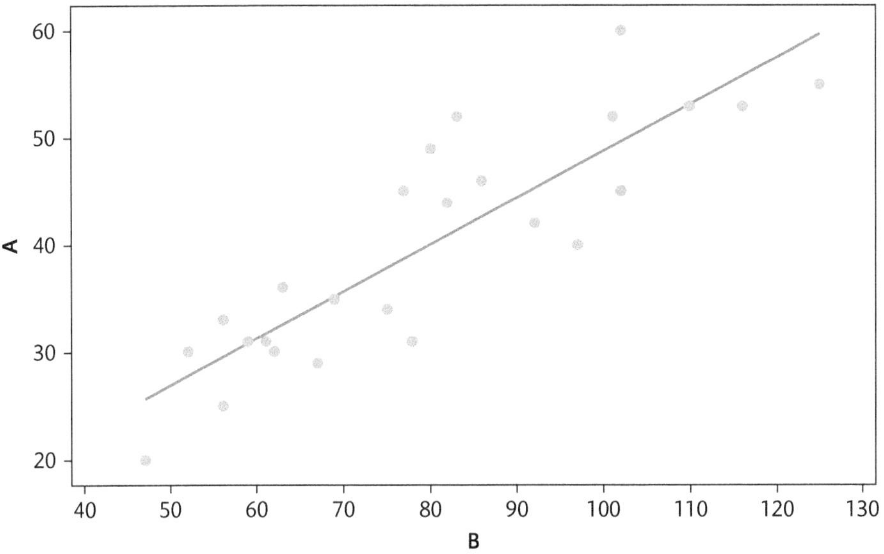

FIGURE 8.1
Pearson correlation between A and B (Graphic from Minitab software).

although the mean value of A is almost exactly one half of the mean value of B, the gradient of the straight line is also far from 0.5, being instead 0.44. This means that there are other factors present in the relationship between A and B, one of which is that the straight line does not pass through zero, but cuts the A axis where A = 4.

We can bring these numbers together in a regression analysis to determine more precisely the extent to which the variability in index A is related to the variability in index B. This index is called r^2, and in this case it is 75%. This value tells us that 75% of the variability of B can be accounted for by its relationship with A. Conversely, 25% of the variability in B must be due to something else apart from A—in this case, random chance.

This regression analysis is clearly only between two indices (A and B) and so is described as univariate, as we are performing only one analysis. When there are two separate but linked analyses (such as A with B and A with C), it is described as bivariate. When there are three or more analyses, the analysis is multivariate, and consequently a multivariate regression analysis is performed.

SELF-CHECK 8.1

Explain the components of the equation of a straight line. Which are the dependent and independent variables?

Two equations of the straight line

Now let us consider two equations of the straight line, plotting the relationship between a marker of renal damage, serum creatinine, and two other indices, systolic blood pressure (SBP) and levels of a new molecule in the blood, Substance X. We might hypothesize that Substance X is a better marker of renal damage (via a relationship with creatinine) than is SBP. We can test this hypothesis by collecting data from selected patients, and then analysing that data. Table 8.2 shows the descriptive statistics of these groups.

Each of these three data sets has a normal distribution, though note that Substance X has been measured in only 19 of the subjects.

If we express the relationship between creatinine and SBP in the form $y = mx + c$ we find that:

$$\text{Creatinine} = 1.86\ \text{SBP} - 128$$

By contrast, the relationship between creatinine and Substance X is:

$$\text{Creatinine} = 0.73\ \text{Substance X} + 33$$

We can combine these data sets into a single analysis, and express this in a graph that also shows the two lines of best fit (Figure 8.2).

TABLE 8.2 **Descriptive statistics of data sets A and B**

Variable	N	Mean	SD	Minimum	Q1	Median	Q3	Maximum
Creatinine	24	125	57	36	4	124	178	192
SBP	24	136	31	90	108	136	165	186
Substance X	19	121	58	24	72	124	172	210

SD = standard deviation; Q1 = 25th (lower) quartile; Q3 = 75th (higher) quartile; SBP = systolic blood pressure

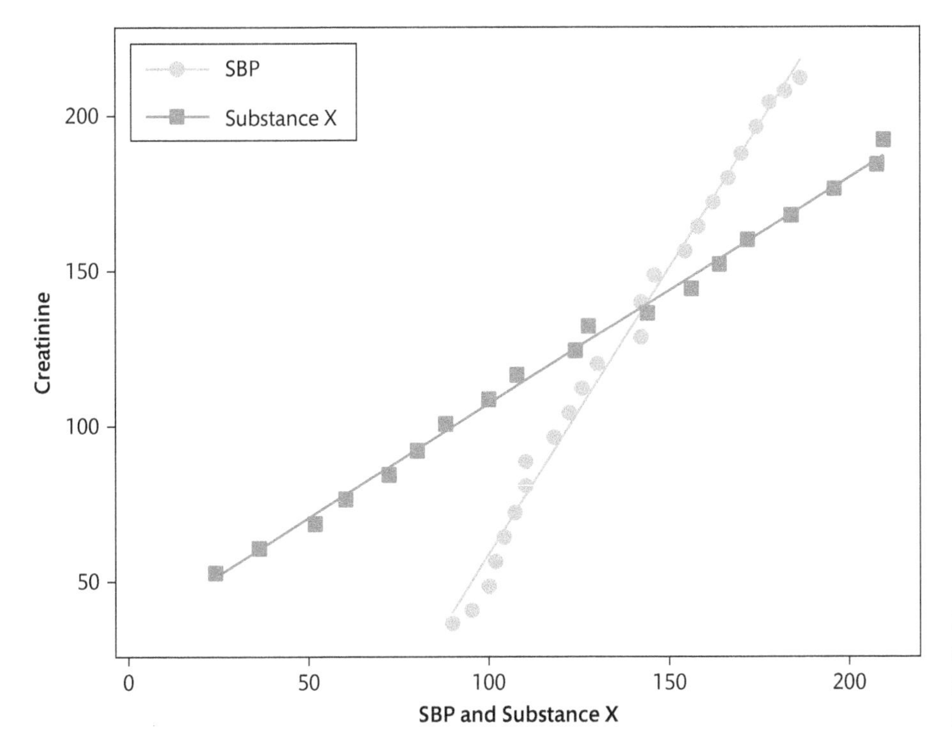

FIGURE 8.2
Scatterplot of creatinine versus SBP and Substance X (Graphic from Minitab software).

One line of best fit represents the equation (*creatinine*) = 1.86 (*SBP*) – 128; the other represents the equation (*creatinine*) = 0.73 (*Substance X*) + 33.

It is clear that the two relationships are very different. The relationship between creatinine and SBP is considerably 'sharper', with a higher gradient, than that between creatinine and Substance X, which is more 'gentle', with a lower gradient. But is this difference statistically significant? We can determine this formally using a **bivariate regression analysis**.

Bivariate regression analysis the determination of which of two factors is independently related to a third factor

Bivariate regression analysis

In this example we look at the potential interaction between two indices on the x axis (Substance B and SBP) and their relationship with the index on the y axis (creatinine). Because we are determining the response of creatinine to Substance B and/or SBP, it is called the response, or responding variable. Similarly, because we seek to determine if we can predict levels of creatinine by looking at levels of Substance B and/or SBP, the latter may be described as the predictor variable. And because Substance X or SBP may influence or predict levels of creatinine independently of each other, then these two are independent variables: they may act on the dependent variable of creatinine.

If we simply merge the SBP data and Substance X data into a combined data set we can take an overview. The equation of the straight line for the merged set of data is:

$$\text{Creatinine} = 1.0\ \text{Combined data} - 6$$

Note from this equation that the gradient of this merged line is 1, so for each increase of one unit of the combined data, the creatinine also increases by one. The line of best fit for this analysis is shown in Figure 8.3.

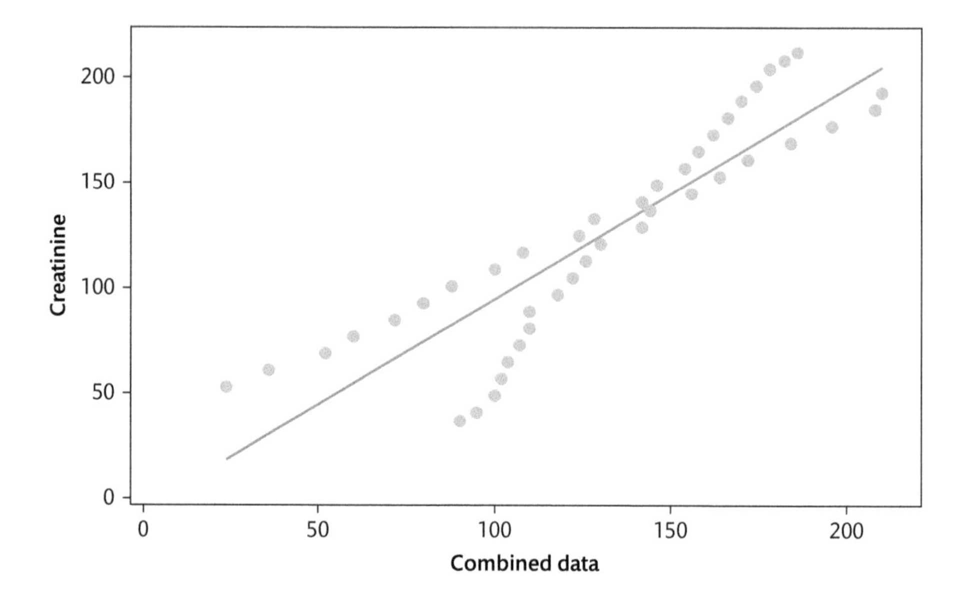

FIGURE 8.3
Scatterplot of creatinine versus the combined set of SBP and Substance X (Graphic from Minitab software).

By a simple 'eyeball' test of the graph, it appears that the Substance X/creatinine data lies closer to the overall line of best fit than does the SBP/creatinine line. This suggests that the relationship between Substance X and creatinine is stronger than that between SBP and creatinine. However, we must test this suggestion formally with a linear regression analysis, where the statistical software provides a more complex equation of the straight line:

$$Creatinine = 0.76\,(Substance\ X) - 0.06 + (SBP) + 37.5$$

Here, there are two 'm' values (0.76 linked to Substance X and -0.06 linked to SBP) and one 'c' value (37.5). The regression analysis dissects the relative importance of the two independent variables (SBP and Substance X) and their ability to predict the variation in the dependent variable of creatinine (Table 8.3).

The first row in this table (Constant) presents data regarding the point where the straight line crosses the y axis (the coefficient, which the regression equation evaluates as 37.5). This value is important in more complex analyses, but in our case we note that the p value is significant (that is, $p = 0.02$). This simply tells us that this point is significantly different from zero.

The two coefficient figures for Substance X and SBP simply repeat the respective gradients (the m value in $y = mx + c$) of each line of best fit. However, the SE (standard error) of the coefficient tells us how accurate each data set actually is. The fourth column—the t value—is the coefficient

TABLE 8.3 **Regression data for Substance X and SBP**

Predictor	Coefficient	SE of the coefficient	t value	P value
Constant	37.5	14.5	2.58	0.02
Substance X	0.76	0.08	9.66	< 0.001
SBP	-0.06	0.19	-0.31	0.763

SE = standard error, SBP = systolic blood pressure

divided by its SE, and is allied to the t in the Student's t test. Finally, in column 5, the significance of this t value is presented as a probability.

We can interpret this data as follows. The column on the far right of Table 8.3 tells us that the probability that the relationship between Substance X and creatinine is very strong at $p<0.001$, whilst that between SBP and creatinine is not significant at $p=0.763$. This does not mean that SBP can no longer be regarded as a marker of renal damage, merely that Substance X is more closely linked to creatinine than is SBP. Furthermore, the conclusions cannot be extrapolated very far outside this setting as the sample size (at most, only 24 people) is so small. Nevertheless, this information is excellent pilot data that should prompt additional studies.

Multiple equations of the straight line

By extending the same mathematical principles in bivariate analyses to that of several sets of data, we move to **multivariate linear regression analysis** (MVLRA), with the fusion of a series of different equations of the straight line. To illustrate this, let us consider which of four independent variables (IV) (A, B, C, and D) has a significant relationship (if at all) with a dependent variable (DV). The four equations of the straight line of these variables are:

Multivariate linear regression analysis
the determination of which of three or more factors is, or are, independently related to a dependent factor

$$DV = (0.42\ IV\ A) - 3$$

$$DV = -(0.66\ IV\ B) + 76$$

$$DV = (0.56\ IV\ C)\ C - 4$$

$$DV = (0.03\ IV\ D) - 39$$

The mathematics of the MVLRA combines these four separate equations into one, which gives ...

$$DV = (0.19\ IV\ A)\ - (0.32\ IV\ B) + (0.078\ IV\ C) - (0.014\ IV\ D) + 32$$

Note that the gradient (m) components for each variable in the combined equation are different from those in each single equation. In the case of Variable A the m value has reduced from 0.42 to 0.19. This is because some of the effect of A on the dependent variable can be accounted for by each of the relationships that A has with B, C, and D. Similarly, by itself, B has quite a marked influence on the dependent variable, with an m value of -0.66. However, when adjusting this effect of B for any effects of A, C, and D, then the influence of B on the dependent variable falls by over half to -0.32.

The relationships between the dependent and independent variables (A, B, C, and D) and their respective equations of the straight line are shown graphically in Figure 8.4, and illustrate the difficulty in trying to determine precise relationships between the indices without the help of statistical software. In highly complex analyses involving thousands of patients and dozens of independent variables, use of a powerful statistical package is mandatory.

The results of the regression analysis are shown in Table 8.4, and demonstrate that variables A and B are both independently related to the dependent variable, with probabilities of $p<0.001$ and $p=0.002$ respectively. The failure of the analysis to find a relationship between variable D and the dependent variable is perhaps not surprising, judging by the almost horizontal line of best fit on Figure 8.4. However, we may have expected a relationship between variable C and the dependent variable, as the slopes of the lines of best fit for variables A and C are reasonably close.

The complex mathematics of the MVLRA consider the relationship between variables A and C to be so close that it cannot be genuine, and so is likely to be due to random chance. Indeed,

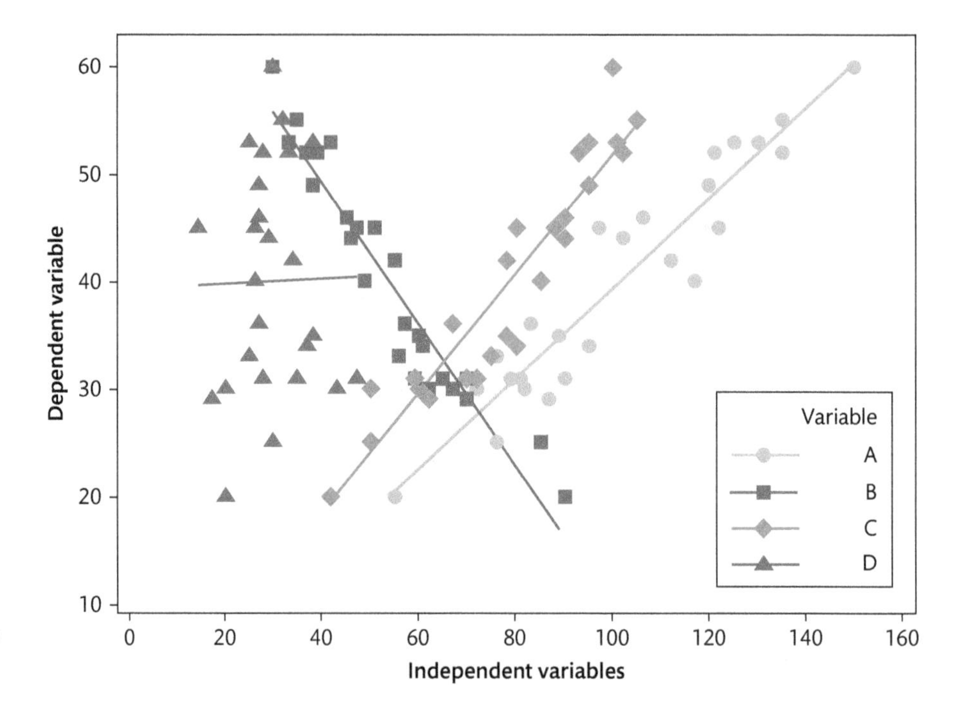

FIGURE 8.4
Scatterplot of the dependent variable and four independent variables (Graphic from Minitab software).

TABLE 8.4 **Regression data for the independent variables**

Predictor	Coefficient	SE of the coefficient	t value	P value
Constant	32.2	11.0	2.92	0.009
Variable A	0.19	0.04	4.45	<0.001
Variable B	−0.32	0.09	−3.58	0.002
Variable C	0.08	0.08	1.00	0.994
Variable D	−0.01	0.06	−0.24	0.810

the correlation coefficient between variables A and C is 0.89, which is exceptionally strong. Consequently, the analysis has determined that variable A is more strongly related to the dependent variable, and that any effect of variable C on the dependent variable is in fact due to its relationship with variable A.

Renal function in diabetes

Now that we have explored the principles of MVLRA with some hypothetical models, we can return to the real-life hypothesis that serum albumin, serum cholesterol, SBP, and HbA1c all contribute significantly to the variation in eGFR. The raw data for this study is summarized in Table 8.5.

We seek relationships between these indices by correlation. As all indices have a normal distribution, then Pearson's correlation is again appropriate, as is shown in Table 8.6. Although some of the correlations between albumin, SBP, cholesterol, and HbA1c on the right-hand

TABLE 8.5 **Descriptive statistics of renal function data**

Variable	N	Mean	SD	Minimum	Q1	Median	Q3	Maximum
eGFR	80	59	17	23	46	57	73	90
Albumin	80	37	7	24	32	37	41	53
SBP	80	141	10	122	134	139	145	160
Cholesterol	80	5.0	1.1	2.3	4.2	5.2	5.8	7.5
HbA1c	80	58	11	34	50	59	65	83

SD = standard deviation; Q1 = 25th (lower) quartile; Q3 = 75th (higher) quartile.
eGFR = estimate glomerular filtration rate, SBP = systolic blood pressure, HbA1c = glycated haemoglobin

side of Table 8.6 are of interest, we are focusing on their relationship with the eGFR on the left. Of these, the relationship between cholesterol and eGFR is not significant (p=0.317), and will therefore be removed from further analysis. With one less index to consider, the analysis becomes simpler and more powerful. We therefore focus on the remaining three indices–the relationships that eGFR has with HbA1c (which is weak: p<0.05), with albumin (which is moderate: p<0.01), and with SBP (which is strong: p<0.001). So we have already achieved part of the answer to our hypothesis: there is no clear relationship between cholesterol and eGFR.

Although we have found that HbA1c, SBP, and albumin are significantly related to eGFR, the relationship between albumin and SBP is also significant, as is that between HbA1c and albumin. This brings confusion to the general picture as it may be that some part of the relationship between SBP and eGFR may be because of the relationships between SBP and albumin, or between eGFR and albumin. To apportion weight to the different relationships we use MVLRA.

Regression analysis

In this analysis, eGFR is the dependent variable (on the y axis), whilst the other indices (those on the x axis: HbA1c, SBP, and albumin) are the independent or predictor variables. This is because our broad hypothesis asks which of HbA1c, SBP, and albumin predict the

TABLE 8.6 **A matrix of Pearson correlations**

	eGFR	Albumin	SBP	Cholesterol
HbA1c	r=–0.23, p=0.036	r=-0.25 p=0.026	r=–0.02, p=0.846	r=–0.05, p=0.66
Cholesterol	r=0.11 p=0.317	r=0.02 p=0.861	r=-0.17 p=0.116	
SBP	r=–0.53 p<0.001	r=–0.35 p=0.001		
Albumin	r=0.45 p=0.002			

eGFR = estimated glomerular filtration rate, SBP = systolic blood pressure, HbA1c = glycated haemoglobin eGFR. We have already ruled out cholesterol as a predictor of eGFR as the two indices have no clear relationship (a correlation coefficient of r=0.11). The three individual y = mx + c equations are:

- $\text{eGFR} = m_1\ \text{HbA1c} + c_1$
- $\text{eGFR} = m_2\ \text{SBP} + c_2$, and
- $\text{eGFR} = m_3\ \text{Albumin} + c_3$

where m_1, m_2, and m_3 are the gradients of the three straight lines and c_1, c_2, and c_3 are the intercepts of the straight line on the y axis (which denotes the value of eGFR). These three separate equations can be merged into one:

$$\text{eGFR} = m_1\ \text{HbA1c} + m_2\ \text{SBP} + m_3\ \text{Albumin} + c_1 + c_2 + c_3$$

The statistical software used to perform the MVLRA can then be directed to fill in the missing parts of the equation and determine the individual m and c values. The software has formed the equation with the independent variables in order of their m values:

$$\text{eGFR} = 0.85\ \text{Albumin} - 0.69\ \text{SBP} + 0.15\ \text{HbA1c} + 116$$

Note that the m value of 0.69 associated with the SBP is negative: this is because the relationship between eGFR and SBP is inverse, with a correlation coefficient of -0.53 (Table 8.6).

So we find that the largest 'm' values are those that are linked to albumin (0.85) and SBP (0.69), with that of HbA1c being much smaller (0.15). With these values in mind, we could perhaps assume that the strongest influence on the variability of the eGFR is albumin, because it has the m value of 0.85. But we would be wrong. This is because we need to factor in the 'tightness', or 'accuracy' of each m value, which can be estimated by determining the standard error (SE) of each value, as shown in Table 8.7.

It is important to understand that the regression software has determined which of the three factors are independently associated with the dependent variable of eGFR. Both albumin and SBP are significantly linked with the eGFR independently of each other and of HbA1c. However, the relationship between HbA1c and eGFR, although significant in the univariate Pearson correlation, was not independent of albumin and SBP. It is very likely that some of the relationship between HbA1c and eGFR would have been due to its relationship with albumin and SBP.

Albumin

Table 8.7 shows that the SE of the m value of the albumin data is 0.26, which is far less than that of the m value itself (0.85). So in a manner similar to the relationship between mean and

TABLE 8.7 **Regression data of eGFR and the independent variables**

Factor	m (gradient) coefficient	SE of the coefficient	t value	p value
Constant	116	30.4	3.81	<0.001
Albumin	0.85	0.26	3.27	0.002
SBP	–0.69	0.16	–4.31	<0.001
HbA1c	0.15	0.14	1.1	0.275

SBP = systolic blood pressure, HbA1c = glycated haemoglobin.

SD that helps to determine distribution, dividing the m value by its SE gives a 't' value (that being 3.27). And in a similar manner to how we combine the sample size with the t value in the Student's t test to derive probability, we do the same thing in a MVLRA, and so determine the extent to which this data is significant. In this case we have $p=0.002$, telling us that the relationship between eGFR and albumin is highly significant.

A note on the terminology of regression analysis. Many statisticians and types of statistical software describe the m value coefficient (the gradient) as beta (β). We will revisit this in Section 8.2.

SBP

In applying the same mathematics to the SBP data, the m value of 0.69 divided by its SE gives a t value of 4.31, which is significant at $p<0.001$. So although the m value of the albumin data is greater than that of the SBP data, the SBP data is 'tighter', with less variability, and has a much smaller SE, leading to a larger t value and smaller p value. Note also that the coefficient and the t value are both negative. This means that as SBP increases, then eGFR decreases: the two variables have an inverse relationship.

HbA1c

Similarly, the relationship between the m value and its SE (which is relatively large) for the HbA1c data is such that the t value is small and so not significant. Although the relationship between eGFR and HbA1c was mathematically significant in the univariate correlation analysis ($r=-0.23$, $p=0.036$), it was not sufficiently strong when matched with the much stronger relationships that the eGFR data has with SBP and albumin in a multivariate analysis. Therefore, we can say that in a multivariate analysis, HbA1c is not a predictor of eGFR.

On the whole, the significance of these three data sets with regard to the eGFR to some extent mirrors those of the correlation coefficients in Table 8.6.

The constant

At the top of Table 8.7 is the data for the combined c values (the constant), where the merged equation of the straight line crosses the y axis. This p value is simply emphasizing the point that 116 is significantly distant from zero.

Attributable proportion

Finally, as mentioned earlier, the regression analysis gives a very useful index, the r^2. This tells us of the extent to which the variability in the y axis (that is, the eGFR) can be accounted for by all the x axis indices in the model (that is, the SBP and albumin). In this case, our model gives an r^2 of 35.7%, telling us that slightly over a third of the variability in the eGFR can be attributed to the combination of SBP and albumin. On the other hand, it also tells us that almost two-thirds (64.3%) of the variability in the eGFR must be attributed to other factors. What the other factors are, we cannot say, but may speculate that some could reflect inflammation, so that an index such as CRP may be useful.

SELF-CHECK 8.2

What is the importance of the coefficient, its standard error, and the t value in linear regression analysis?

Stepwise analysis

We have clarified the relationship between eGFR and the four possible predictors by removing cholesterol and HbA1c. But of the remainder, which is more important? It appears that SBP is the more important as it has p<0.001, whereas that of albumin is less, at p=0.002. However, we must determine whether this is the case in a formal analysis. To do so we can use a sub-analysis of MVLRA called **stepwise analysis**. The mathematics of this approach determines which of the two indices contributes most to the variability in eGFR. It then recalculates this figure in terms of the second most important index, and so on; although in this case there are only two steps. This is summarized in Table 8.8.

TABLE 8.8 **Stepwise regression analysis**

Independent variable	Step 1	Step 2
SBP	t=−5.6; p<0.001	t=−4.4; p=0.001
Albumin		t=3.1; p=0.003
r^2	22.2%	35.7%

SBP = systolic blood pressure

Stepwise analysis **the determination of the order in which a number of independent factors are related to a dependent factor**

The first step has identified SBP as the primary influence on eGFR. The software then adjusts this data for any potential effect of albumin. Note that in the second step, the t value has fallen from −5.6 to −4.4, and the p value has also become larger, indicating that some part of the effect of SBP was due to an effect of albumin. Nevertheless, the effect of albumin is still strong (with a t value of 3.1) and independent of SBP.

An additional bonus is that the software gives us more information about the attributable proportion, r^2. We have already noted that the regression model accounts for 35.7% of the variability in eGFR. The stepwise model tells us that SBP itself accounts for 22.2% of this variability, and so albumin by itself must account for 13.5% of the variability in eGFR.

Interpretation

All our analyses have done is to point out associations and relationships. It is up to us to interpret. The stepwise analysis indicates that increasing blood pressure and falling serum albumin are related to deteriorating renal function; but are either of these mechanistic? It is difficult to conceive of a pathophysiology where albumin somehow causes renal failure, but there is ample evidence that hypertension is directly responsible for kidney disease.

In this particular risk factor the kidney disease is called diabetic nephropathy, and is an important predictor of large artery disease that leads to myocardial infarction and stroke. The link with albumin is likely to be the result of renal disease, as damage to the glomerulus may well lead to the loss of albumin into the urine (that is, albuminuria), and so falling blood levels of this important protein. As hinted at previously, a criticism of this analysis is its failure to include the marker of inflammation, and cigarette smoking may well have an effect.

Problems and assumptions in linear regression analysis

Perhaps the most obvious assumption about linear analysis is that all of the indices have a linear (straight line) relationship with each other. However, this is not always the case: some may have exponential, curvilinear, or sigmoidal relationships. Consequently, the assumption of a linear relationship may therefore lead to error. Some complex statistical software packages have sub-programs to detect non-linear relationships, and, if present, will make an adjustment.

Another possible problem is the variability in each individual data set, which is assumed to be constant throughout the full range of that data set. For example, as regards age, a well-balanced data set includes roughly the same number of people in each decade: a data set with poor balance has more 50 to 60-year-olds and 70 to 80-year-olds than 60 to 70-year-olds.

In the examples used in this chapter, all of the indices—both dependent and independent—have had a normal distribution, and analytical software generally assumes this to be the case. However, consider the following data set from a group of patients with atherosclerosis (Table 8.9), and their possible relationships with a new molecule, Substance Y.

TABLE 8.9 **Descriptive statistics of atherosclerosis patients**

Variable	N	Mean	SD	Minimum	Q1	Median	Q3	Maximum
Cholesterol	200	5.1	0.9	2.5	4.6	5.1	5.8	8.0
HDL	168	1.4	0.4	0.6	1.1	1.4	1.6	2.4
LDL	154	3.2	0.8	1.0	2.6	3.2	3.8	4.7
Glucose	200	4.1	0.9	2.8	3.5	4.0	4.6	7.0
eGFR	200	63.8	18.5	23.0	50.0	62.0	74.0	90.0

SD = standard deviation; Q1 = 25th (lower) quartile; Q3 = 75th (higher) quartile. HDL = high density lipoprotein, LDL = low density lipoprotein

These five indices all have a normal distribution. In hypothesizing the ability of these indices to independently predict levels of the dependent variable, Substance Y, a MVLRA was performed, the outcome being shown in Table 8.10. The column on the far right gives the probability that the particular index is related to Substance Y. With probabilities very much less than 0.05, this analysis clearly identifies cholesterol (p=0.008) and glucose (p=0.006) as predictors of levels of Substance Y, although LDL has only just missed statistical significance (p=0.06).

TABLE 8.10 **MVLRA for Substance Y**

Predictor	Coefficient	SE of coefficient	t value	p value
Constant	132.2	39.0	3.39	0.001
Cholesterol	0.19	7.0	2.7	0.008
HDL	−16.8	12.4	−1.36	0.175
LDL	15.3	8.1	−1.89	0.06
Glucose	13.4	4.8	2.79	0.006
eGFR	−0.001	0.245	−0.00	0.997

HDL = high density lipoprotein, LDL = low density lipoprotein, eGFR = estimated glomerular filtration rate

However, we have made the assumption that Substance Y has a normal distribution. But in fact the descriptive statistics for this index are mean 53, standard deviation 56.8, minimum 5, lower quartile 10, median 18, upper quartile 83.5, and maximum 250, which indicates a non-normal distribution. Accordingly, this analysis may be flawed.

Cross reference

Chapter 7 discussed the value of transforming data to allow an analysis of variance

Transformation

One way around the problem of different types of distribution is to transform that data in order to give it a normal distribution. The forms of transformation include taking a logarithm, the square, or the square root of the particular index.

If we consider the analysis in Table 8.10, we need to transform the Substance Y data into a form that has a normal distribution, commonly achieved by taking the log of the data. In doing so, the descriptive statistics of Substance Y become mean 3.33, standard deviation 1.17, minimum 1.61, lower quartile 2.3, median 2.89, upper quartile 4.42, maximum 5.52, which has a normal distribution. Using these data we obtain a different regression analysis, as shown in Table 8.11.

TABLE 8.11 **MVLRA for log Substance Y**

Predictor	Coefficient	SE of coefficient	t value	p value
Constant	5.12	0.82	6.25	< 0.001
Cholesterol	0.36	0.15	2.41	0.017
HDL	–0.30	0.26	–1.14	0.255
LDL	–0.29	0.17	–1.7	0.092
Glucose	0.31	0.1	3.07	0.003
eGFR	–0.01	0.01	1.0	0.912

HDL = high density lipoprotein, LDL = low density lipoprotein, eGFR = estimated glomerular filtration rate, SE = standard error

The effect of transforming Substance Y into an index with a normal distribution is to change its relationships with several of the independent variables. The relationship with cholesterol has become less significant by about a factor of two, but the relationship with glucose has become stronger by a factor of two; the relationship with LDL has also become less significant. These findings may have implications for both public health and the treatment of diabetes.

This analysis makes several other assumptions, including that the data is collected from a random population (that is, it is unbiased). If it is possible that such a bias is present, this is acknowledged. However, if present, these anomalies only become significant when the sample size is small.

Logistic regression

a method for determining which of a number of independent factors are related to a categorical dependent factor

8.2 Logistic regression analysis

The second type of regression analysis has a key feature in common with linear regression: in both types of analysis we use a number of independent variables to help us to predict a single dependent variable. However, in **logistic regression** the dependent variable is not continuously variable (as in the example of creatinine we have already looked at), but is a categorical variable (such as sex or the presence or absence of cancer).

The initial aspect of the mathematics of logistic regression is different from linear regression as there is no clear equation of a straight line in the former. Instead, the data relies on the probability that one or more of the independent variables has a significant influence on whether an individual is associated with one of the dependent variables. This probability can be described as an **odds ratio** odds ratio $= \frac{ad}{bc}$. It is important to state now that an odds ratio is not a hazard ratio. The latter will be addressed in Section 8.3.

Odds ratio

a simple method for assessing the likelihood of a factor being related to a particular category

Calculating an odds ratio

In a group of individuals, some may have cancer, whilst others will be free of cancer (a typical binary [two options] classification). If p is the probability of having cancer, then 1-p is the probability of not having cancer. We can use the ratio between p and 1-p to estimate the overall probability of having cancer or not having cancer. By extending this principle to two groups, such as smokers and non-smokers, we can estimate the relative risks that each group has of also having cancer.

Let us consider an example. Suppose that of 567 non-smokers, 40 have some form of lung disease such as cancer, chronic obstructive pulmonary disease, bronchitis, shortness of breath, asthma, and emphysema. Therefore, the probability of having lung disease is 40/567 = 0.07055 (about 7%), and the probability of not having lung disease is 1-0.07055 = 0.92945 (about 93%). Hence the overall probability, or likelihood, of having lung disease for a non-smoker (in this population) is 0.0705/0.92945 = 0.076. This probability is also described as the odds (a term adopted from gambling) of a particular outcome—in this case, of having lung disease.

Now suppose that of 234 smokers, 65 have lung disease. These data give a probability of having lung disease as 65/234 = 0.27777 (about 28%), and of not having lung disease as 1-0.278 = 0.72222 (about 72%); so that for a smoker the odds of having lung disease is 0.27777/0.72222 = 0.385.

Given these two odds—of having lung disease as a non-smoker or as a smoker—we can then combine them to determine the risk of lung disease due to smoking as an odds ratio:

$$\text{Odds ratio of lung disease due to smoking} = 0.385/0.076 = 5.07$$

This tells us that smokers have over a five-fold risk of lung disease compared with non-smokers. However, as with all analyses, there is an inherent likelihood of error, and we should state how confident we are that our data is robust, and not due to error, in the form of a confidence interval (CI) around this odds ratio. A 95% CI can be determined from the SE in the same way that it can be with other methods—in this case, the odds ratio ± the standard error (SE) × 1.96. Hence the odds ratio (95% CI) = 5.13 (3.26 – 7.85). As the odds ratio is far from 1, and the 95% confidence intervals fail to include 1, we can be assured that the data is highly significant, with $p<0.001$.

Cross reference

Confidence intervals: The method of defining assurance is explained in more detail in Chapter 6

The odds ratio calculation can be summarized as a 2×2 table, as shown in Table 8.12. Here, the number of subjects with or without a particular disease, and with or without a particular index such as a risk factor, are placed in one of four boxes. The odds ratio for one of the subjects with the risk factor having the disease, compared to one of those free of the risk factor not having the disease is given by the solution to the summary (a × d)/(b × c). Substituting the smoking and lung disease figures cited above also give an odds ratio of lung disease if a smoker as 5.06. Incidentally, this format is of course exactly the same as a Chi-squared analysis, which gives $p<0.001$ for the hypothesis that there is a difference in the proportions.

TABLE 8.12 **Template for an odds ratio**

	Risk factor present	Risk factor absent	Total
Disease present	a	b	a+b
Disease absent	c	d	c+d
Total	a+c	d	a+b+c+d

Relative risk

Relative risk

the ratio of the probability of an event occurring in a group exposed to factor compared to the probability of the same event occurring in a group not exposed to that factor

This common term frequently arises when statistics are discussed, and now is a good time to address the problem, which is that some incorrectly use it interchangeably with odds ratio. The **relative risk** (sometimes described as risk ratio, hence the difficulty) is a tool used to estimate the risk of the development of disease in a group exposed to a factor, compared to the estimated risk of the development of the disease in the unexposed group. This sounds very much like the definition of an odds ratio, except the relative risk is used in a cohort study where subjects are followed over a long period, perhaps years, hence *development* of the disease. An odds ratio can be generated from data collected at a single time point. Relative risk should not be used for case-control studies.

Confusion is compounded by the use of the same 2×2 in Table 8.12, with its a, b, c, and d groups. But the mathematics are slightly different, in that relative risk $= \frac{\frac{a}{a+c}}{\frac{b}{b+d}}$. If we incorrectly analysed our smoking/lung data with this equation it would give 3.94. A further confusion is that where a and b are very small (as in rare diseases), the relative risk equation approximates to that of an odds ratio. Should it be necessary, a 95% confidence interval can also be calculated for this statistic, which calls for determination of the standard error of the logarithm of the relative risk.

Note These data (and others in the chapter) have been constructed simply as examples to illustrate a point. They must not be taken as firm public health data. Cigarette smoking is a powerful cause of lung cancer, but the exact risk varies with age, the number of years smoking, the number of cigarettes smoked daily, and other factors.

Types of logistic regression

In logistic regression analysis, the dependent variable is a category. These categories can themselves be binary (where there are two categories) or multinomial (where there are three or more categories). Examples of binary categories include being male or female, having blue eyes or brown eyes, or being left-handed or right-handed. There is no natural order of one category being considered before or after the other.

Multinomial categories can be classified as one of two types. Categories such as mild, moderate, or severe pain are in an order, where the pain gets worse. These categories are described as ordinal. Other categories have no clear order. The blood groups A, B, AB, and O are an example of this. There is no natural order—there is no scientific reason for group B to be placed between group A and group AB. These categories are called nominal. This is important because in ordinal analysis the software assumes that pain progresses through the three stages. We will now explore the three types of logistic regression with worked examples.

TABLE 8.13 **Descriptive statistics of data in a binary logistic regression**

Variable	N	Mean	SD	Minimum	Q1	Median	Q3	Maximum
Age	100	56	9	25	51	57	62	74
BMI	100	28.8	7.5	15	23	26	36.7	45
Calcium	100	2.44	0.24	1.96	2.26	2.41	2.59	3.13
D-dimers	100	309	88	135	241	297	375	500
eGFR	100	76	16	43	63	73	88	90

eGFR = estimated glomerular filtration rate, BMI = body mass index

Binary logistic regression

In seeking to test the hypothesis that age, body mass index (BMI), serum calcium, plasma D-dimers, and the eGFR (the independent variables) all predict the presence or absence of cancer (the dependent variable), we would first need to define our sample size via a power calculation. The mathematics of the power calculation in this setting is more demanding than in a MVLRA, so that a sample size of 100 would be appropriate, of whom at least 20 would need to have an endpoint (in this case the presence of cancer). Once the data has been collected, it can be entered into a statistical spreadsheet for analysis. The descriptive statistics are shown in Table 8.13.

Of these 100 subjects, we find that 23 have cancer and 77 are free of the disease. The next step is to perform a univariate analysis of the data according to the presence or absence of disease (Table 8.14). As all the indices have a normal distribution, then the univariate analysis consists of a series of t tests.

These find that the group of patients with cancer have a significantly higher BMI, higher calcium, and a lower eGFR than those without cancer. By contrast, the differences in age and D-dimers are not statistically significant, and so are unlikely to be useful in determining the likelihood of the presence or absence of cancer. Consequently, we can eliminate these from further analysis, which will therefore be simpler and so more powerful. We are now ready to perform a binary logistic regression analysis, hypothesizing that BMI, calcium, and eGFR

TABLE 8.14 **Analysis according to disease status**

	Presence of cancer (n=23)	Absence of cancer (n=77)	p value
Age	58 (8)	55 (9)	0.145
BMI	31.8 (7.3)	27.9 (7.4)	0.031
Calcium	2.61 (0.26)	2.39 (0.21)	0.001
D-dimers	300 (84)	312 (90)	0.58
eGFR	68 (11)	78 (17)	0.001

eGFR = estimated glomerular filtration rate, BMI = body mass index. Data are mean and standard deviation

TABLE 8.15 **Binary logistic regression analysis**

Predictor	Beta coefficient	SE of the coefficient	Z value	p	Odds ratio	95% CI
Constant	-8.66	3.27	-2.65	0.008		
BMI	0.042	0.036	1.15	0.251	1.04	0.97–1.12
Calcium	3.836	1.255	3.06	0.002	1.15	1.08–1.29
eGFR	-0.0457	0.0189	-2.41	0.016	0.92	0.87–0.95

eGFR = estimated glomerular filtration rate, BMI = body mass index, SE = standard error. Modified from a Minitab printout

independently predict the dependent variable, cancer or no cancer. It is likely that the statistician, when entering the data into the spreadsheet, will code the presence of cancer as 1 and the absence of cancer as 0. This analysis is set out in Table 8.15.

The table summarizing the logistic analysis shows similarities with that of a linear regression analysis (Table 8.10). The principal similarity is the (beta) coefficient, which in linear analysis is the gradient of the equation of the straight line; the SE of the coefficient is also common between logistic and linear analyses. However, in this logistic regression the coefficient refers to the probability of an event occurring (that is, of an individual being either in the 'cancer' or the 'not cancer' group) when all of the dependent variables have been accounted for.

The next step in logistic regression is also similar to that in linear regression, which is to calculate the ratio between the beta coefficient and its SE. This value, Z, is called the Wald statistic, and is the equivalent of the t statistic in linear regression. With both the Student's t and the Wald statistic, the larger the value, the more significant the particular relationship.

Finally, the statistical software combines the Z value with the sample size (the number of subjects in the study) to give the probability, and then the odds ratio with its 95% CI. This is analogous to the simple determination of a difference between two sets of normally distributed data in a t test, in which the probability also depends on the difference between the means of the two groups and the sample sizes.

Note that the Z values of BMI and calcium are positive, whereas the Z value for eGFR is negative. This means that cancer is associated with an increase in BMI and calcium but a decrease in the eGFR. The positive or negative values depend entirely on whether or not the cancer/not cancer binary notations are entered by the statistician as 0/1 or as 1/0. But whatever the 0/1 combination, BMI and calcium act in the opposite direction from the eGFR. Note that the Constant value is markedly significant at p=0.008. This tells us that there is a great deal of the probability of being in a cancer/not cancer group that is not accounted for by all the independent variables. This is hardly surprising—it would be astonishing if we could account for all cancer with just a handful of predictors!

BMI

This analysis has found that the significance that BMI had in univariate analysis has become non-significant in multivariate analysis. Note that the odds ratio is 1.04, telling us that obesity increases the risk of having cancer by 4%. However, the 95% CIs for this lie either side of 1 (that is, 0.97 and 1.12), meaning the reliability of this 4% is unacceptable. This is not so surprising as the univariate association between cancer and BMI (as shown in Table 8.14) was weak (p=0.031), and so bringing in other more powerful indices (calcium, eGFR) weakened any effect of BMI so much so that it became not significant. Put another way, the reason for BMI losing significance may be that some of the variability of the BMI data could have been accounted for by relationships with calcium and the eGFR. Indeed, the Pearson correlation between calcium and BMI was r=0.24, giving p=0.017.

Calcium and eGFR

The key result of this analysis is that both calcium and eGFR have remained as independent significant predictors of cancer. The independence of each index is demonstrated by the Pearson correlation between calcium and eGFR giving a value of r=–0.057, giving p=0.576.

Note from Table 8.15 that the odds ratio and its 95% CI for calcium are greater than 1, that is, 1.15 (1.08–1.29). This means that raised calcium increases the risk of cancer by some 15%. This is in contrast with the odds ratio and 95% CI for eGFR, all of which are less than 1, being 0.92 (0.87–0.95). This means that a reduction in the eGFR is associated with cancer. In fact, by subtracting 0.92 from 1 we get 0.08, and can translate this as meaning that a higher eGFR brings an 8% protection from the risk of being in the cancer group.

Interpretation

In this particular example, of the five indices of age, BMI, calcium, D-dimers, and eGFR, only calcium and eGFR are independently associated with cancer. It is unlikely that either of these indices is causal of cancer, but raised calcium may be the product of cancer within bone. A falling eGFR reflects deteriorating renal function, but we cannot determine whether this is due to primary renal cancer, or another factor. But we must note that the study is small, some would say very small, and others too small to warrant analysis—only 23 cases are being compared to 77 controls.

Nominal and ordinal logistic regression

Nominal and ordinal logistic regression analyses consider three or more categories. In the case of nominal regression, the categories are in no clear order (such as lung cancer, breast cancer, and prostate cancer), but in ordinal regression, there is a clear order (such as increasingly bad pain: mild, moderate, and severe pain). In our worked example, we will use the same five independent variables as in the binary logistic regression (Table 8.16). We will first make an initial univariate analysis of this data in terms of three categories that we will call Groups 1,

TABLE 8.16 **Univariate analysis of the five independent variables**

	Group 1 (n=94)	Group 2 (n=102)	Group 3 (n=104)	p value
Age	54 (11)	56 (8)	60 (7)	<0.001
BMI	32.4 (7.5)	31.1 (7.5)	29.3 (7.2)	0.011
Calcium	2.1 (0.25)	2.09 (0.19)	2.11 (0.24)	0.923
D-dimers	281 (93)	296 (62)	327 (90)	<0.001
eGFR	71 (19)	70 (20)	73 (21)	0.578

eGFR = estimated glomerular filtration rate, BMI = body mass index. Data are mean and standard deviation (SD).

2, and 3, which may be three nominal categories (the three different cancers) or three ordinal categories (the three progressive stages of pain).

The p values have been determined by ANOVA, as all indices have a normal distribution. If any index followed a non-normal distribution, then the Kruskal–Wallis test would need to be applied. Tukey's post-hoc test could be used to determine whether there are differences between the groups where the ANOVA gives p<0.05. This is possible for the age, BMI, and D-dimers data, and the results showed that levels are significantly different between Groups 1 and 3 with a probability of p<0.05.

Cross reference

We discuss the Kruskal-Wallis test and Tukey's post-hoc test in Chapter 7

We can then proceed with ordinal and nominal regression, having refined the hypothesis of determining which of age, BMI, calcium, D-dimers, and eGFR are predictive of the groups, by eliminating potential roles for calcium and eGFR.

Ordinal regression

In this analysis, the three groups must be linked in ordered steps, such as stages of increasing pain: Group 1 = mild pain, Group 2 = moderate pain, Group 3 = severe pain. This is reminiscent of the ordered groups analyses of the stages of pregnancy in Section 7.6. In Table 8.16, there are clear trends in age (increasing), BMI (decreasing), and D-dimers (increasing) that change stepwise across the three groups. However, we must determine formally if this is a statistically significant trend with the appropriate software. The analysis is presented in Table 8.17.

If we look at Table 8.17 we notice that there are two values for the constant. As in binary logistic regression, the constant provides information regarding the probability of there being an ordered trend between the groups once all the independent predictors have been accounted for. There is no constant 0 because this is the reference group, Group 1, where the pain is mild.

The major point about ordinal regression is that we are telling the statistical software that the three indices are in a strict order.

The result that constant 1 is significant (p=0.02) tells us that age, BMI, and D-dimers do not account for all of the differences between Groups 1 and 2 (that is, the transition from mild to moderate pain), so there must be some other as yet undetermined factors. Similarly, the high significance of constant 2 (p<0.001) tells, once more, that age, BMI, and D-dimers do not account for all the differences between Groups 2 and 3 (that is, the transition between moderate to severe pain).

TABLE 8.17 **Ordinal logistic regression of three pain categories versus age, BMI, D-dimers**

Predictor	Coefficient	SE of the coefficient	Z	p	Odds ratio	95% CI
Constant 1	2.108	0.909	2.32	0.02		
Constant 2	3.693	0.927	3.98	<0.001		
Age	0.047	0.012	3.78	<0.001	0.92	0.88–0.96
BMI	-0.046	-0.015	-3.11	0.002	1.05	1.02–1.08
D-dimers	0.006	0.001	6.0	<0.001	0.90	0.82–0.94

BMI = body mass index, SE = standard error. Modified from a Minitab printout

We would conclude that age, BMI, and D-dimers are each independently associated with increasing pain. We cannot say that any of these relationships are causal, although we may speculate on these points. We also conclude that calcium and the eGFR are unrelated to this trend.

SELF-CHECK 8.3

What do you deduce from an odds ratio of 1.16 (95% CI 0.95–1.46) and another of 0.85 (95% CI 0.71–0.96)?

Nominal regression

The basis of this analysis is that there is no natural trend between the three groups, such as three different types of cancer—none is intrinsically worse than another. The analysis is shown in Table 8.18, and again consists of three sets of data, where prostate is compared with lung, then prostate is compared with breast, and then breast compared with lung.

- The uppermost part of Table 8.18 considers the difference between prostate and lung cancer. None of the indices significantly predicts a difference between these two groups.

TABLE 8.18 **Nominal logistic regression of three types of cancer versus age, BMI, D-dimers**

Predictor	Coefficient	SE of the coefficient	Z	p	Odds ratio	95% CI
Groups 1/2: prostate and lung						
Constant	0.975	1.154	0.84	0.398		
Age	0.02	0.015	1.33	0.191	1.02	0.99–1.05
BMI	0.026	0.019	1.37	0.178	0.97	0.94–1.01
D-dimers	0.003	0.002	1.50	0.168	1.01	0.98–1.03
Groups 1/3: prostate and breast						
Constant	-2.821	1.311	2.15	0.031		
Age	0.067	0.023	2.91	0.011	1.07	1.03–1.11
BMI	0.067	0.021	3.19	0.008	0.94	0.90–0.97
D-dimers	0.007	0.002	3.5	0.001	1.09	1.05–1.18
Groups 2/3: breast and lung						
Constant	2.845	1.265	2.25	0.024		
Age	0.047	0.017	2.76	0.019	0.95	0.92–0.99
BMI	0.041	0.02	2.05	0.042	1.03	1.00–1.07
D-dimers	0.005	0.002	2.5	0.015	1.06	1.03–1.12

SE = standard error, BMI = body mass index

- The middle part of the table considers differences between prostate and breast cancer. Each of the three variables significantly predicts the difference between these two groups. However, the constant is also significant, so these three indices in combination do not explain in its entirety the difference between the groups.
- The lower part of the table considers differences between breast and lung cancer. Each of the three variables again significantly predicts the difference between these two groups. However, once more, the constant is also significant, so these three indices in combination do not explain all the differences between the groups.

The layout of the data is almost the same in nominal regression (Table 8.18) as in ordinal regression (Table 8.17). The principal difference between the methods is that there are separate constant indices for each pair of cancers. Neither age, BMI, nor D-dimers can differentiate prostate from lung cancer: all p values are >0.05, and the 95% CIs of each of the odds ratios lie either side of 1. However, all three indices are significant in the comparison of breast and prostate cancer. Note that the odds ratios and 95% CIs for age and D-dimers are all greater than 1, indicating an increase in these factors in prostate cancer. In contrast, the odds ratio and 95% CI for BMI are lower than 1 (the reverse of the age and D-dimers data), indicating that lower BMI is associated with breast cancer.

In the third analysis, breast with lung cancer, all three indices are significantly different once more. However, the pattern is different from that when prostate and breast cancers were compared. The lung cancer patients are younger than those with breast cancer, whilst BMI and D-dimers are both higher in lung cancer.

As with all regression analyses, we must be cautious in our interpretation, and must not assume causation. However, those results with the smallest p values may prompt additional studies. We must also be aware of confounders, such as that different drug treatments or lifestyle may influence the BMI, and so introduce a source of bias and, hence, error.

8.3 Survival analysis

Survival analysis focuses on the time taken for a particular event to occur. Although described as survival analysis, the event need not be death: it may be any precise and exact event. In cardiovascular disease, the event may be a heart attack or a stroke; in transplantation, it may be the rejection of an organ. The endpoint of a particular study may be the proportion of subjects having or surviving an event by a certain time, or perhaps the length of time that a patient can survive whilst being free of the event.

In clinical research, survival analysis may well be called upon to test the efficacy of a new drug or drug regime when compared with the existing drug regime (often called standard care). At the onset of a study, subjects (often patients) may be randomized to one of two or more different treatments. As the study progresses, the patients are regularly reviewed (perhaps every three months) to determine which of them has suffered an endpoint. At the conclusion of the study, the outcomes of patients on the different treatments are compared, and conclusions are drawn.

As one of the outcomes is an exact endpoint (it is either present or not) there are similarities with binary logistic analysis. One of these is that the calculation of the risk of an individual suffering an endpoint is allied to that of the odds ratio. In survival analysis, however, this risk is a **hazard ratio**.

Hazard ratio
a technique for determining the relative event rates in two groups

Clearly, if we describe something as a hazard, it must be worth avoiding, and the factor worth avoiding is often an adverse endpoint such as death. However, an endpoint may also be another aspect of health care, such as being discharged from hospital so many days after surgery. The key aspect about a hazard ratio is that the numbers of endpoints in two or more different groups are collected over a set period of time, such as two years. Although the rate at which hazardous endpoints develop is reasonably easy to determine (e.g. 54 events from 463 = 11.7%), this does not take into account the time over which the data is collected. So if the observation period is, say, 700 days, then the rate becomes 11.7/700 = 0.0167% per day, which converts to 6.1% per year. Compare this rate with that for another group (perhaps with a different disease or being treated in a different way) and you have the hazard ratio.

However, this explanation is too simple, as the correct analysis of survival data is more complex and demands certain statistical software (such as that of Cox, hence Cox regression analysis) that takes account of the length of time for each endpoint to appear, and, if they are provided, other features such as the age of the subjects. The same software will also provide the 95% confidence intervals of the ratio.

Worked example

Let us consider the discovery of a new molecule in the blood, Substance X. Studies have found it to be raised in the blood of patients with cancer. A null hypothesis states that levels of this molecule will not differ according to the long-term outcome of patients with this disease. An alternative hypothesis states that high levels are found in patients who succumb earlier to an adverse event, such as death or the requirement for surgery. Another way of putting this is that increased levels are a predictor of poor outcome.

Following ethics committee approval, Substance X is measured in several hundred patients with the particular cancer, and their disease is monitored monthly over the next two to three years. As each particular patient suffers an endpoint, the date of this event is plotted according to whether the level of Substance X was above or below a particular cut-off point. In this case the cut-off point of 0.24 ng/mL would be arrived at by sensitivity, specificity, positive and negative predictive values, and an ROC analysis (Section 3.4). The graph is called a Kaplan–Meier plot (Figure 8.5).

Implications

It appears that high levels of Substance X have some kind of protective role in survival. This may lead to a different type of treatment in those with low levels of Substance X compared with those with a high level.

Note that the two lines are roughly straight: the rate of the development of endpoints is constant throughout the whole 800 days of the study. Endpoints appear at a rate of about 1.5% per 200 days when Substance X <0.34 ng/mL and at about 2.6% per 200 days when Substance X >0.34 ng/mL. The ratio of the rate at which these hazardous endpoints accrue is therefore in the order of 1.5/2.6 = 0.58. However, by adjusting for the differences in the times to each endpoint, the final hazard ratio may well be a little different, such as 0.55.

The probability of there being a meaningful overall difference between these lines is $p<0.01$, but at what exact time point does this difference become apparent? By doing individual analyses at regular intervals, perhaps every 200 days, the slow increase in significance can be determined. This is relatively easy to do as long as the sample size is known. If there were 100 patients in each group at baseline, the difference in the accrual of endpoints could be plotted in a number of ways, such as by the Chi-squared test (Table 8.19).

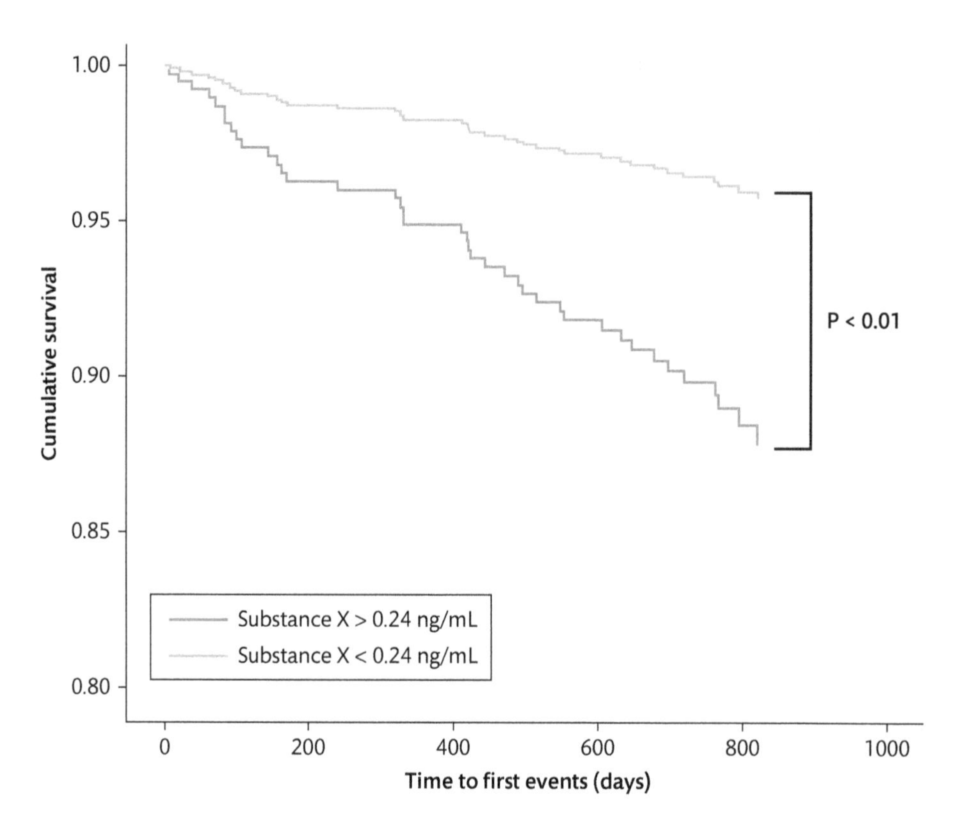

FIGURE 8.5
Kaplan-Meier plot of survival according to levels of Substance X. The two lines show the slow and steady reduction in survival over an 800-day period. The upper line is those patients whose baseline Substance X levels (at day zero) are less than 0.24 ng/mL; the lower line, those whose baseline levels are higher than the cut-off point. The p value gives the probability of a genuine overall difference between these two lines, and clearly (p<0.01) this difference is significant.

TABLE 8.19 **Trend of significance of endpoints**

	% Patients free of an endpoint		
Day	Substance X < 0.34 ng/mL	Substance X > 0.34 ng/mL	p value*
200	99%	97%	0.312
400	98%	95%	0.248
600	97%	92%	0.121
800	96%	88%	0.037

*Chi-squared test based on 100 subjects per group

This indicates that the difference between the two groups became significant at around day 700. However, there are several other ways of seeking a difference between the groups, such as sensitivity, specificity, and odds ratios, etc.

Assumptions

The analysis makes many assumptions, the main one being that there are no major differences in standard clinical, demographic, and laboratory indices between those with a high or low level of Substance X at baseline. These indices include age, sex, body mass index, smoking, the stage of the disease, and many other factors that vary according to the nature of the disease.

For example, if this were a study regarding cardiovascular disease, all groups would also need to be matched for factors such as hypertension, diabetes, and serum lipoproteins. Accordingly, there must be data which shows that Substance X does not vary with these factors.

A second assumption is that there are no changes in the treatment of the cancer (probably surgery, radiotherapy, and chemotherapy) in the two groups over the 800 days. In an ideal world, all those who started the study will remain contactable for the entire 800 days. Regrettably, some patients may be lost to follow-up or decline to take further part, and it is conceivable that this sub-group may be different from those who complete the study, leading to difficulties in interpretation.

8.4 Epidemiology

Epidemiology is the study of trends and causes of disease in populations. As a science it grew from the study of diseases that appeared to come and go with no clear cause, and so caused an epidemic. Diseases that are present at a low level all the time, and that do not appear to fluctuate, are called endemic. In the Victorian era, many epidemics turned out to be caused by infectious microorganisms: such as the outbreak of cholera in Soho (part of inner-city London) in the mid-nineteenth century, which involved hundreds of people. As those who studied epidemiology (that is, epidemiologists) developed their craft, studies became larger and larger, often involving thousands and then tens of thousands of people.

There is no set point where 'large' studies become epidemiology. However, it is perhaps the purpose and emphasis of the study that makes it epidemiological. For example, epidemiology studies can test large and more complex hypotheses in broad populations. One frequent theme is that data is often merged or pooled from several smaller and similar studies, which brings additional power. These studies are described as **meta-analyses**, and collect together information originally published by other researchers, and so are described as retrospective. Studies that collect new data from scratch are called prospective. We will consider statistical aspects of epidemiology in three case studies.

Meta-analysis
the fusion of several separate small studies into a larger study, enabling analyses that were not possible in each of the smaller studies

Case study 1: Diabetes and heart disease in thalassaemia: Pepe et al. (2013)

The most severe form of thalassaemia major produces such a profound anaemia that regular blood transfusions are called for. Whilst this is beneficial in the short term, it can cause problems. Iron is a valuable resource, and the body is programmed to retain and store this mineral. When red blood cells die, the iron that each cell has is retained in organs such as the liver and bone marrow. This is fine in health, but if there are excess red blood cells (such as those that are transfused), the body retains all the excess iron. This iron is deposited in various organs, where it causes damage that can in turn lead to additional disease, such as of the pancreas (causing diabetes), and in the muscle of the heart, causing heart disease. Manifestations of this heart disease include heart failure, arrhythmia, fibrosis, and ventricular dysfunction.

The researchers tested the hypothesis that diabetes is associated with a higher risk of heart complications in patients with thalassaemia major. A sub-hypothesis was that the extent of damage was related to the degree of myocardial iron overload. The hypothesis was tested by probing an existing database of 957 patients with thalassaemia major, gathered from 68 thalassaemia centres. Of these patients, 86 have an additional diagnosis of diabetes, allowing a case-control study, selected results of which are shown in Table 8.20.

TABLE 8.20 **Comparison of selected demographic and clinical data in thalassaemia patients with and without diabetes**

	Patients with diabetes (n=86)	Patients free of diabetes (n=709)	p value
Age (years)	37.4 (6.2)	32.0 (6.7)	<0.001
Sex (male/female)	35/51	350/359	0.129
Splenectomy (%)	81.4	54.6	0.001
Hypertension (%)	3.8	2.2	0.416
Hypogonadism (%)	41.9	19.5	<0.001
Hepatitis C virus RNA positive (%)	54.4	40	0.01
Myocardial iron overload (%)	66.3	53.6	0.026

Data are mean (standard deviation), absolute numbers, or percentages. Analysis by t test or Chi-squared test

Case-control analysis

According to this data, thalassaemic patients with diabetes are more likely to be older, to have had their spleens removed, to have underactive gonads, and to be positive for hepatitis C virus RNA, implying infection. One could therefore say that these are all risk factors for the presence of diabetes, but it cannot be said that they predict the development of diabetes. However, note that the frequency of hypertension in those with diabetes is 3.8%, and in those free of diabetes is 2.2%—a relative increase of over 72% in the former. This difference is not significant (p=0.416).

Now consider the frequencies for splenectomy: 81.4% in diabetes and 54.6% in non-diabetes, which is significant (p=0.001). In fact, the relative rate of splenectomy is 49% higher in diabetes when compared to non-diabetes. The paradox of a smaller relative difference of splenectomy (49%) being significant where a larger relative difference in hypertension (72%) is not significant can be explained in terms of the sample size and power—there are far fewer subjects with hypertension than have had their spleens removed.

SELF-CHECK 8.4

Which of the factors in Table 8.20 are likely to be predictors of diabetes in a multivariate regression analysis?

Logistic regression analysis

The researchers subsequently evaluated the impact of diabetes on heart disease, and the extent to which it was related to the burden of iron within the heart; which was assessed as myocardial iron overload (Table 8.21). The data in the middle of the table indicate that all five types of heart disease are independently related to diabetes once non-myocardial iron overload has been adjusted for.

The most powerful findings are that diabetes is most strongly related to cardiac complications and to arrhythmias. Both of these have an odds ratio of over 4, which can be translated as

TABLE 8.21 **Logistic regression analysis of diabetes and heart disease in thalassaemia**

	Adjusted for non-myocardial iron overload		Adjusted for non-myocardial iron overload, age, and other endocrine disease	
	Odds ratio (95% CI)	**p value**	**Odds ratio (95% CI)**	**p value**
Cardiac complications	4.23 (2.65–6.76)	<0.001	2.84 (1.71–4.69)	<0.001
Arrhythmias	4.09 (2.16–7.74)	<0.001	2.21 (1.12–4.37)	0.023
Heart failure	3.14 (1.87–5.26)	<0.001	2.33 (1.33–4.06)	0.003
Myocardial fibrosis	2.12 (1.24–3.63)	0.006	1.91 (1.11–3.29)	0.021
Right ventricle dysfunction	1.82 (1.01–3.30)	0.048	1.33 (0.71–2.49)	0.366

that both carry over a four-fold risk (over 400%). The probability of the presence of heart failure and myocardial fibrosis is less (over three-fold [actually 314%] and over two-fold [actually 212%] respectively), whilst the presence of diabetes increases the risk of having right ventricular dysfunction by 'only' about 82%.

However, many of these different types of heart disease are likely to share the same pathophysiology, and may also be related to age and other endocrine disease. The right side of Table 8.21 shows analyses that have been adjusted for these confounders (also called covariates). Although cardiac complications and arrhythmias are still related to diabetes, this risk has fallen to between two- and three-fold. The risk for the other types of heart disease has also fallen, and in the case of right ventricular dysfunction, this risk has become not significant. This finding implies that some of the risk of the different types of heart disease was linked to the covariates, and in the case of right ventricular dysfunction, this link is so strong that it can no longer be considered as an independent link with diabetes.

It is very easy to assume that diabetes directly causes these types of heart disease. Whilst this assumption is very persuasive, the type of study and the analysis cannot confirm this association. Nevertheless, studies of this type, although retrospective, are very useful in refining pathophysiology and are likely to prompt other investigations.

Case study 2: Haematological malignancy and ethnic groups in England: Shirley et al. (2013)

The haematological malignancies (principally of white blood cells: leukaemia, lymphoma, and myeloma) comprise about 7% of all cancers, although this figure varies markedly throughout the world, possibly as a result of racial, ethnic, and geographical factors. The latter can be controlled for by studying these diseases in a single location, therefore leaving only racial (genetic) and ethnic factors.

The researchers tested the hypothesis that there are differences in the frequencies of the various white blood cell cancers, by probing a database of cancer registrations in England collected over the years 2001 to 2007. The self-defined ethnic/racial groups were White, Indian, Pakistani, Bangladeshi, Black African, Black Caribbean, and Chinese. Taking National Census data, the frequencies of these seven groups in England were taken to be 93%, 2.2%, 1.5%, 0.6%, 1%, 1.2%, and 0.5% respectively. The rate of cancers in White race/ethnicity was taken as a reference point to which the other groups were compared.

In broad terms, the analysis for cancer in these groups is relatively simple. For example, assuming no effect of race between the Whites and the Chinese, if there are 930 cancers in the former then there should be five in the latter. If follows that finding 10 cancers in the Chinese group implies a doubling of the rate ratio, which may be statistically and pathogenically significant in that being of Chinese origin may (in this setting alone) be a risk factor for cancer. Conversely, finding only one cancer implies that being of Chinese origin is protective against cancer.

However, as in other analyses, we must consider how confident we can be about this data, and because of multiple testing, the researchers opted for a more demanding confidence interval than usual—that of 99% instead of 95%. If present, this brings a probability that any difference is meaningful at $p<0.01$, as opposed to the standard probability level of 0.05. Similarly, because of the potential confounding covariates, the raw data was adjusted for age, sex, and income, and was presented in terms of 100 000 people. The results of the analysis of cancer of mature B-lymphocytes are shown in Table 8.22.

In interpreting this data, we consider whether the rate ratios are different and if the confidence intervals (CIs) overlap. For example, the 99% CIs of the rate ratio for these cancers in Indians of both sexes are less than 1, being 0.70 (99% CI 0.63–0.78) and 0.75 (99% CI 0.64–0.89) in men and women respectively. These data suggest that Indian ethnicity protects against cancer as compared with White ethnicity. Pakistani ethnicity seems to carry no such advantage, whereas only Bangladeshi men had a reduced rate of cancer compared with White men. Being of Black African origin carries an increased risk in both sexes (by 26% in men and 65% in women), whilst Chinese race carries a reduced risk in both cases (by 43% in men and 34% in women). In pooling the three South Asian ethnicities and the two Black ethnicities, relative risks of the cancers differed markedly (higher in the Black group, $p<0.001$).

In many cases, the reasons for these differences are unclear, but some may be related to under-reporting of their disease by certain groups. Alternatively, increased rates may be due to other

TABLE 8.22 Frequency of mature B-lymphocyte cancers according to race/ethnicity

	Men			Women		
Race/ethnicity	Number of cases	Incidence of cancer*	Rate ratio (99% CI)	Number of cases	Incidence of cancer*	Rate ratio (99% CI)
White	32 330	12.5	1.00 (0.98–1.02)	25 511	8.1	1.00 (0.98–1.03)
Indian	334	7.0	0.70 (0.63–0.78)	254	5.9	0.75 (0.64–0.89)
Pakistani	262	12.5	1.09 (0.93–1.27)	136	7.1	0.96 (0.77–1.20)
Bangladeshi	49	5.8	0.58 (0.4–0.84)	34	6.3	0.74 (0.48–1.16)
Black African	179	15.5	1.26 (1.04–1.53)	155	11.6	1.65 (1.34–2.03)
Black Caribbean	325	11.9	1.01 (0.87–1.16)	263	8.9	1.13 (0.97–1.33)
Chinese	56	7.1	0.57 (0.40–0.80)	49	5.0	0.66 (0.45–0.95)

*Adjusted to 100 000 subjects. CI = confidence interval

disease such as infection with the human immunodeficiency virus, increased rates of smoking, and the consumption of high-meat and high-fat diets by different groups. The impressive sample size (almost 60 000 people) provides considerable power and the authors can be very sure of the conclusions.

Case study 3: Long-term effect of aspirin on colorectal cancer: Rothwell et al. (2010)

Colorectal cancer (CRC) is the second most common cancer, and carries a lifetime risk of 5%. It has long been noted that high doses of aspirin (at or greater than 500 mg daily) reduce the risk of developing CRC, but the long-term effectiveness of lower doses (75–300 mg daily) is unknown. Given the free availability of aspirin, a major step forward in the public health of CRC could be achieved if it could be shown that small doses of this drug are beneficial. On the other hand, excessive aspirin use can lead to haemorrhage (bruising and bleeding). It is therefore a question (as it is with almost all treatments) of a balance being struck between benefit (protection from CRC) and unwanted side effects (haemorrhage). However, there are no formal drug trials of aspirin in CRC, but there are numerous trials of this drug in the prevention of cardiovascular disease (principally heart attack and stroke).

Therefore, the null and alternative hypotheses that aspirin does not/does protect against CRC can be asked indirectly by probing databases of trials of this drug in cardiovascular disease. The researchers performed a meta-analysis by combining data from a series of other trials that sought to prevent heart attack and stroke with different doses of aspirin. This much larger database provides the power to test hypotheses that could not be achieved in each of the smaller trials by themselves. For example, the researchers were able to show a reduced risk of mortality due to CRC in those taking 75–300 mg aspirin over a 20-year period (Figure 8.6).

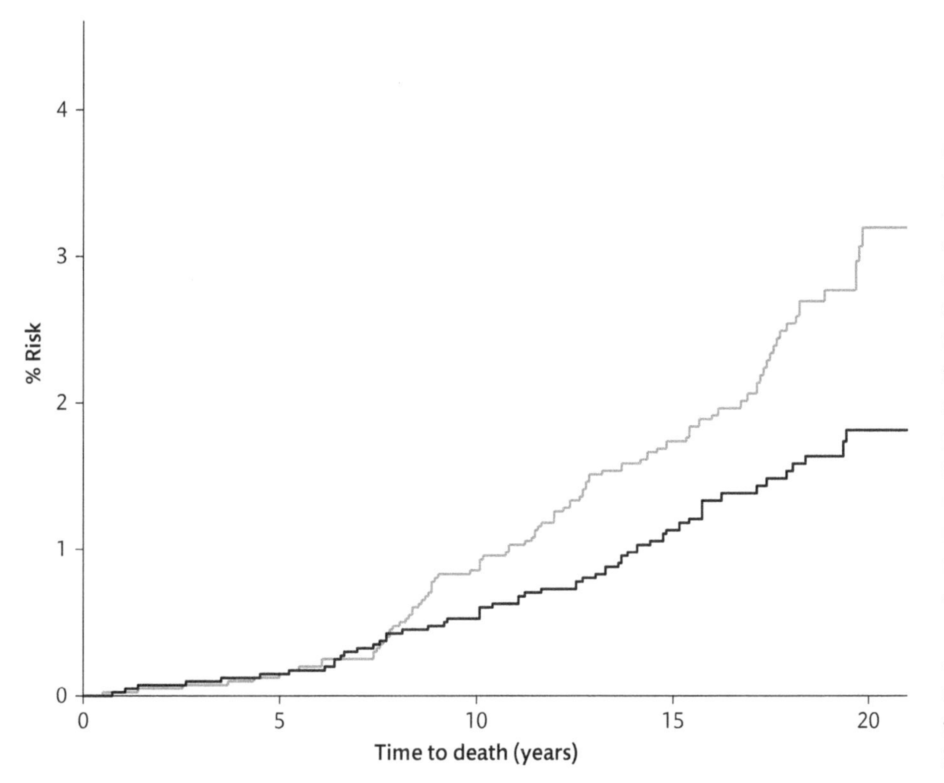

FIGURE 8.6

Kaplan-Meier plot of the rate of death due to CRC. The effect of low dose (75–300 mg daily) aspirin (black line) versus controls not taking aspirin (blue line) on the subsequent incidence and mortality due to CRC. The two lines diverge after about 7 years, at which point those taking aspirin have a reduced risk of death. Reproduced from Rothwell, P.M., Wilson, M., Elwin, C.E., Norrving, B., Algra, A., Warlow, C.P., and Meade, T.W. (2010) Long-term effect of aspirin on colorectal cancer incidence and mortality: 20-year follow-up of five randomised trials. *Lancet* 376,1741–50.

Note that this graphic is essentially the reverse of Figure 8.5, where the survival data start at 100% and fall over time. In the Kaplan–Meier plot of deaths due to CRC, the rate of course starts at zero (as everyone is alive) and slowly increases.

The slowly rising lines represent the steady increase in the appearance of events (deaths from CRC) over the 20 years of the study. The upper (blue) line plots the increasing risk of death for those not taking aspirin; the lower (black) line plots the increased risk in those taking aspirin. The two lines start to diverge at around seven years, and after about 17 years the two lines diverge markedly, and this is significant at p=0.006.

Forest plot
a method for graphically representing odds ratios and the relative sample size of different studies

A further aspect of this research is illustrated by Figure 8.7, which is described as a **forest plot**. This shows data from the different trials and the extent to which these data can be merged. The data sets in Figure 8.7(a) show two trials that used 500–1300 mg of aspirin daily. In the British Doctors Study, although 27% fewer subjects on aspirin died (the odds ratio being 0.73), this was not significant, as shown by the 95% confidence interval (CI) which ranges from 0.49 to 1.10, and so which straddles 1. It is also shown by the graphic in Figure 8.7(b), where both lines (or whiskers) from the central orange squares cross the line where the odds ratio is 1.

This pattern is partially mirrored by the second set of data, from the UK-TIA study (1200 mg daily), which reported a better odds ratio (0.53) in favour of a beneficial effect of aspirin. However, because this study recruited fewer participants, the reduction in the proportion of cancers by a seemingly impressive 47% is in fact not significant, as is demonstrated by the wide 95% CI of 0.23–1.25. It could be argued that this is an example of a false negative due to a small sample size. That the UK-TIA study is much smaller (1379 subjects) compared with the British Doctors Study (5238 participants) is reflected by the smaller black box in Figure 8.7(b).

By combining these two studies (the sub-total row, so the sample size is over 6600), the additional power means that the 95% confidence interval is tighter (0.48–0.98) and so now does

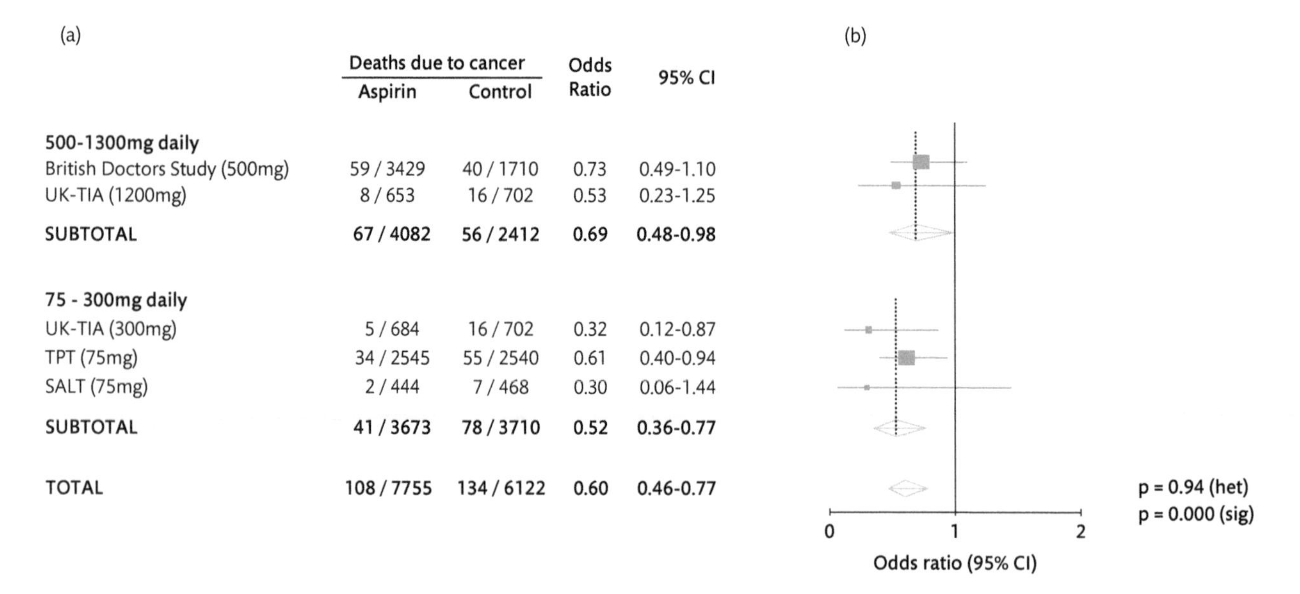

(a)

	Deaths due to cancer		Odds Ratio	95% CI
	Aspirin	Control		
500-1300mg daily				
British Doctors Study (500mg)	59 / 3429	40 / 1710	0.73	0.49-1.10
UK-TIA (1200mg)	8 / 653	16 / 702	0.53	0.23-1.25
SUBTOTAL	**67 / 4082**	**56 / 2412**	**0.69**	**0.48-0.98**
75 - 300mg daily				
UK-TIA (300mg)	5 / 684	16 / 702	0.32	0.12-0.87
TPT (75mg)	34 / 2545	55 / 2540	0.61	0.40-0.94
SALT (75mg)	2 / 444	7 / 468	0.30	0.06-1.44
SUBTOTAL	**41 / 3673**	**78 / 3710**	**0.52**	**0.36-0.77**
TOTAL	**108 / 7755**	**134 / 6122**	**0.60**	**0.46-0.77**

FIGURE 8.7

Meta-analysis of the effect of aspirin on long-term risk of death due to colorectal cancer in randomized trials of aspirin versus control. Analysis includes all events from time of randomization. Each box and its whiskers represent the odds ratio and 95% confidence interval. Note how in some cases the right-hand whisker crosses the line of odds ratio =1, and so this result is not significant.

not include 1. Consequently, it is significant, as represented by the compressed diamond shape on the right. It may be concluded that this dose of aspirin brings an odds ratio of 0.69, and so a reduction in the risk of CRC of 31%.

The same principle broadly applies to the merging of three trials that used 75–300 mg of aspirin daily, although two of these trials showed a beneficial effect of aspirin (data sets (b)). Combining these three studies into one with a sample size of 7502 subjects tightens the 95% CI compared with each of the three individual studies, and shows that use of this dose of aspirin brings an odds ratio of 0.52, and so a reduction in the risk of CRC of 48%. Note that the different sizes of the studies are reflected by the sizes of the black boxes of the graphic on the right, and that the compressed parallelogram becomes smaller with an increasing sample size.

Finally, combining the studies of high and low doses of aspirin, with over 13 000 participants, brings an odds ratio of 0.6, indicating a 40% reduction in the risk of CRC. The 95% CI of this risk ranges from 23% to 54% (derived from the 95% CI of the odds ratio, being 0.46–0.77). On the lower right of the figure are two probability values. The p=0.000 (sig) tells us that the reduction in a risk of CRC due to cancer is extremely unlikely to be spurious. The second figure, p=0.94 (het), is a test of the heterogeneity of the data set. This statistic confirms that there are no gross differences in the design of the individual studies, and so the meta-analysis is reliable. Many statisticians would feel uncomfortable seeing p=0.000, as this implies infinite probability, which is impossible. A more appropriate value would be p<0.001.

Chapter summary

- Large data sets, as may be present in epidemiological studies, enable complex techniques using methods such as regression analysis and survival analysis.
- Multivariate regression analyses determine which of a number of independent variables are related to a single dependent variable.
- In linear regression analyses the dependent variable has a continuous distribution; in logistic analysis the dependent variable has a categorical distribution.
- In binary logistic analysis, there are two dependent variables. In ordinal logistic analysis, three or more dependent variables have a natural order. In nominal logistic analysis, three or more dependent variables have no natural order. The outcome is reported as an odds ratio.
- Kaplan–Meier plots show survival analysis over time, but also the acquisition of endpoints.
- Hazard ratios give an estimation of the likelihood of the development or presence of an endpoint, focusing on survival over time.
- A forest plot provides a visual representation of the odds ratios and their 95% confidence intervals.

Suggested reading

- **Christensen, E., Neuberger, J., Crowe, J., Altman, D.G., Popper, H., Portmann, B., Doniach, D., Ranek, L., Tygstrup, N., and Williams, R. (1985). Beneficial effect of azathioprine and prediction of prognosis in primary biliary cirrhosis.** *Gastroenterology* **89, 1084–91.**
- **Pepe, A., Meloni, A., Rossi, G., Caruso, V., Cuccia, L., Spasiano, A., Gerardi, C., Zuccarelli, A., D'Ascola, D.G., Grimaldi, S., Santodirocco, M., Campisi, S., Lai, M.E., Piraino, B., Chiodi, E., Ascioti, C., Gulino, L., Positano, V., Lombardi, M., and Gamberini, M.R. (2013). Cardiac complications and diabetes in thalassaemia major: a large historical multicentre study.** *Br J Haematol* **163, 520–27.**
- **Rothwell, P.M., Wilson, M., Elwin, C.E., Norrving, B., Algra, A., Warlow, C.P., and Meade, T.W. (2010). Long-term effect of aspirin on colorectal cancer incidence and mortality: 20-year follow-up of five randomised trials.** *Lancet* **376, 1741–50.**
- **Shirley, M.H., Sayeed, S., Barnes, I., Finlayson, A., and Ali, R. (2013). Incidence of haematological malignancies by ethnic groups in England, 2001-7.** *Br J Haematol* **163, 465–77.**
- **Scott, I. (2008). Interpreting risks and ratios in therapy trials.** *Australian Prescriber* **31, 12–16.**

Questions

This final research chapter brings together many of the aspects of research and statistics that we have been addressing over the last three chapters, so we provide an extended activity to work through.

Introduction

A group of researchers wish to test the hypothesis that a new form of chemotherapy reduces the progression of rheumatoid arthritis (RA). The definition of this progression is any one of a series of well-defined endpoints—death, any hospital admission (such as for stroke, diagnosis of pulmonary oedema, surgery, perhaps to correct or replace a deformed joint), an increase in the disease activity score (DAS), the appearance of new symptoms (such as the development of rheumatoid nodules, anaemia, or scleritis), or more severe disease (examples of which include vasculitis and nephropathy). However, the researchers are aware of the likelihood that other factors (confounders, or covariates) may influence an endpoint, and so prospectively collect data on C-reactive protein (CRP, a marker of inflammation), age, the duration of the disease, rheumatoid factor (RhF), the DAS itself, and sex (as women are known to be at a greater risk of RA than are men). These data are collected before the patients are started on the new therapy.

Sample size determination

If this were a simple multivariate analysis, then the power calculation to determine the sample size would call for 12–15 patients per variable, giving a recruitment target of around 100 patients. However, the alternative hypothesis states that those on the new therapy will have fewer endpoints—but how many fewer? Here the researchers need to decide on the degree of improvement in outcome, and this to some degree is guided by other data.

Supposing that over a five-year period, it is reasonably well established that 8% of subjects will suffer an endpoint each year, so that after five years this rate will be 40%. Let us hypothesize that a new treatment brings this rate down to 5% per year, so that after five years the rate is 25%. For this to be significant at $p<0.05$, 200 subjects are needed, with 100 being in each group (standard treatment and the new treatment). However, for extra confidence, the researchers decide to recruit 10% extra, so set their sample size at 220. However, this is 220 people completing the trial, and many will drop out part of the way through, or may simply be lost to follow-up. Accordingly, it may be that 250 or even 300 people will need to start the study to ensure that enough complete the trial.

Management of the study

Once regulatory authority approval (Medicines and Healthcare Products Regulatory Agency, Local Research Ethics Committee, and each of the Hospital Trusts) has been obtained, recruitment can begin, and, after informed consent, patients randomized to standard care or to the new drug. If the patient is unaware to which arm of the trial they have been allocated, the trial is called 'single blind'. If both the patient and the physician caring for the patient are unaware of the regime, the trial is said to be 'double blind'. In the case of the latter, a third party independent of the physician is responsible for allocating the patients. In many cases, the 'blinding' aspect is helped if the patient is given a tablet containing no active ingredient, which is called a placebo. Hence the best trials are double blind, randomized, and placebo controlled, as possible bias by the patient or their physician may influence the outcome. This is less likely if the patient and the physician are unaware if the former is taking an active drug or a placebo.

The patients are reviewed at regular periods (perhaps every two to three months), and the appearance of an endpoint is noted. After five years, the spreadsheet is analysed. One of the initial problems is to determine that people in the two groups (standard treatment and new drug) are balanced for major clinical and demographic factors. If not, and perhaps there are more women in one group than the other, then any resulting difference between the groups may have been due to sex, not to the new drug. To get around this, statisticians are often given the responsibility of balancing potential confounders, and determining which patients are randomized to each group. This process is called stratification.

Analysis

Step one is often to ensure that the two groups are balanced for major factors at the outset. These are shown in Table 8.23. These indicate that there are no major differences in the possible confounders that may influence the results.

Question 8.1

Name and justify the statistical tests used to determine differences in each index between the two groups in Table 8.24.

Of the 220 people completing the study, 65 had suffered an endpoint and 155 were free of an endpoint. Of those 110 randomized to standard care, 40 (36.4%) suffered an endpoint. Of those 110 randomized to the new drug, 25 (22.7%) suffered an endpoint. One of the simplest methods for comparing this difference is with the Chi-squared test, which reports a 97.3%

TABLE 8.23 Clinical demographic, and laboratory factors at the outset

	Group 1: Standard care	Group 2: New treatment	p value
Age (years)	56 (10)	54 (11)	0.295
Sex (men/women)	25/85	23/87	0.295
Disease activity score (units)	3.4 (1.1)	3.3 (1.1)	0.763
C-reactive protein (mg/mL)	3.3 (1.6-6.0)	3.5 (1.9-6.5)	0.972
Rheumatoid factor (units/mL)	61 (47-69)	52 (43-68)	0.110
Disease duration (years)	12 (8-19)	9 (5-17)	0.456

Data are mean and standard deviation, median and inter-quartile range, or absolute numbers of subjects

likelihood that the difference is genuine, and only a 2.7% likelihood that the difference is spurious, with p=0.027.

Having established that there is a difference in endpoints between the two treatment groups, we now move to investigate which of the six research indices could have predicted the appearance of an endpoint. This is done by comparing the indices between the two groups, as in Table 8.25.

This analysis indicates that levels of C-reactive protein, the disease activity score, and whether or not the subject was taking the new drug or a placebo (that is, was in the standard care group) were all potential predictors of the possibility of suffering an outcome. However, this univariate analysis must only be preliminary as confounders may be present. For example, it is possible that patients with the highest disease activity score might well also have a high C-reactive protein. In order to address this issue we need to perform a multivariate analysis, and this should be logistic.

TABLE 8.24 Analysis of research indices according to outcome

	Subjects suffering an endpoint (n=65)	Subjects free of an endpoint (n=155)	p value
Age (years)	55 (10)	56 (10)	0.358
Sex (men/women)	15/50	33/122	0.770
Disease activity score (units)	3.7 (1.0)	3.2 (1.2)	0.006
C-reactive protein (mg/mL)	4.0 (1.8-6.8)	3.0 (1.6-5.5)	0.013
Rheumatoid factor (units/mL)	62 (44-69)	54 (45-68)	0.305
Disease duration (years)	10.5 (7-15.5)	12.5 (8.5-18.0)	0.456
New treatment/standard care	25/40	85/70	0.027

Data are mean and standard deviation, median and inter-quartile range, or absolute numbers of subjects

TABLE 8.25 **Logistic regression analysis**

	Coefficient	Standard of the coefficient	Z	Odds ratio (95% CI)	p value
Constant	1.325	1.123	1.18		0.238
C-reactive protein	-0.489	0.227	−2.15	0.61 (0.39–0.96)	0.031
Age	0.012	0.01	1.20	1.01 (0.97–1.04)	0.991
Rheumatoid factor	0.024	0.020	1.20	1.00 (0.98–1.02)	0.865
Disease activity score	−0.313	0.134	-2.34	0.73 (0.56–0.93)	0.019
Sex	0.106	0.391	0.27	1.11 (0.52–2.39)	0.786
Disease duration (years)	1.96	2.11	0.928	1.08 (0.90–1.28)	0.615
New treatment/ standard care	0.691	0.312	2.21	2.00 (1.08–3.69)	0.012

95% CI = 95% confidence interval. Modified from a Minitab printout

Question 8.2

Why not use multivariate linear regression analysis?

The results of the logistic regression analysis are shown in Table 8.25. Each of the three indices that were significant in the univariate analysis of Table 8.24 have been retained as independent influences on the likelihood of being in the group suffering an endpoint compared with the group free of an endpoint.

Question 8.3

The coefficient and Z value of the C-reactive protein and disease activity score are negative, whereas those for age and disease duration are positive. What does this mean?

Interpretation

The researchers have successfully tested their hypothesis, in that they have established that C-reactive protein, the disease activity score, and being on the new treatment or on usual care all influence outcome. Therefore, as a clinical drug trial, the study has succeeded in that being in the group on the new drugs protects against the development of an endpoint. This represents an advance in our ability to care for these patients.

A note of caution

All of the examples in this chapter are extremely simplified versions of real clinical research, and are put together purely to explain principles of analysis. In the real world, hypotheses are considerably more complex and there are inevitably more confounders (other drugs, other clinical syndromes), and this is certainly true for this end-of-chapter activity. Accordingly, clinical trials of new drugs often recruit thousands, perhaps tens of thousands, of subjects and call for international cooperation.

Cox regression analysis

Much of the development of survival analysis is due to the work of David Cox, after whom the method is named. The survival analysis in Section 8.4 looks at the effect of only one index—Substance X. Similarly, the survival plot of Figure 8.6 considers only the use or not of aspirin. So both of these are examples of univariate analysis. This is in contrast to the example we have just been looking at, where we asked which of seven indices have an influence on outcome, and so is multivariate. If we had data of the time point at which an individual suffered an endpoint, we could have used Cox's method to work out which of the variables had the most influence on whether someone suffered an endpoint. The method itself is very complex and its use by non-statisticians is not recommended.

9

Communication

Learning objectives

After studying this chapter, you should be able to ...

- Understand the need to communicate
- Explain the importance of good communication with colleagues within and beyond the laboratory
- Describe the key aspects of the layout of a PowerPoint presentation
- Outline the major sections of a written communication
- Recognize the significant elements of an abstract
- Identify leading features of a research paper

Communication is often a much-overlooked skill, and in the context of biomedical science practice ranges from the everyday interactions between colleagues regarding clinical matters to the dissemination of the results of research, either as a publication or as a presentation to an international meeting such as the IBMS Congress. In this final chapter we will look at how to communicate effectively with various colleagues—presenting findings of studies internally (Sections 9.1 and 9.2) or externally as abstracts to learned societies (Section 9.3), writing formal documents (Section 9.4), and preparing data for publication in a journal (Section 9.5).

9.1 Routine communication with colleagues

The correct passage of information is crucial for the safe and efficient working of a routine laboratory and is a key part of good laboratory practice.

Communication within the laboratory

In many cases, intra-laboratory communication is often seen as informal (such as the oral statement that a batch of samples is ready for, or has completed, analysis), but nevertheless must be focused and acted upon professionally. An example of developing practice is the increasing passage of information between computers, many of which are linked to analysers, the use of which requires specialized training. In other settings, information and communications may be on paper (such as on noticeboards). Senior staff will regularly hold updates and team-briefs with their staff on a variety of topics.

A crucial medium of communication is the noticeboard, which can present formal information such as the composition of team members, details of quality control, quality assurance, and audit, and of upcoming scientific meetings. The departmental noticeboard is also an important source of non-professional information such as welcome and leaving social events. Although not the absolute requirement of a good department, social events are an excellent opportunity to get to know one's colleagues in a less formal setting.

Communication beyond the laboratory

More important is the passage of information outside the laboratory to service users. Many of our colleagues within and beyond the laboratory (such as in other parts of a hospital or in an industrial setting) seek the results of tests, and technical or clinical advice. In a hospital or GP practice this information may change a treatment, and so could be a factor in whether or not the patient lives or dies. Accordingly, in many establishments, clinical information of a factual nature is passed beyond the laboratory only by staff of a minimum grade. However, advice stemming from the interpretation of that result will only be made by staff of a higher grade, and who therefore possess the appropriate clinical knowledge. If any such communications are by telephone, the passage of that information is often logged, along with the date, time, nature of the information, and details of the recipient.

Whoever communicates outside the laboratory must be prepared to stand by that communication, be it oral or written (generally e-mail), and so must ensure the information is correct, unambiguous, timely, and concise. Every good laboratory will have a communications standard operating procedure (SOP) that staff will be trained in and expected to adhere to. Communications both within and beyond the laboratory should be regularly audited.

Very rarely, scientists will directly communicate with patients, perhaps in an out-patient setting such as an oral anticoagulant clinic, or in administering an oral glucose tolerance test, or in venepuncture. In all cases the scientist will formally confirm the identity of the patient, use their professional manner to seek to reassure the (possibly anxious) patient, and answer all questions (wherever possible). But the scientist should be guarded about disclosing certain clinical facts that may come better from a colleague who is medically qualified.

SELF-CHECK 9.1

Why is good communication within the laboratory so important?

9.2 Focused communication with colleagues

Although not essential, good communication skills are a valuable part of a scientist's professional portfolio. As a scientist progresses in his or her career, they can expect to have these skills developed and tested. In many cases, staff of certain grades may be 'invited' to give an oral presentation to their colleagues, perhaps as part of a rota. It is likely these skills will have been developed whilst in education, and many will have had experience of presenting whilst at university. The particular topic may be of the scientist's choice, or it may be suggested by a more senior colleague or lecturer. An informal lunchtime meeting provides an ideal opportunity for staff to acquire and develop their communication skills.

The exact nature of the communication may take many forms, such as:

- Reporting back key points from an external meeting
- An update on a new disease or method
- An audit
- A case report of an interesting patient.

Presentation skills are worthy of development, as it could be that giving a presentation may be part of the interview process. Oral communication—in the form of a presentation—is often delivered in conjunction with some kind of visual component, typically PowerPoint slides.

A PowerPoint presentation

The gold standard method for an oral presentation in front of an audience (of any size) is to use a PowerPoint presentation. PowerPoint (or equivalent software package) is widely available and very user friendly.

Whatever the topic, arguably the most important element of planning for and delivering an oral presentation is the length of time available to deliver the talk. A broad rule of thumb is that there should be one slide for each minute of the presentation (e.g. 10 slides in a 10-minute talk), although the number of slides per minute can sometimes be increased as does experience of presenting.

A second rule of thumb is that the presentation adopts a standard pattern, which is generally the same regardless of the audience. The first slide typically gives the title and details of the presenter, followed by an introduction and objectives, the information being delivered, and a summary/conclusion. Further details are presented in Section 9.3. The most successful presentations are those where the presenter broadly follows a script, and that it has been rehearsed, probably at home. A further valuable tip is to record one of these rehearsals on a dictaphone-like device or phone, so that blockage points, inappropriate pauses, and other features that interrupt a good flow of words can be detected and addressed.

9.3 Presentations outside the laboratory

The professional development of a mid-grade scientist typically involves presentations to colleagues outside the laboratory, most commonly at local, national, or international scientific meetings. The model is essentially the same as for an internal presentation, but the content is very different, and is markedly more formal. The speaker is representing not only him/herself, but also the department and hospital/university/commercial enterprise. Most presentations take one of two forms—that where there is original data, and that of an educational update.

Presenting original research

The presentation of original research must show new information that advances our knowledge or practice of a particular aspect of biomedical science. As indicated, if presented to one's own local laboratory colleagues, this can be reasonably informal. However, a presentation will be more structured if presented elsewhere, such as if part of a project undertaken to fulfil the requirements of a higher degree at an institute of higher education. In this situation the audience is likely to be one's peers and lecturers. However, the ultimate form of a presentation is to an external scientific meeting, inevitably organized by a learned society.

In most cases, the organizing committee of a society will circulate to its members details of a forthcoming meeting, and with it an invitation to present data. Those wishing to submit data have to do so in the form of an abstract, which is an extended paragraph summarizing the project. It is often limited to perhaps 250 words, and accordingly, putting an abstract together is an exercise in succinct writing. The abstract should include the key features of the project, and should be written to a common format. These are separate sections, consisting of the introduction, an objective or hypothesis, details of methods and subjects (if appropriate, and the sample size), the result, and conclusion. The most successful abstracts also present some data and p values. An example of an abstract is shown in Box 9.1: it has 246 words. An abstract is not usually called for in internal meetings or in the setting of higher education.

BOX 9.1 Example of an abstract presenting research data

Value of CD789 as a marker of leukaemia

A Scientist, Department of Blood Science, Your Hospital, Postcode

> The introduction should give an overview of the project, and ideally state why the research is justified

Introduction: There are many types of leukaemia, each defined by factors such as morphology, intracellular markers, and by the presence of CD molecules at the cell surface. Despite the large number of these CD markers, there is still often difficulty in precisely identifying the characteristics of a particular leukaemia, leaving room for additional potential markers.

> The objective tells the purpose of the work

Objective: To provide pilot data on the potential of CD789 as (a) a general marker of leukaemia and (b) a marker of acute versus chronic leukaemia.

Setting: An inner-city University Teaching Hospital.

> Methods and subjects tell us of the laboratory techniques and the patients and controls

Methods: Expression of CD789 on white blood cells (defined by CD45) using fluorescence flow cytometry.

Subjects: Fifty patients with leukaemia (white blood cells count > 25×10^6 cells/mL, 25 with acute leukaemia, 25 with chronic leukaemia), and fifty patients with anaemia or healthy laboratory staff as controls (white blood cell count < 11×10^6).

> The results tells us what happened, and is best with data and p values

Results: Mean fluorescence intensity (MFI) expression of CD789 was median 234 units (inter-quartile range 157-415 units) in the patients with leukaemia compared with 193 (157-238) in the controls ($p = 0.0025$, Mann-Whitney U test). MFI was higher than the 95th centile of the control group in 17 of the 50 patients (34%) with leukaemia. In those 25 patients with acute leukaemia, MFI was 205 (170-376) units compared with 249 (188-448) in those 25 with chronic leukaemia ($p = 0.269$).

> The final section summarizes the project

Conclusions: Expression of CD789 was higher in 34% of patients with leukaemia, suggesting it may have potential as a marker of this disease. However, CD789 was unable to differentiate acute from chronic leukaemia.

BOX 9.2 Example of an abstract for an update presentation

Recent developments in leukaemia diagnosis

A Scientist, Department of Blood Science, Your Hospital, Postcode

Despite marked advances in the diagnosis and treatment, leukaemia continues to be a challenge for scientists and clinicians. This presentation will review recent developments in the diagnosis of this disease, and will focus on cell surface markers (such as CD789) and changes in the expression of certain genes (such as *HifN*) as are defined by methods such as flow cytometry and rtPCR/hybrid mapping respectively. The importance of these new laboratory techniques, and others such as automated robotic cell and gene typing, for diagnosis of different types of leukaemia and in the detection of minimal residual disease, will be discussed.

The abstract opens with a statement of why the topic is worthy of attention

This is followed by details of those features of leukaemia that will be addressed

The abstract concludes with a note of the practical value of these methods

The organizers of the meeting will judge which of the submitted abstracts will be most interesting and relevant to the meeting attendees. Accordingly, the abstract must be written in such a way that will make it attractive to the organizers, and then, if successful, attractive to the conference attendees. The abstract may be published by the organizers, allowing those not able to attend the meeting to have access to the information.

An update presentation

Meetings of learned societies often have defined educational sessions that cater for those seeking to improve or refresh their knowledge of different aspects of biomedical science. Indeed, it is the exceptional scientist who can maintain their knowledge of all aspects of their practice by themselves.

In most cases, the presenters of update lectures are invited by the organizing committee. There will still be a summarizing abstract, but it will lack the overview of results included in the abstract for a research presentation. Nonetheless, there must still be sufficient information to enable potential attenders to decide whether or not they choose to attend. An example of the abstract for an update presentation is shown in Box 9.2.

Should the abstract prove to be sufficiently worthy to the organizing committee, they are likely to invite the author to make an oral presentation (typically, a series of PowerPoint slides) or to present a poster.

An oral PowerPoint presentation

Once the organizing committee of the particular meeting has invited the author to make an oral presentation, he or she can then put together the slides. The number of slides in a presentation may vary, but their order will follow a pattern very much like that of the abstract. A common fault in inexperienced presenters is the desire to cram too much information into a slide.

Simplicity, and so clarity, is preferred. The most common format for a presentation of research data is as follows:

1. **Title of the presentation**, with name(s) and work address(es) of the presenter (and co-authors, if any).
2. **Introduction**. The exact number of slides needed to set out the introduction varies with the nature of the presentation, and the time allowed, but would generally include a description of why the particular project is worthy of investigation, and then a description of the 'knowledge gap', which can be summarized as 'what it is we don't know', and 'why we need to know it'.
3. **Statement of the hypothesis**. As discussed in other chapters, all good research tests a specific alternative hypothesis against its null hypothesis. However, in an abstract there is usually not enough space to show the null and alternative hypotheses, so that one simple overall hypothesis will suffice. As rigorous and quantitative scientists we are not moved to simply 'investigate', or have an 'objective' or an 'aim' regarding the research—these words are far too vague and non-specific. We must state the research question as precisely and exactly as possible.
4. **Plan of investigation**. Having defined the hypothesis, the next slide shows the plan of investigation—that is, how the hypothesis is to be tested. The method adopted depends on the nature of the investigation. If the project is clinical research, then the plan may be to recruit a certain number of patients and controls with or without a particular disease. In a laboratory setting, the nature of the samples (blood, tissues, reagents) and their origins (patients with particular diseases) will be described.
5. **Materials, methods, and subjects**. It is very likely that there will be one (or several) technical slides showing the analyses to be performed. These slides may set out the methods used to determine levels of certain cells or molecules in the blood or within the tissues of certain organs. There may also be a slide of the details of the patients and controls (their age, sex, and often other factors such as blood pressure, medications, and other disease).
6. **Statistics and analysis**. The last slide of this opening section will be a calculation to provide justification of the sample size (the number of subjects and/or the number of analyses to be performed) and a description of the statistical methods that will be used to analyse the results.
7. **Results**. The presentation then moves to the results, which may be presented as tables and/or figures. These should not be too complex as the presenter will be able to run through the salient features of the data in each slide, and perhaps orally offer some interpretation. Depending on the size of the study, there may be several slides showing results, and some may be shown graphically (as outlined in Chapter 4).
8. Many presenters have a slide that summarizes the key aspects of the data.
9. **Conclusion**. The last slides provide a conclusion, and perhaps an acknowledgement of those who have provided support.

Once the format of the presentation has been fixed, the best presenters will practise giving their talk time and time again, in order to become completely familiar with their slides. Presenting informally in front of laboratory colleagues is also recommended.

The session of the particular meeting will be chaired, and he or she will generally invite the audience to ask questions. A good presenter will be fully aware of their data and its weaknesses, and so is likely to have prepared answers to potential questions. This can only be achieved by a thorough understanding of all aspects of the project.

Figure 9.1 outlines the format of slides in a general topic.

Presenting data to colleagues

A Scientist

Department of Blood Science
Your Hospital

Introduction

(Statement of the current position: e.g.)
Different types of leukaemia can be defined by the expression of certain CD molecules

(Statement of the problem: e.g.)
Despite this, there are still some types of leukaemia that are not easily classified.

(Statement of need, e.g.):
Therefore, there is a need for additional CD markers that may be able to help classify certain leukaemias

(Possible solution to the problem, e.g.):
A new antibody, directed towards CD789, has been developed that has shown potential in leukaemia classification

Hypotheses

1. Expression of CD789 is increased in leukaemia compared to controls
2. Expression of CD789 is increased in acute leukaemia compared to chronic leukaemia

Plan of investigation

Use flow cytometry to measure the expression of CD789 on the surface of white blood cells from...

(a) 50 patients with leukaemia and 50 subjects free of leukocyte neoplasia

(b) 25 patients with acute leukaemia and 25 with chronic leukaemia

Analysis by Student's t test or the Mann-Whitney U test, depending on distribution

Results

Hypothesis 1
Controls: 193 MFI units (157-238)
Leukaemia: 234 MFI units (177-415): p=0.0025

The 95th centile of MFI expression in controls was 295 units. Levels were higher than this in 17 (34%) of the patients with leukaemia

Hypothesis 2
Acute: 205 MFI units (170-376)
Chronic: 249 MFI units (188-448): p=0.269

Data: median and interquartile range of the mean fluorescence intensity (MFI) expression of CD789. Analysis: Mann-Whitney U test

Graphic of hypothesis 1

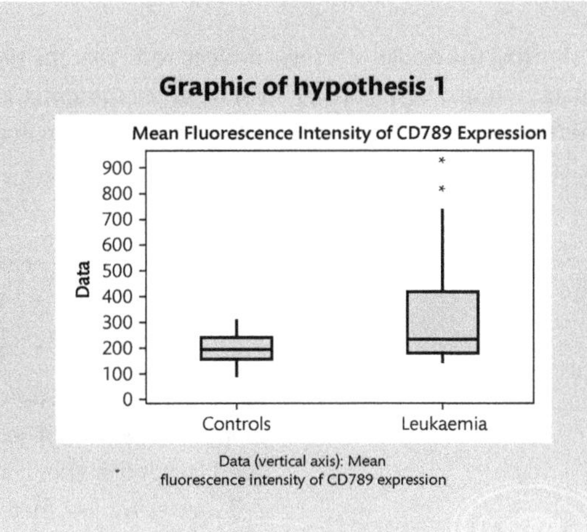

Data (vertical axis): Mean fluorescence intensity of CD789 expression

FIGURE 9.1 *(CONTINUED)*

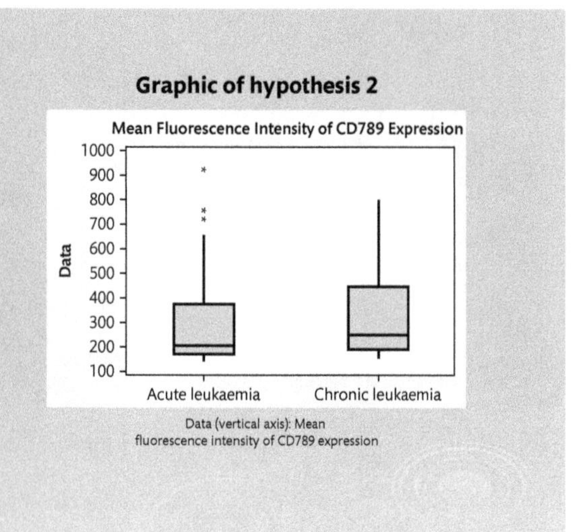

Summary

Hypothesis 1
Our data support the hypothesis that the expression of CD789 is higher in leukaemia

Hypothesis 2
Our data does not support the hypothesis that the expression of CD789 is higher in acute leukaemia than in chronic leukaemia

Conclusion

CD789 may be useful in screening for leukaemia

It does not seem to be able to differentiate acute from chronic leukaemia

Additional studies are warranted as it may have a role in myeloid versus lymphocytic leukaemia

Acknowledgements

With thanks to Dr Jekyll for help with obtaining clinical material, and to Mr Hyde for help with statistics.

The project was part of an MSc at the University of Somewhere

FIGURE 9.1

Slides showing the results of a small project. Note that the slides are formatted to a common form with a regular margin. Information should be kept to a minimum as the presenter can give further details orally. All the font styles and sizes are consistent, and there are no more than four points per slide. The emphasis is on simplicity: cluttered or busy slides are counter-productive (graphics from Minitab software).

Presentation as a poster

Posters are an essential component of scientific meetings, and represent an important opportunity for the author to make his or her mark publicly. To compile a poster, the author will need to put together a story, much like a PowerPoint presentation, on a large board, typically 2–3 metres by 1–2 metres. The poster itself may comprise a series of smaller cards, each one focusing on one or more of the sections of a PowerPoint presentation, i.e. the title of the presentation, introduction, statement of the hypothesis, plan of investigation, materials, methods and subjects, statistics and analysis, results, and conclusion. The size of each card will vary according to the amount of information the author wishes to present. An alternative format is one very large sheet of heavy paper, upon which the same sections are printed.

At the meeting, the poster will be placed on its board, generally for half a day, so it can be viewed informally by delegates. The author will be required to stand by their poster, generally for an hour, at times set by the organizers, giving interested delegates an opportunity to ask questions of the presenter. Accordingly, the presenter has the opportunity to develop their communication skills, and has a useful opportunity to defend their data.

SELF-CHECK 9.2

What are the four key sections of a structured abstract?

9.4 Writing for the laboratory

Oral communication, whether by PowerPoint or a poster, is clearly important, but is not a permanent record of the information covered (unless, of course, the oral communication was recorded in some way for future reference). By contrast, the written word has a greater degree of permanence—with journal articles being archived for access many years into the future. As such, more care must be put into written communication as it cannot be retracted.

Writing for the laboratory is a further skill, as it will impact on the practice of all laboratory scientists. Documents written for internal viewing only include those for audit, good laboratory practice, health and safety, COSHH etc. However, some of these documents may also be required by external assessors of the working of the laboratory. Probably the most important internal document is the standard operating procedure.

The standard operating procedure

All work in the laboratory is performed according to a series of steps set out in a document called a standard operating procedure (SOP). This document enables any practitioner, once trained, to successfully complete a particular task. The SOP will be written by senior staff, often to a set pattern. Indeed, there is even an SOP on how to write an SOP!

Each individual step in an SOP must be clear, focused, and unambiguous. The document itself will have a defined lifespan—it will be renewed by a date that will be stated on the SOP itself. This allows new (hopefully improved) practice points to be introduced. SOPs can be large documents, extending to several pages. An example of a section of an SOP is presented as Box 9.3. This SOP is for a research procedure (that is, not a routinely measured analyte): SOPs in routine practice will be very different, and (as patient care is involved) are likely to be more demanding and with greater referral to quality control. However, regardless of the nature of the laboratory, the steps in each section of any SOP are ordered to enable the method to proceed as easily as possible. Note also, that there is no right or wrong format for an SOP: the only criterion is that it works to its required specification.

SELF-CHECK 9.3

Compare the SOP in Box 9.3 with an SOP in your own laboratory.

BOX 9.3 Standard Operating Procedure 63: ELISA for Soluble E Selectin

Name of author, date

Procedure only to be performed by staff holding appropriate competency

Contents

1. INTRODUCTION

Synopsis:

This is the standard ELISA for measurement of endothelial cell marker E-Selectin\CD62E using the duoset from R&D catalogue number DY724. The method and quantities described are for two plates (~75 samples). You should allow a whole day to do the assay.

Brief method:

1. Coat the microtitre plates with 112 μL of capture primary antibody in PBS buffer in the fridge (4°C) overnight or 1.5 hours at room temperature.
2. Wash, add 100 μL serum/plasma (diluted 1/5 [20 plasma plus 80 blue or PBS-Tween buffer] on the plate) or standards (top 'prepared' 50 ng/mL) for 1.5 hours at room temperature. ALSO, prepare a sample of the 'Universal' plasma.
3. Wash, add 112 μL detection antibody (one vial in 20 mL 1% BSA PBS for two plates) for 1.5 hours at room temperature.
4. Wash, add 100 μL Streptavidin-HRP conjugate (diluted 1/200 in PBS, i.e. 100 μL plus 20 mL) for a minimum of 20 minutes at room temperature in the dark.
5. Wash, add 100 μL substrate (made up from equal volumes of reagents A and B). It should go blue. Wait 3–5 minutes. The key definition is a clear gradation of blue colour from the top to the blank.
6. Stop with 75 μL Acid. It will go yellow. Read at 450 nm.

Expected Values:

About 20-40ng/mL. The Universal standard is coming in at about 40 ng/mL.

2. MATERIALS AND EQUIPMENT

2.1 Microtitre plates Flat-bottomed 96 well microtitre plates (Immunlon 2). *Supplier, storage and location*: Thermo Life Sciences, Telephone 01256 817 282. Stored at room temperature (RT) in cupboard 3. If you open the last case, inform the laboratory manager.

2.2 Yellow and Clear Tips *Supplier;* Alpha Laboratories Limited Tel–02380483000 Fax–02380643701

For 0–200 μL use Yellow tips Cat: no FR1200. For 200–1200 μL use Clear tips Cat: no FR1250. Tips racked and stored at RT in special boxes on the desk in the main lab. Use fresh tips at each stage of the procedure.

2.3 Micropipettes 0–10 μL, 0–200μL and 0–1000μL single channel, 0–250 μL 8 channel micropipettes.

2.4 Wash Buffer *To make this up, you need ...*

2.4.1 Phosphate buffer saline (PBS) tablets. Sigma catalogue number P 4417. Buy in lots of 50 tablets. They are kept on the shelves above the right-hand side of the plate reader.

2.4.2 Tween 20. Sigma Catalogue number P 1379. Order in 100 mL lots. *Supplier:* Sigma–Aldrich, 0800 717181 or website.

2.4.3 Distilled water.

Method: To one litre of water (plastic measuring cylinder), add 5 PBS tablets and 0.5 mL Tween (use a pastette) into a large conical flask. Place on rotamixer with stir bar, wait until tablets have dissolved. Store this at RT on the rota mixer. N.B. Generally, make up 2, 3 or more litres in one batch. Write the date on the flask.

2.5 Reagents:

N.B. Scientific Staff are responsible for preparing antibodies and standards from scratch.

2.5.1 Primary capture mAb: For assay, defrost 1 aliquot labelled as 112 μL 1° Ab, and dilute this in total volume 20 mL PBS buffer to make a working solution for 2 plates.

2.5.2 Recombinant standard: Each vial in the DuoSet contains 190 ng/mL when reconstituted with 0.5 mL of 1% BSA PBS. Add 500 μL of 1% BSA PBS to the vial and vortex and incubate at room temperature for 15 minutes. The required top standard for an ideal standard curve is 10 ng/mL. To prepare the top standard take 52.6 μL from the standard vial and add 947.4 μL of 1% BSA PBS. The standard is to be double diluted in 1% BSA PBS in the wells of the plate as follows ...

(i) Place 200 μL (of 10 ng/mL) in each of wells 11 and 12 of row A.

(ii) Place 100 μL in wells of rows B to H of columns 11 and 12.

(iii) Double dilute down the plate from A to G. Discard the leftover 100 μL standard.

(iv) It follows that 11H and 12H are blanks.

2.5.3 Controls

Internal control (Universal Plasma): This is a pool of plasma aliquoted into 60 μL and 500 μL volumes. Each ELISA (not simply every sEsel ELISA but EVERY departmental ELISA) is to have one of these samples analysed. If in doubt, check with Scientific Staff. Treat this as one of the plasmas, i.e. dilute it 20 μL plus 80 μL in the wells as the other samples you have.

Control 1 (8 ng/mL): This is prepared afresh from the standard vial to check the standard curve obtained from double dilution done on the plate. This will provide an alternative to check the accuracy of our standard curves at different points of dilution. Take 42.1 μL from the standard vial and add 957.9 μL of 1% BSA PBS.

Control 2 (1 ng/mL): Take 5.2 μL from the standard vial and add 994.8 μL of 1% BSA PBS.

Control 3 (0.5 ng/mL): Take 2.6 μL from the standard vial and add 997.4 μL of 1% BSA PBS.

Place these controls above the blank. The universal control must be placed below the blank.

2.5.4 Secondary detection mAb: For the assay, defrost one aliquot and make to 20 mL in 1% BSA PBS for two plates.

2.5.5 Streptavidin HRP conjugate: This reagent lives at 4 °C in a small white cardboard box in refrigerator B. Dilute this 1/1000 in PBS for use, e.g. 100 μL to 20 mL PBS for 2 plates.

2.5.6 Substrate: From R&D Systems, Substrate reagent pack, Catalogue number DY999. Store at 4 °C in refrigerator B. Substrate reagent pack contains two different reagents in glass bottles, reagents A and B, which should be mixed in equal amounts immediately before use.

2.5.7 Stop Solution: 1 mol/L hydrochloric acid. No need to dilute. *Supplier/Storage* Sigma, Catalogue number 920-1. Order in lots of 1 gallon (3.8 L). The acid is stored in the red metal safety cupboard.

Biohazard/COSHH: (N.B. this is not an official assessment!) **Considerable**. Do not get on skin/clothes. Wash off spills with water. Otherwise maintain good laboratory practice.

2.6 Disposable gloves.

3. METHOD

3.1 Become fully familiar with this protocol and identify all reagents and buffers in advance of the procedure. Check expiry dates of relevant reagents (if supplied).

3.2 Using a micropipette and a yellow tip, add one aliquot of defrosted primary antibody to approximately 20 mL PBS buffer and, using an eight-channel micropipette with yellow tips, add 100 μL to each well of two microtitre plates. Place in lunch box with lid and label, then to the fridge overnight OR incubate on the bench for two hours.

3.3 Using the eight-channel washing manifold (Sigma), wash out the unbound antiserum with three lots of >250 μL PBS/Tween (2.5). Blot out on tissue paper between each step.

3.4 Defrost 34 plasma samples per plate if doing in duplicates and leave in fridge overnight (your own and the "Universal" [see 2.5.3]). Ensure well mixed on the day of the assay.

3.5 Place 80 μL of blue PBS buffer into the microtitre plate wells. To this add 20 μL of plasma.

3.6 Top Standard rEsel should be prepared as in 2.5.2. Add 200 μL standard to wells of columns 11 and 12. Add 100 μL buffer to wells B–H columns 11 and 12. Double dilute in 100 μL volumes down the plate leaving wells 11H and 12 H as blanks.

3.7 Place the plate on the plate reader and shake for a minute to enable homogenization. Place plates back in the lunch box and incubate for 2 hours at RT.

3.8 Wash again as in 3.3.

3.9 Defrost one aliquot of secondary detection mAb and make to 20 mL in 1% BSA PBS for two plates. Using the 8-channel micropipette, add 100 µL of this detection mAb to each well.

3.10 Incubate for 2 hours at room temperature in the lunch box.

3.11 Wash again as in 3.3.

3.12 Add 100 µL streptavidin-HRP diluted 1/200 in PBS to each well. Generally, this would be 100 µL + 20 mL buffer. Get reagents A and B out of the fridge to warm up on the bench.

3.13 Incubate for 30 minutes at room temperature but in the dark. This can be inside a cardboard box, but take care it is not disturbed.

3.14 Wash off excess enzyme with THREE cycles of 250 µL PBS/Tween. Be thorough.

3.15 Add equal volumes of substrate components A+B in an appropriate plastic tray. With the 8-channel micropipette, add 100 µL of the mixture to each well from left to right or vice versa, depending on location of the standards. A blue colour will develop quite rapidly.

3.16 After 3–5 minutes, when there is a large and clear gradient between the top and bottom standards, add 75 µL acid per well from left to right at the same speed as the substrate A+B. The blue colour will go lemon yellow. Read using the ELISA plate reader (instructions on wall) at 450 nm. See SOP E15 for operation of the plate reader.

3.17 Dispose of the plates and tips as appropriate. Clean up the bench and wipe it. See GLP document about biohazards. All tips and soiled tissues, leftover whole plasma to go in yellow bins. Microtitre plates to be emptied down a sink, then washed out with tap water, then put in yellow bins. Switch off the plate reader and complete usage log.

3.18 Draw a standard curve on three-log graph paper (optical density Y axis, standard X axis) and consult with senior staff for reliability of results obtained if in doubt. Ensure the internal and external quality control samples are within range. Write up in log book. Store plate reader printout and graph in appropriate location.

SOP 63 SIGNED OFF BY (.....Laboratory manager .....) (.....Date .....)

This SOP in the SOP file in the laboratory

Copy in Laboratory Manager's office and on computer.

9.5 Writing for external publication

New knowledge (such as the findings of a research project) must be disseminated beyond the lecture hall, in the form of a written article. Vehicles for these articles are broadly of two types: those that purport to be academic, formal, and peer-reviewed (mostly journals), being recognized by Pub-Med (the leading search engine for original research), and those whose standards are more informal (such as magazines and trade updates or reports). These are exemplified by two publications of the Institute of Biomedical Science—the quarterly formal research-orientated *British Journal of Biomedical Science* (for which the accepted abbreviation is *Br J Biomed Sci*), and *The Biomedical Scientist*, with monthly issues. The latter often contains excellent articles on subjects relevant to biomedical science (perhaps an update, or with an educational aspect), whilst the former publishes only new data that represents an advance in the practice of biomedical science. There are marked differences in the layout and writing style of these two types of vehicle, and we shall focus on writing an article for a formal journal.

The layout of an academic research article seeking to be published in a Pub-Med recognized vehicle roughly follows the format of an oral presentation, but considerably more care is required, as the information will go into the public domain, and will remain there, possibly forever, for public and international scrutiny. The approach to writing a review article, which by definition lacks data, is markedly different from that of a research paper. However, when preparing a written article of either format, it is a good idea to read widely; these papers will provide not only facts and figures, but will demonstrate the acceptable quality of work that is expected, and provide general ideas on layout.

But before embarking on the writing process, a key question must be addressed—"is the data worthy of publication?"

Quality of data for publication

Academic publication is fiercely competitive—some journals reject 90% of submitted manuscripts. One way in which journals are ranked is according to a feature called 'Impact Factor', a number that, as expected, quantifies the impact of the articles in the journal in terms of how many articles are cited as references in other articles. Those vehicles with the highest impact factors (such as the weekly *The Lancet* [impact factor around 44] and *The New England Journal of Medicine*: [impact factor 72]) demand only the best material that is likely to describe only the most important advances in medicine. Many journals published monthly (such as the *British Journal of Haematology* [4.7] and the *Journal of Clinical Pathology* [2.9]) also have high standards, reflecting their impact factor, whilst some journals (such as the *British Journal of Biomedical Science* [1.9] and the *Quarterly Journal of Medicine* [2.8]) publish quarterly. Many journals have a mission statement, that of the *British Journal of Biomedical Science* being:

> The ***British Journal of Biomedical Science*** is committed to publishing high quality original research that represents a clear advance in the practice of biomedical science, and reviews that summarize recent advances in the field of biomedical science. The overall aim of the Journal is to provide a platform for the dissemination of new and innovative information on the diagnosis and management of disease that is valuable to the practicing laboratory scientist.

Some journals also offer guidance on the quality of potentially publishable data. That of *British Journal of Biomedical Science* is shown in Box 9.4, key features being the originality of the data, the power of the study (in terms of, for example, the number of subjects recruited), and its likely impact on the day-to-day work of a laboratory scientist. The choice of vehicle is also important: the subject of the paper must be appropriate for the particular journal—a manuscript on prostate cancer submitted to the *Journal of Breast Cancer* will be returned immediately! But some choices can be difficult—if there is data on renal disease in cancer should the paper go to a journal specializing in renal disease or a journal specializing in cancer? Answer—get advice from an experienced colleague. An additional point about the choice of a publishing vehicle is to be realistic about the nature of the data and importance of the finding—is it really good enough for the *British Medical Journal*?

In order to be published, manuscripts must be submitted to a journal, the editor of which will demand certain features. These are laid down in 'instructions to authors', a set of rules that the manuscript must adhere to, which will be available on the journal's website. Typical instructions may include, for example, that the report must be written as a portrait-style manuscript, often with one-inch margins on all four sides, and with the text typed in 12 point font and double-spaced. For example, the leading pathology laboratory journal for biomedical and clinical scientists is the *British Journal of Biomedical Science*. Its instructions to authors can be accessed via its website: www.bjbs-online.org/authors.asp., and must be followed exactly. Failure to do so results in the manuscript being returned to the corresponding author for changes.

The manuscript will be laid out in a number of sections, generally as follows, but with major differences dependent on which of the two types it is: that which is presenting research data

BOX 9.4 Major features in submission outcome

The Journal is published by the Institute of Biomedical Science, a UK body that speaks for the biomedical and clinical scientists that work predominantly in routine National Health Service pathology laboratories. Accordingly, published material will reflect their day-to-day working and experience. Features likely to lead to acceptance or rejection are as follows ...

Original Article	In Brief	Rejection
Reports results that represent a clear advance in the practice of biomedical science	Reports results that represent a clear advance in the practice of biomedical science	Results do not represent a clear advance in the practice of biomedical science
Well powered, with a good, clear original hypothesis	Sufficiently powered with clear aim/objective/hypothesis	Under-powered (i.e. small numbers), no clear objective/aim
Several research indices (blood cell or molecule, microbe, gene polymorphism, tissue expression of a gene or molecule, etc.), with perspective from several routine markers/processes, possibly across disciplines (e.g. haematology and histology)	Few, perhaps only one or two research indices (blood cell or molecule, microbe, gene polymorphism, tissue expression of a gene or molecule, etc.) with perspective from only a few routine markers/processes	Few, perhaps only one research index, with little or no perspective with routine laboratory markers/processes. Results perhaps too preliminary. Disease or condition is rare and so unlikely to be seen in a routine laboratory
Follow-up data (e.g. survival outcome) or serial data (results on several different time points, e.g. days/weeks/months)		Poor formatting and English grammar that the message is unintelligible. Failure to adhere to instructions for authors
Additional cell biology/gene data (e.g. tissue culture, flow cytometry, SNPs, etc.) that support the original hypothesis		Not original, poor literature search, local audit of practice, confirming other findings

N.B. These criteria (which are not exhaustive) are not necessarily or always present in a particular article, an overview is taken on the entire manuscript by Editor and referees.

The sample size (number of patients, tissues, experiments, etc.) is a key factor, and varies according to the type of work (see below). It may be defined by a formal power calculation. Many papers are rejected because the sample size is too small.

Clinical science describes work directly relevant to patient care, or to disease diagnosis or management. An example of this would be a case control study, perhaps linked to disease stage or other subgroups, and including outcome. Recently published papers have a minimum of 100 recruits, but better papers have a median of 162 recruits. This is likely to increase with the number of analyses and sub-groups. The latter are likely to have a minimum of 30 per sub-group.

Laboratory development describes new methods, techniques, approaches, and links between biomedical science and basic biology of the cell or gene. Papers reporting new methods will have validation data, and a perspective of testing the new method/technique on clinical samples (serum, urine, sputum, microbes, etc.) typical of those undergoing routine analysis. Recently published papers have a minimum of 50 such clinical samples, better papers being validated by data from a median of 100 samples, which will again rise with the number of additional analyses.

In both types of work, the higher the sample size, the more robust and so more believable will be the conclusions, and so the greater the likelihood of publication.

(as a full original article, or a short *'In Brief'* report), or that of a review article. Other journals will have their own particular and specific instructions. For example, the *British Journal of Biomedical Science* allows up to four tables or figures in full original articles—other journals may demand fewer, or have no restriction.

An original article

A manuscript with a comprehensive mass of research data is likely to adopt the following format.

Title page

This is a single page, stating (in order, from the top of the page), the title, the authors, and their address(es), often where the work is performed. The title must capture the attention of the reader, and will ideally contain several key words, as these may be picked up by computer search engines. Some authors prefer the title to be open and descriptive, such as 'The role of gene X in breast cancer', whereas other authors may phrase the title to state a result, such as 'Gene X is overexpressed in breast cancer'. A third option is to pose a question, such as 'Is there a place for gene X in breast cancer?'.

Towards the bottom of the page there will be details of which of the authors will interact with the editors (that is, the corresponding author). This is generally whoever writes the paper, but it may be a more senior colleague. There must be full contact details of the correspondent, including postal address, telephone and fax numbers, and e-mail address. These details will be published as part of the final article so that readers can get in touch with the researchers.

In almost all cases the instructions to authors call for three to six key words. These are provided for indexing, and to aid computer searching (and, hence, the enhanced dissemination of the research).

The abstract

The second page of the article is the abstract. This is where the work must be summarized, usually in a single paragraph of up to 250 words, as per the example in Box 9.1. Depending on the instructions to authors, this paragraph may be 'structured' into sections, following the same sub-headings as those used in the rest of the manuscript. However, there may be other requirements, such as 'setting' (perhaps it was conducted in primary or secondary care, or it may be results of a questionnaire). All abstracts must provide the sample sizes (such as the number of cases and controls), and have some data and p values.

Introduction

This section provides the reader with the background information that justifies the study, and must include a review of the pertinent literature. In this literature review, the author(s) will cite key references that describe why the study is important and therefore why the particular work needs to be done. It may point out a gap in knowledge that the project seeks to fill. The penultimate part of the introduction is the firm original hypothesis that the authors seek to test; an aim or an objective is often too non-specific. This is often followed by a brief plan of investigation, which explains how the hypothesis is to be tested. A good introduction rarely exceeds two pages (any longer betrays lack of focus), and may include 10 to 20 references.

Subjects, materials, and methods

In this section the authors must clearly describe the investigations that were performed, and (if clinical) on which subjects. This description must be to a level of detail that gives the reader all they need in order to reproduce the work, and it is likely to include details of the source of the reagents and equipment (name of company, city, country). If laboratory methods are non-standard, then intra- and inter-assay coefficients of variation must be provided. If ethical committee approval was required, this must be stated, alongside a note that the written informed consent from all subjects was obtained.

Some editors require a statement that the Declaration of Helsinki was adhered to. This document, published by the World Medical Association, consists of a code of conduct, and a number of points that should be adhered to in order to ensure that a research study has been undertaken ethically. One such point is that each participant must be able (physically and psychologically) to give written informed consent.

This section concludes with a statement of which statistical tests will be used to analyse the data, and a justification of the number of subjects recruited and/or the number of experiments or repetitions performed (that is, the sample size and power calculation). It is likely that a statistician will have been consulted when the study was being designed, to ensure that the sample size is correct. The length of this section depends on the nature of the research, and may be four to six pages long.

Results

Data can be presented in the text, or as separate tables. Some data may also be shown graphically. If the study is clinical (perhaps with patients and healthy controls), Table 1 is often the clinical data (such as blood pressure, body mass index), laboratory data (such as serum potassium), and demographic data (age, sex, race/ethnicity). It may also list other diseases that the subjects have, and their medications (but not usually with the doses of the particular drug). Purely laboratory studies are likely to have tables showing relevant aspects of the particular indices (molecules, cells, etc.) being studied. Tables should not be so large that they are spread over two pages.

In a clinical paper, Table 2 and subsequent tables will show the analyses of the raw data. There will also be a section of the text that will summarize the results in terms of the original hypotheses. Data is presented only once, either in the text, or in tables. In many cases the explanation of some (or all) of the data is enhanced if presented in graphical form; such graphs appear as numbered figures (as per the PowerPoint slides of Figure 9.1). Examples of how data can be presented are shown in Chapter 4.

Results sections are as long as they need to be. However, if much of the data is presented in tables, then the text of a results section may be as short as one or two pages. This section is rarely referenced, but if so it will cite papers explaining methodology or statistics.

Discussion

This section also follows a set pattern:

- The first paragraph is a brief summary of the major findings of the study.
- The second often compares these results with those of other researchers, whose work may have already been cited in the introduction.
- The third paragraph often discusses the implications of the study upon the laboratory or clinical medicine as a field, as the nature of the study demands.

Towards the end of the discussion there is often an acknowledgement of the limitations of the study—for example, that relatively small numbers do not allow extensive interpretations, or that the exclusion of certain groups of subjects means that the results cannot be extrapolated too widely. Finally, there are two or three concluding sentences, and perhaps some mild (but guarded) speculations.

A good discussion is generally about three pages; more than this again betrays a lack of focus. In some cases there may be acknowledgement of the help provided by colleagues, and perhaps the source of the funds that allowed the study to proceed. There may also be a declaration of interests—for example, that funds were obtained from a commercial organization.

References

The introduction section of an article will cite key works that set the scene and justify the research. These 'cited papers' are the references. As the article proceeds, methods and data from other workers will be referred to in the materials and methods, and the discussion sections respectively. It is likely that key papers will be referred to in the introduction and again in the discussion.

References are typically cited (that is, their relevance to a particular passage of text flagged) in one of several ways. For example:

- The *Harvard* system is also called the author-date system; the citation in the text takes the form of the author surname and year of publication: (Smith, 2014).
- The *Vancouver* system is also called the author-number system: the citation takes the form of a number, which can appear in the text in various forms—for example, in brackets, such as [5] or (5), or as a superscript number 5.

The reference list then gives full details of each article cited in the text. The order in which articles are listed depends on the reference system being used: if using the Harvard system, articles are listed in alphabetical order according to the surname of the (leading) author; if using the Vancouver system, the articles are listed numerically, according to the order in which they are cited in the text.

The information provided for each article referenced may be presented in a number of forms, the most common being to list the authors (surname and initials), the title of the research, the journal, its year, volume, and pages. For example:

> Goon, P.K., Lip, G.Y., Stonelake, P.S., Blann, A.D. Circulating endothelial cells and circulating -progenitor cells in breast cancer: relationship to endothelial damage/dysfunction/apoptosis, clinicopathologic factors, and the Nottingham Prognostic Index. *Neoplasia*. 2009;11:771–9.

However, the exact form will be described in the instructions for authors. In some cases the instructions may be to include the monthly issue number, such as in *Neoplasia. 2009;11(4):771–9*, where the (4) will most likely be the April issue (the fourth month). In other cases the reference may be followed by a code, often placed there by the search engine (which in this instance is Pub Med), such as:

> Blann, A.D. Soluble P-selectin: the next step. *Thromb Res*. 2014 Jan;133(1):3–4. doi: 10.1016/j.thromres.2013.10.030. PMID 4216322

Overall, a research paper may have 2500–5000 words, with perhaps 30–50 references. The *British Journal of Biomedical Science* sets a limit of 40 references. An example of an original article is shown in Figure 9.2.

British Journal of Biomedical Science

ISSN: 0967-4845 (Print) (Online) Journal homepage: http://www.tandfonline.com/loi/tbbs20

Serum miR-210 and miR-155 expression levels as novel biomarkers for rheumatoid arthritis diagnosis

RS Abdul-Maksoud, AM Sediq, AAA Kattaia, WSH Elsayed, N Ezzeldin, SM Abdel Galil & RA Ibrahem

To cite this article: RS Abdul-Maksoud, AM Sediq, AAA Kattaia, WSH Elsayed, N Ezzeldin, SM Abdel Galil & RA Ibrahem (2017) Serum miR-210 and miR-155 expression levels as novel biomarkers for rheumatoid arthritis diagnosis, British Journal of Biomedical Science, 74:4, 209-213, DOI: 10.1080/09674845.2017.1343545

To link to this article: http://dx.doi.org/10.1080/09674845.2017.1343545

Accepted author version posted online: 15 Jun 2017.
Published online: 07 Aug 2017.

FIGURE 9.2 (*CONTINUED*)

BRITISH JOURNAL OF BIOMEDICAL SCIENCE, 2017
VOL. 74, NO. 4, 209–213
https://doi.org/10.1080/09674845.2017.1343545

Taylor & Francis
Taylor & Francis Group

Check for updates

Serum miR-210 and miR-155 expression levels as novel biomarkers for rheumatoid arthritis diagnosis ①

RS Abdul-Maksoud[a], AM Sediq[b], AAA Kattaia[c], WSH Elsayed[d], N Ezzeldin[e], SM Abdel Galil[e‡] and RA Ibrahem[f] ②

[a]Faculty of Medicine, Medical Biochemistry Department, Zagazig University, Zagazig, Egypt; [b]Faculty of Medicine, Clinical and Chemical Pathology Department, Zagazig University, Zagazig, Egypt; [c]Faculty of Medicine, Histology and Cell Biology Department, Zagazig University, Zagazig, Egypt; [d]Faculty of Medicine, Pathology Department, Zagazig University, Zagazig, Egypt; [e]Faculty of Medicine, Rheumatology and Rehabilitation Department, Zagazig University, Zagazig, Egypt; [f]Faculty of Medicine, Public Health and Community Medicine Department, Menoufia University, Menoufia, Egypt

ABSTRACT

Background: MicroRNAs play a crucial role in the regulation of immune response. We hypothesised roles for serum miR-210 and miR-155 in the diagnosis of rheumatoid arthritis (RA) and relationships with the clinical and laboratory variables including erythrocyte sedimentation rate (ESR), C-reactive protein (CRP), rheumatoid factor (RF), anti-cyclic citrullinated peptide (CCP) antibodies, tumour necrosis factor-alpha (TNF-α) and interleukin-1β (IL-1β).
Methods: MiR-210 and miR-155 levels were identified by real-time polymerase chain reaction (PCR). TNF-α and IL-1β were measured by enzyme-linked immunosorbent assay and routine markers by standard techniques in 100 patients with RA and 100 individuals as healthy controls. Disease activity in the patients was assessed by DA-S28.
Results: MiR-210 was lower in RA compared to controls [median/IQR 0.96 (0.8–1.24) vs. 4 (1.28–3.93), $p < 0.001$]. miR-210 correlated inversely with clinical and laboratory variables including TNF-α and IL-1β (both $r = -0.96$, $p < 0.001$). MiR-155 expression was increased in RA compared to controls [median/IQR 6 (3.5–8.1) vs. 1.0 (0.95–1.6), $p < 0.001$] and correlated with TNF-α and IL-1β (both $r = 0.94$, $p < 0.001$). In multivariate analysis, miR-210 and miR-155 were both independent diagnostic markers for RA, and both were associated with RA disease activity.
Conclusion: Serum miR-210 and miR-155 levels are independent diagnostic markers for RA, out-performing several routine indices and reflect disease activity. Thus, miR-210 and miR-155 might serve as non-invasive biomarkers for the diagnosis of RA.

③

ARTICLE HISTORY
Received 24 January 2017
Accepted 6 June 2017

KEYWORDS ④
miR-210; miR-155; rheumatoid arthritis; TNF alpha; IL-1β

Introduction ⑤

Rheumatoid arthritis (RA) is a systemic, chronic inflammatory, autoimmune disease that affects primarily the articular cartilage and bone. The most important features of RA are persistent inflammation, cartilage erosion with subsequent joint destruction and deformity [1]. Untreated patients have a progressive course resulting in short- and long-term disability, so early treatment can prevent severe disability and leads to patient benefits [2,3]. Therefore, biomarkers for the diagnosis and prediction of therapeutic outcomes are needed, enabling clinicians to treat patients as early as possible with the best therapeutic approach. Pro-inflammatory cytokines play an important role in the pathophysiology of RA: the release of TNF-α and IL-1β causes synovial inflammation, and cytokines promote the production of acute-phase proteins (such as CRP) and development of cardiovascular disease and anaemia [4].

MicroRNAs are small (18–22 nucleotides) non-coding RNA molecules that have an important role in the regulation of gene expression. It has been reported that around 30% of human genes are regulated by miRNAs [5]. They mediate this regulation either by mRNA degradation or by translational repression and possess diverse functions in regulating some cellular processes, including cellular differentiation, proliferation, apoptosis as well as cancer development [6–8]. Moreover, miRNAs have important roles in regulating the immune responses and the development of autoimmunity [9,10].

Increasing evidence has linked circulating miRNAs with the diagnosis and progression of a variety of diseases, including cardiovascular disease, infectious diseases and some cancers [11–13]. Biochemical and genetic studies revealed the physiological functions of individual miRNAs in immunity [14]. Altered expression of miRNAs in synovial tissue, synovial fibroblasts, peripheral blood mononuclear cells, osteoclasts and isolated

⑥

CONTACT RS Abdul-Maksoud rehabshaaban2014@gmail.com
‡Medicine Department, Faculty of Medicine, Umm Al-Qura University, Makkah, Saudi Arabia

FIGURE 9.2 (CONTINUED)

T lymphocytes results in the degradation of extracellular matrix, inflammation and invasive behaviour of resident cells [15–17]. Furthermore, several studies have supported an aberrant miRNA expression in different cell types in RA, thus regulating specific pathways involved in pathogenesis [18–20]. It has been reported that miR-210 inhibit the activation of NF-$_{\kappa}$ B pathway, an important regulator of the immune response, inflammation as well as cell survival [21,22]. In clinical and experimental arthritis models, Kurowska-Stolarska et al. [23] showed that miR-155 plays an important role in the proinflammatory activation of myeloid cells and the development of antigen-derived inflammatory arthritis.

⑦ We hypothesise that miR-210 and miR-155 are more effective in the diagnosis of RA than other laboratory markers and are linked to disease activity. Moreover, we hypothesise that these miRNAs could play an important role in the pathogenesis of RA through a link with TNF-α and IL-1β.

Materials and methods ⑧

The study was carried out on 200 individuals: 100 patients with RA (disease duration 8.0 ± 3.0 years) and 100 healthy age matched as controls. Patients were diagnosed according to 1987 revised American association rheumatism criteria for the classification of rheumatoid arthritis [24]. They were recruited from the outpatient clinics and inpatient units of Rheumatology and Rehabilitation Department, Zagazig University Hospitals, Faculty of Medicine, Zagazig, Egypt. We excluded patients with hepatic, diabetes, renal or malignant diseases. Patients were subjected to full history taking and thorough clinical examination. Patients medical records were reviewed for the documentation of clinical and laboratory data. All patients received non-steroidal anti-inflammatory drugs (NSAIDs) daily and 68% took steroids. DMARDs alone or in combination with other drugs were administered in 80% of subjects and 15% of patients were treated with biologics (infliximab and humira). RA disease activity was assessed using the Disease Activity Score (DAS-28 score). DAS-28 depends on the following parameters (swollen joint count (SJC), tender joint count (TJC), erythrocyte sedimentation rate (ESR)). For DAS-28 score, low disease activity was defined as DAS-28 ≤ 3.2, moderate as 3.2 < DAS-28 ≤ 5.1, high as DAS-28 > 5.1 and remission as DAS-28 < 2.6 [25]. The median/IQR of SJC, TJC and DAS-28 were 12 (8–15), 16 (11–20) and 4 (3–6), respectively. The control group had no family history of RA, did not show any clinical or laboratory signs of autoimmune diseases and they were free of any treatment. This study was approved by the Ethics Committee of Zagazig University and all subjects were included in the study after giving their informed consent after explanation of the purpose and procedures of the study.

After an overnight fast, venous blood samples were collected by venipuncture under complete aseptic conditions and divided into two portions: 5 ml of blood was collected into specific tubes for obtaining serum. Serum was stored at −80 °C till analysis. Another 1.6 ml was collected in tubes contain 0.4 ml trisodium citrate (109 mmol/l) and used to measure ESR according to Westergren [26]. Levels of serum CRP and rheumatiod factor (RF) were measured by immunturbidimetric assay on a Roche/Hitachi cobas c system (c501) autoanalyzer (Roche Diagnostics, Mannheim, Germany) using CRPL3 and RFII reagents respectively. Anti-CCP antibodies were measured on a Cobas e601 immuno-autoanalyzer (Roche Diagnostics) that adopts an electrochemiluminescence immunoassay (ECLIA) technique using dedicated reagents. TNF-α and IL-1β were measured by ELISA using Quantikine Kits (R&D Systems, Minneapolis, MI, USA).

MiRNAs were extracted from 400 µl of serum using the miRNeasy kit (Qiagen, Valencia, CA, USA), quantified spectrophotometrically at 260 nm and stored at −80°C. RNA was reverse transcribed to cDNA with miScript II RT Kit (Qiagen, CA). For real-time PCR of miR-210 and miR-155, the amplification was performed using StepOnePlus™ RealTime PCR system (Applied Biosystems Inc., Foster, CA). Syn-cel-miR-39 miScript miRNA Mimic was used as an internal control for data normalisation. Real-time PCR was performed using miScript Primer Assays [containing miRNA-specific forward primers: miRNA-210 (Cat. No. MS00003801); miRNA-155 (Cat. No MS00031486); Syn-cel-miR-39 (Cat. No MSY0000010)] and miScript SYBR Green PCR kit (Qiagen, Valencia, CA, USA) according to manufacturer's protocol with the manufacture-provided miScript Universal primer (reverse primer) and QuantiTect SYBR Green PCR Master Mix. PCR was performed in a mixture containing 10 µl 2× QuantiTect SYBR Green PCR Master Mix (Qiagen, USA), 0.5 µl of each primer (100 pmol/ µl), 5 µl cDNA and water to adjust the final volume to 20 µl. PCR was performed under the following cycling conditions: denaturation at 95 °C for 15 min, followed by 40 cycles at 94 °C for 15s; 55 °C for 30s and at 70 °C for 30 s. Each measurement was performed in duplicate. Melting curves were performed to confirm the specificity of PCR products. MiRNAs expression levels were determined by using ΔΔCT method where the Ct of the target miRNAs were normalised to the Ct of Syn-cel-miR-39 (internal control). Relative changes of gene expression were calculated from the equation $2^{-\Delta\Delta Ct}$.

The intra-assay CV was 0.7% and 0.8% for miR-210 and miR-155, respectively. The interassay CV of miR-210 and miR-155 was 2.3% and 2.5%, respectively. These values indicate that the techniques used in this study (RNA isolation and qRT-PCR) are reproducible. Data were analysed using SPSS statistical package version 17 (SPSS Inc., Chicago, IL, USA). Student's *t*-test and the chi square (x^2) test were used for comparison between groups. Mann–Whitney U-tests were used for non-normally distributed data. Correlations were assessed by Spearman's rank correlation. Contrast hypothesis test [27] was used to test the trend in continuous quantitative data. A p-value < 0.05 was considered significant. ⑨

FIGURE 9.2 (*CONTINUED*)

Results

Clinical, demographic and laboratory data are presented in Table 1. Patients and controls were age and sex matched. Serum miR-210 levels were significantly lower in patients with RA compared to healthy control subjects, whilst miR-155 was higher in RA compared to controls. Unsurprisingly, all other laboratory indices were higher in RA than in controls. Multivariate regression analysis revealed that both miR-210 and miR-155, but not any other laboratory index, were independent diagnostic markers for RA (Table 2). Correlation of miRNAs with clinical and laboratory data in RA patients are shown in Table 3. Serum miR-210 correlated negatively, and miR-155 positively with ESR, CRP, SJC, TJC, DAS-28, RF, anti-CCP, TNF-α and IL-1β levels. No correlation was found between miRNAs expressions and other studied parameters including age and disease duration.

When stratifying RA patients according to disease activity (DAS-28 score) (Table 4), levels of miR-210 fell, whilst levels of miR-155, RF and IL-1β increased with increased disease activity. There was no significant difference in anti-CCP and TNF-α levels between different subgroups. When stratifying RA patients according to the type of treatment, there was no significant difference in miR-210 levels between patients treated with DMARDs or biologics (0.9 ± 0.3 vs. 1.0 ± 0.2, $p = 0.63$). Furthermore, miR-155 levels showed no significant difference between patients treated with DMARDs or biologics (6.7 ± 2.4 and 6.3 ± 2.5; $p = 0.51$).

Table 1. Demographic and laboratory data.

Variable	RA *N* = 100	Control *N* = 100	*P* value
Age (years)	51.0 ± 7.0	51.0 ± 6.0	0.51
Sex male/female	26 (26)/74 (74)	34 (34)/66 (66)	0.22
ESR (mm/h)	52 (36–64)	13 (10–13)	<0.001
CRP (mg/l)	34 (22–41)	10 (7–11)	<0.001
RF (IU/ml)	150 (99–179)	23 (14–56)	<0.001
Anti-CCP (U/ml)	120 (102–242)	10 (8–42)	<0.001
TNF-α (pg/ml)	35 (21–43)	12 (9–15)	<0.001
IL-1β (pg/ml)	38 (24–46)	18 (15–23)	<0.001
MiR-210	0.96 (0.8–1.2)	4 (1.2–3.9)	<0.001
MiR-155	6 (3.5–8.1)	1.0 (0.9–1.6)	<0.001

Notes: Data presented as mean ±SD, median (IQR) or n (%). ESR, erythrocyte sedimentation rate; CRP, C-reactive protein; DAS28, 28-joint Disease Activity Score; RF, rheumatoid factor; Anti CCP, anti-cyclic citrullinated peptide antibodies; TNF, tumour necrosis factor; IL-1β, interleukin 1β.

Table 2. Multivariate regression analysis for independent factors for detecting RA.

Parameters	Odds ratio	95%CI	*P* value
ESR	1.19	0.53–3.42	0.43
CRP	0.99	0.63–5.66	0.21
RF	0.97	0.32–12.33	0.45
Anti-CCP	1.12	0.23–7.88	0.17
TNF-α	1.06	0.73–5.66	0.11
IL-1β	1.10	0.02–11.45	0.09
miR-210	0.31	0.001–0.56	0.001
miR-155	2.30	1.5–7.8	0.009

Note: CI = confidence interval, see Table1 for other abbreviations.

Table 3. Spearman rank correlations (r) between serum miRNAs and clinical and laboratory data in RA patients.

Variable	miR-210	miR-155
TNF-α	$r = -0.96$: $p < 0.001$	$r = 0.94$: $p < 0.001$
IL-1β	$r = -0.96$: $p < 0.001$	$r = 0.94$: $p < 0.001$
Anti-CCP	$r = -0.95$: $p < 0.001$	$r = 0.94$: $p < 0.001$
SJC	$r = -0.93$: $p < 0.001$	$r = 0.90$: $p < 0.001$
TJC	$r = -0.94$: $p < 0.001$	$r = 0.91$: $p < 0.001$
DAS28	$r = -0.75$: $p < 0.001$	$r = 0.70$: $p < 0.001$
RF	$r = -0.54$: $p < 0.001$	$r = 0.48$: $p < 0.001$
CRP	$r = -0.35$: $p < 0.001$	$r = 0.32$: $p = 0.001$
ESR	$r = -0.28$: $p = 0.005$	$r = 0.27$: $p = 0.006$
Age	$r = 0.10$: $p = 0.33$	$r = -0.01$: $p = 0.90$
Disease duration	$r = 0.07$: $p = 0.50$	$r = -0.08$: $p = 0.86$

Notes: CI = confidence interval, SJC = swollen joint count, TJC = tender joint count. See Table 1 for other abbreviations.

Discussion

Rheumatoid arthritis (RA) is one of the leading causes of chronic morbidity in the developed world. Often seen as a minor health problem despite potentially fatal systemic manifestations, if untreated, it can lead to extensive damage of the cartilage and bone, causing deformity and disability [1,2,28]. Therefore, new biomarkers for RA diagnosis are urgently needed.

Recently, several miRNAs have been identified as playing important roles in the regulation of immune responses and the development of autoimmune disorders including RA [29,30]. In contrast to cellular or tissue miRNAs, the expression profile of circulatory miRNAs in RA has not been fully investigated. Therefore, we investigated levels of serum miR-210 and miR-155 in 100 patients with RA and 100 healthy controls. We also investigated the association of miR-210 and miR-155 with proinflammatory cytokines (TNF-α and IL-1β) as the precise role of miRNA regulation of proinflammatory cytokines in RA needs to be explored. Our principle observation is lower serum levels of miR-210 and higher levels of serum miR-155 in

Table 4. Change in levels of studied parameters in the sera of patients with different disease activities.

	DAS-28 score				
	Remission *N* = 14	Low disease activity *N* = 20	Moderate disease activity *N* = 38	High disease activity *N* = 28	*P* value
RF (IU/ml)	123 (56–149)	145 (112–176)	165 (89–180)	167 (110–187)	0.04
Anti-CCP (U/ml)	129 (67–204)	148 (98–204)	175 (85–285)	205 (159–246)	0.89
TNF-α (pg/ml)	26 (15–35)	30 (12–35)	35 (25–43)	43 (26–49)	0.33
IL-1β (pg/ml)	29 (22–45)	35 (23–42)	38 (24–42)	45 (27–50)	0.02
MiR-210	1.2 (1.15–1.67)	1.0 (0.89–1.17)	0.97 (0.89–1.19)	0.75 (0.65–0.85)	<0.001
MiR-155	3.9 (1.07–5.3)	5.8 (4.0–7.2)	6.1 (3.9–7.7)	9.0 (7.6–9.8)	<0.001

Note: *P* value calculated using contrast hypothesis test. Data presented as median (IQR). See Table 1 for abbreviations.

FIGURE 9.2 (CONTINUED)

RA patients compared to healthy controls. These results are in line with reports that miR-210 may be associated with osteoarthritis (OA) [31,32]. In articular cartilage of OA rats, Zhang et al. [22] analysed miR-210 expression and showed that it was much lower than that of normal rats. Regarding miR-155, several studies have focused on its role in autoimmune diseases, such as RA. Stanczyk et al. [14] were first to report increased miR-155 expression in RA synovial fibroblasts compared to osteoarthritis synovial fibroblasts. Others observed that miR-155 expression is increased in RA peripheral blood mononuclear cells compared to normal controls [15,33].

Our multivariate regression analysis revealed that miR-210 and miR-155 are independent factors for detecting RA and can diagnose RA more accurately than existing laboratory markers. We also observed that low levels of miR-210 and increased miR-155 were associated with increased levels of TNF-α and IL-1β. This supports data from Qi et al. [34] who reported that overexpression of miR-210 inhibits expression of proinflammatory cytokines (IL-6 and TNF-α) induced by TLR4 in murine macrophages. They also noted that MiR-210 has anti-inflammatory and anti-apoptotic effects in LPS-induced chondrocytes. Moreover, our data extend that of Li et al. [33], who demonstrated that miR-155 in peripheral blood mononuclear cells of 45 RA patients was associated positively with increased TNF-α, IL-1β, CRP, RF, ESR, SJC, TJC and DAS-28. We predict that the down regulation of miR-210 and the upregulation of miR-155 are potential markers of RA disease activity and may have an important role in the pathogenesis of the disease. When analysing the influence of treatment on miR-210 and miR-155 expression, there was no significant difference between patients treated with DMARDs or biologics. This might be due to the small sample size in our study. Therefore, we recommend our data be confirmed in a larger population.

The major question now is how serum miR-210 and miR-155 relate to RA. Recently, some miRNAs have emerged to play an important role in the control of toll-like receptor (TLR) pathways [35]. These pathways can activate NF-κB and induce the expression of many genes responsible for the inflammatory response in OA and for the production of the proinflammatory cytokines. MiR-210 could regulate cell-cycle E2F3, cell differentiation, proliferation and apoptosis. Moreover, it stimulates the proliferation of fibroblast growth factor receptor protein 1 and homeobox A1 [36]. MiR-155 could increase TNF-α levels directly by binding to its 3'UTR or targeting gene transcripts coding for the repressor of TNF-α translation [37]. In germinal centres, miR-155 plays an important role in the development of B cells and mediates regulatory roles in T-cell homeostasis [38]. Moreover, miR-155 can contribute to RA pathogenesis by inducing chemokine production and downregulating chemokine receptor, which may lead to monocyte retention at the sites of inflammation in RA [39]. ⑯

In conclusion, low levels of MiR-210 and high miR-155 may be useful markers of RA disease activity. Due to their correlations with serum TNF-α and IL-1β, both serum miR-210 and miR-155 may be implicated in the pathogenesis of the disease. This work represents an advance in biomedical science because it shows that serum miR-210 and miR-155 levels independently differentiate RA patients and healthy controls and are better than existing biomarkers for the diagnosis of RA. ⑬

Summary table

What is known about this subject

- Several miRNAs have been identified to play important roles in the regulation and the development of rheumatoid arthritis (RA)
- In contrast to cellular or tissue miRNAs, the expression profile of circulatory miRNAs in RA has not been fully investigated
- The role of serum MiR-210 and miR-155 in the diagnosis of RA has not been established yet

What this paper adds ⑭

- Downregulation of serum miR-210 and upregulation of miR-155 expression serve as non-invasive independent biomarkers for the diagnosis of RA
- *Serum miR-210 and miR-155 are potential markers of RA disease activity and may be implicated in the pathogenesis of the disease*

Disclosure statement

No potential conflict of interest was reported by the authors.

ORCID

AM Sediq http://orcid.org/0000-0003-1371-2595
AAA Kattaia http://orcid.org/0000-0002-7188-0100

References

[1] Lee DM, Weinblatt ME. Rheumatoid arthritis. Lancet. 2001;358:903–911.
[2] Goekoop-Ruiterman YP, De Vries-Bouwstra JK, Allaart CF, et al. Clinical and radiographic outcomes of four different treatment strategies in patients with early rheumatoid arthritis (the BeSt study): a randomized, controlled trial. Arthritis Rheum. 2005;52:3381–3390.
[3] Breedveld FC, Weisman MH, Kavanaugh AF, et al. A multicenter, randomized, double-blind clinical trial of combination therapy with adalimumab plus methotrexate versus methotrexate alone or adalimumab alone in patients with early, aggressive rheumatoid arthritis who had not had previous methotrexate treatment. Arthritis Rheum. 2006;54:26–37.
[4] Choy E. Understanding the dynamics: pathways involved in the pathogenesis of rheumatoid arthritis. Rheumatology. 2012;51:v3–v11.
[5] Sun B, Liu X, Gao Y, et al. Downregulation of miR-124 predicts poor prognosis in pancreatic ductal adenocarcinoma patients. Br J Biomed Sci. 2016;73(4):152–157.
[6] Yadegaria ZS, Akramia H, Hosseinib SY, et al. miR-146a gene polymorphism and susceptibility to gastric cancer. Br J Biomed Sci. 2016;73(4):201–203.
[7] Ren X, Shen Y, Zheng S, et al. miR-21 predicts poor prognosis in patients with Osteosarcoma. Br J Biomed Sci. 2016;73(4):158–162.

FIGURE 9.2 (CONTINUED)

[8] Liu C, Xing M, Wang L, et al. miR-199a-3p downregulation in thyroid tissues is associated with invasion and metastasis of papillary thyroid carcinoma. Br J Biomed Sci. 2017;74(2):90–94.
[9] Brooks WH, Le Dantec C, Pers JO, et al. Epigenetics and autoimmunity. J Autoimmun. 2010;34:J207–J219.
[10] Furer V, Greenberg JD, Attur M, et al. The role of microRNA in rheumatoid arthritis and other autoimmune diseases. Clin Immunol. 2010;136:1–15.
[11] Mitchell PS, Parkin RK, Kroh EM, et al. Circulating microRNAs as stable blood-based markers for cancer detection. Proc Nat Acad Sci. 2008;105:10513–10518.
[12] D'Alessandra Y, Devanna P, Limana F, et al. Circulating microRNAs are new and sensitive biomarkers of myocardial infarction. Eur Heart J. 2010;31:2765–2773.
[13] Ji F, Yang B, Peng X, et al. Circulating microRNAs in hepatitis B virus-infected patients. J Viral Hepat. 2011;18:e242–e251.
[14] Stanczyk J, Pedrioli DM, Brentano F, et al. Altered expression of MicroRNA in synovial fibroblasts and synovial tissue in rheumatoid arthritis. Arthritis Rheum. 2008;58(4):1001–1009.
[15] Pauley KM, Satoh M, Chan AL, et al. Upregulated miR-146a expression in peripheral blood mononuclear cells from rheumatoid arthritis patients. Arthritis Res Therapy. 2008;10(4):R101.
[16] Nakamachi Y, Kawano S, Takenokuchi M, et al. MicroRNA-124a is a key regulator of proliferation and monocyte chemoattractant protein 1 secretion in fibroblast-like synoviocytes from patients with rheumatoid arthritis. Arthritis Rheum. 2009;60(5):1294–1304.
[17] Niederer F, Trenkmann M, Ospelt C, et al. Down-regulation of microRNA-34a* in rheumatoid arthritis synovial fibroblasts promotes apoptosis resistance. Arthritis Rheuma. 2012;64(6):1771–1779.
[18] Dong L, Wang X, Tan J, et al. Decreased expression of microRNA-21 correlates with the imbalance of Th17 and Treg cells in patients with rheumatoid arthritis. J Cell Mol Med. 2014;18:2213–2224.
[19] Lin J, Huo R, Xiao L, et al. A novel p53/microRNA-22/Cyr61 axis in synovial cells regulates inflammation in rheumatoid arthritis. Arthritis Rheumatol. 2014;66:49–59.
[20] Xu T, Huang C, Chen Z, et al. MicroRNA-323-3p: a new biomarker and potential therapeutic target for rheumatoid arthritis. Rheumatol Int. 2014;34:721–722.
[21] Olarerin-George AO, Anton L, Hwang YC, et al. A functional genomics screen for microRNA regulators of NF-kappaB signaling. BMC Biol. 2013;11:19–34.
[22] Zhang D, Cao X, Li J, et al. MiR-210 inhibits NF-κB signaling pathway by targeting DR6 in Osteoarthritis. Sci Rep. 2015;5:12775–12782.
[23] Kurowska-Stolarska M, Alivernini S, Ballantine LE, et al. MicroRNA-155 as a proinflammatory regulator in clinical and experimental arthritis. Proc Nat Acad Sci. 2011;108:11193–11198.
[24] Arnett FC, Edworthy SM, Bloch DA, et al. The American rheumatism association 1987 revised criteria for the classification of rheumatoid arthritis. Arthritis Rheum. 1988;31:315–324.
[25] Arnett FC, Edworthy SM, Bloch DA, et al. Modified disease activity scores thatinclude twenty-eight-joint counts. Development and validation in a prospective longitudinal study of patients with rheumatoid arthritis. Arthritis Rheum. 1995;38:44–48.
[26] Westergren A. Studies of the suspension stability of the blood in pulmonary tuberculosis. Acta Med Scand. 1921;54:247–282.
[27] Everitt BS. Contrasts and custom hypothesis. The Cambridge dictionary of statistics. 2nd ed. Cambridge: CUP; 2002;13:329–338. ISBN 0-521-81099-X.
[28] Mody GM. Reflections on rheumatoid arthritis in selected sub-Saharan African countries. East Afr Med J. 2009;86:201–203.
[29] Cobb BS, Hertweck A, Smith J, et al. A role for Dicer in immune regulation. J Exp Med. 2006;203:2519–2527.
[30] Li QJ, Chau J, Ebert PJ, et al. miR-181a is an intrinsic modulator of T cell sensitivity and selection. Cell. 2007;129:147–161.
[31] Yamasaki K, Nakasa T, Miyaki S, et al. Angiogenic microRNA-210 is present in cells surrounding osteonecrosis. J Orthop Res. 2012;30:1263–1270.
[32] Zhang Y, Fei M, Xue G, et al. Elevated levels of hypoxia-inducible microRNA-210 in pre-eclampsia: new insights into molecular mechanisms for the disease. J Cell Mol Med. 2012;16:249–259.
[33] Li X, Tian F, Wang F. Rheumatoid arthritis-associated MicroRNA-155 targets SOCS1 and upregulates TNF-α and IL-1β in PBMCs. Int J Mol Sci. 2013;14:23910–23921.
[34] Qi J, Qiao Y, Wang P, et al. microRNA-210 negatively regulates LPS-induced production of proinflammatory cytokines by targeting NF-κB1 in murine macrophages. FEBS Lett. 2012;586(8):1201–1207.
[35] O'Neill LA, Sheedy FJ, McCoy CE. MicroRNAs: the fine-tuners of Toll-like receptor signalling. Nat Rev Immunol. 2011;11:163–175.
[36] Samaan S, Khella HW, Girgis A, et al. miR-210 is a prognostic marker in clear cell renal cell carcinoma. J Mol Diagn. 2015;17:136–144.
[37] Faraonia I, Antonettib FR, Cardonea J, et al. miR-155 gene: a typical multifunctional. microRNA. Biochem Biophys Acta. 2009;1792(6): 497–505.
[38] Luzi E, Marini F, Sala SC, et al. Osteogenic differentiation of human adipose tissue derived stem cells is modulated by the miR-26a targeting of the SMAD1 transcription factor. J Bone Mineral Res. 2008;23(2):287–295.
[39] Elmesmari A, Fraser AR, Wood C, et al. MicroRNA-155 regulates monocyte chemokine and chemokine receptor expression in Rheumatoid Arthritis. Rheumatology. 2016;55(11):2056–2065.

FIGURE 9.2

An example of an original article. This research paper exhibits most of the features described in the text: (1) the title, (2) the list of authors and their addresses, (3) the abstract, with data and p value, (4) up to six key words, enabling indexing by on-line search engine, (5) the introduction, (6) details of the corresponding author, (7) the final sentence of the introduction has (ideally) a hypothesis, (8) the material and methods (with no separate section for the subjects, the number of which are clearly stated), (9) statistical methods, (10) results, (11) the discussion, which ends (12) with a conclusion, and (13) a statement of how the work represents a significant advance in biomedical science, (14) a summary table, and (15) references, shown in the text as numbers in square brackets/superscripts (16) such as [39]. Note in the references that only the first three authors are named.

In Brief or short reports

As may be expected, these papers carry a short message on perhaps only one or two research outcomes (a new molecule or gene, perhaps). The style and layout of the text will be the same as a full article, but journals generally set additional rules. For example, a 'Biomedical Science In Brief' submitted to the *British Journal of Biomedical Science* has no abstract or heading, but continuous text of up to 1750 words, with two tables/figures and up to 15 references. An example is shown in Figure 9.3.

SELF-CHECK 9.4

What are the most common reasons for the rejection of a manuscript?

Review articles

Update and review articles follow a similar pattern. They begin, like research articles, with a title and abstract. The opening part of the introduction will, as in a research article, set the scene and justify the interest in the subject. However, there will be no original hypothesis, but possibly an aim or objective, the latter possibly being to provide a focused and up-to-date summary of the subject. In place of a hypothesis there is likely to be a statement about how relevant publications were sourced, which will include the key words. This is particularly relevant in a systematic review (one which follows a logical and prescribed course of data collection and reporting) such as that of the Cochrane system.

The key aspect of a review article is its scope: the author(s) must decide on both the breadth and depth of the information they wish to convey. The literature review will be considerably more demanding for a review article than for a research article. There must be decisions regarding which articles to include and which are to be omitted. Having too many will lead to a lack of focus, whereas too few will lead to lack of depth.

One of the most popular search engines is 'Pub Med', part of the US National Institutes of Health (NIH). It can provide the abstracts of perhaps thousands of articles, depending on the general or specific nature of the key words used. A powerful feature of Pub Med is that the search strategy can be modified to include, for example, only review articles, or only articles where the entire text can be downloaded free of charge.

Review articles may have tables, but these will not cite original data; instead, they are likely to summarize key data from other studies. Similarly, if there are figures, they will be of the authors' own making, and will again summarize other work. In bringing together conclusions from a large series of research papers, reviews often exceed 5000 words, and the list of references often exceeds one hundred.

Other articles

Some journals accept guidelines, reports from meetings, etc. that are not research, but in the minds of the Editorial Board warrant publication. In some cases, abstracts from a scientific meeting may be published in a supplement to the journal. Some journals publish an annual update of the key works of the previous year, and those journals published by a learned society may have news and other information of value to their membership.

In some cases, editors are prepared to publish an even shorter version of research data in the form of a 'Letter to the Editor', perhaps one long communication of maybe 1000 words;

British Journal of Biomedical Science

ISSN: 0967-4845 (Print) (Online) Journal homepage: http://www.tandfonline.com/loi/tbbs20

Fasting urinary calcium to creatinine ratio for the evaluation of calcium nephrolithiasis in adults

Lijing Sun, Zhiyong Guo, Jianping Shan & Gengru Jiang

To cite this article: Lijing Sun, Zhiyong Guo, Jianping Shan & Gengru Jiang (2017) Fasting urinary calcium to creatinine ratio for the evaluation of calcium nephrolithiasis in adults, British Journal of Biomedical Science, 74:2, 101-103, DOI: 10.1080/09674845.2016.1264703

To link to this article: http://dx.doi.org/10.1080/09674845.2016.1264703

Full Terms & Conditions of access and use can be found at
http://www.tandfonline.com/action/journalInformation?journalCode=tbbs20

Download by: [Dr Andrew Blann] **Date:** 17 November 2017, At: 04:24

FIGURE 9.3 (CONTINUED)

BRITISH JOURNAL OF BIOMEDICAL SCIENCE, 2017
VOL. 74, NO. 2, 101–103
http://dx.doi.org/10.1080/09674845.2016.1264703

BIOMEDICAL SCIENCE IN BRIEF

Fasting urinary calcium to creatinine ratio for the evaluation of calcium nephrolithiasis in adults ①

Lijing Sun[a,b], Zhiyong Guo[b], Jianping Shan[a] and Gengru Jiang[a] ②

[a]Department of Nephrology, Xinhua Hospital Affiliated to Shanghai Jiaotong University School of Medicine, Shanghai, China; [b]Department of Nephrology, Changhai Hospital Affiliated to Second Military Medicial University, Shanghai, China

ARTICLE HISTORY Received 1 September 2016; Accepted 3 October 2016

③

Calcium has many important roles in physiology and pathology, such as in ensuring bone integrity, in maintaining haemostasis, and as a cofactor for metabolic enzymes. The kidney is the major organ responsible for blood calcium homeostasis and ideally maintains plasma levels under tight control through urinary excretion.[1] Accordingly, changes in urinary calcium excretion are likely to reflect alterations in homoeostasis. Levels of urinary calcium vary widely in different populations as a result of dietary habits, mineral composition of water, climate, genetics and race.[2] Hypercalciuria is a common clinical finding, which increases the risk of renal stones, principally calcium nephrolithiasis. In China, nephrolithiasis is a common condition with a lifetime prevalence of 4.8% in men and 3% in women.[3]

The current gold standard criteria for diagnosing hypercalciuria is urinary calcium excretion >4 mg/kg/24 h or >300 mg/24 h in men and >250 mg/24 h in women and is likely to be a directly relevant diagnostic test for calcium nephrolithiasis. Although urinary calcium excretion is best measured by a 24-h collection, this is not always suitable or practical as a test because of difficulties in sample collection and leading to invalid results. An alternative, the urinary calcium/creatinine (UCa/Cr) ratio is often used in this circumstance instead of a 24-h sample collection. Nordin first proposed the UCa/Cr ratio as a measure of urinary calcium excretion,[4] and although some studies showed a strong correlation between UCa/Cr ratio and 24 h urinary calcium excretion, two studies reported opposite result.[5,6] Consequently, the replacement of the 24-h urinary calcium excretion method with the UCa/Cr ratio is controversial and requires clarification.

Most studies focus on UCa/Cr ratio in paediatric populations: data on adult UCa/Cr ratio (especially in calcium nephrolithiasis) limited. Our study was initiated to determine the value of UCa/Cr ratio in discriminating normal adults from patients with calcium nephrolithiasis, and to determine the effects of factors that may influence the UCa/Cr ratio. We tested the primary hypothesis that the UCa/Cr ratio correlates significantly with 24-h urinary calcium excretion in healthy controls and in patients with nephrolithiasis. ④

This prospective study included 120 healthy volunteers and 120 nephrolithiasis patients. The participants selected were in the period between November 2011 and November 2014. The calcium nephrolithiasis patients with only one stone episode were diagnosed based on the clinical findings (medical history, physical examination, chemical composition analysis of nephrolithiasis, radiography and ultrasound). Exclusion criteria were patients with renal tubular acidosis, congenital bone disease, hyperparathyroidism, inflammatory bowel disease, those who were treated with bisphosphonates, calcium, vitamin D, steroids and diuretics were also excluded. The study performed according to the Declaration of Helsinki and was approved by the ethics committee of our hospital, all of the participants gave informed consent and took a standard Chinese diet. ⑤

First morning fasting urine samples were obtained for calcium, creatinine, albuminuria, pH and density. 24-h urine samples were analysed for urinary calcium excretion. Serum urea, creatinine, electrolytes, urinary calcium and albuminuria were determined by a Hitachi 7600 biochemistry analyser (Hitachi, Japan), urine PH and density were determined by Urisys 2400 analyser (Roche, Switzerland). Body mass index (BMI) was calculated as weight (kg)/height (m^2).

The SPSS 20.0 program was used for the data analyses. Data were expressed as mean with standard deviation, while Pearson's correlation analysis was used to assess the correlation between UCa/Cr ratio and 24-h urinary calcium excretion. Receiver operator characteristic (ROC) curves were constructed to establish the most appropriate cut-off values of UCa/Cr ratio that would be ⑥

CONTACT Gengru Jiang xinhuakidney@163.com ⑦

FIGURE 9.3 (*CONTINUED*)

Table 1. Demographic and laboratory data.

	Controls	Patients	P Value
Male sex (%, n)	73.3, 88	80.3, 97	0.167
Age (years)	51.6 (10.7)	52.8 (9.5)	0.358
BMI (kg/m²)	23.5 (3.0)	23.8 (3.1)	0.406
UCa/Cr(μmol/μmol)	0.15 (0.7)	0.28 (0.11)	<0.001
UAbl/Cr(μmol/μmol)	16.7 (5.8)	40.4 (23.6)	<0.001
Urine pH	6.0 (1.0)	5.8 (0.9)	0.188
Urine density (mg/mL)	1.016 (0.004)	1.015 (0.005)	0.401
24hr urinary calcium excretion (mg)	188 (54)	318 (37)	<0.001
Serum urea mmol/L)	4.3 (0.8)	4.2 (0.8)	0.379
Serum creatinine(μmol/L)	66 (11)	64 (13)	0.110
Serum calcium (mmol/L)	2.32 (0.08)	2.41 (0.83)	<0.001
Serum potassium (mmol/L)	4.0 (0.4)	4.0 (0.4)	0.824
Serum sodium (mmol/L)	140 (3)	140 (3)	0.636
Serum chloride (mmol/L)	106 (4)	106 (4)	0.546

Data mean (SD) or %, n (number of subjects).

indicative of a high probability of calcium nephrolithiasis. Multivariate linear regression analysis was performed to determine whether any of sex, age, BMI, Ualb/Cr, urine pH or density could influence the UCa/Cr ratio. *P* value < 0.05 was considered to be significant.

⑧ The characteristics of the participants involved in our study was shown in Table 1. There was no significant difference in age, gender, BMI, urine PH, urine density, serum urea, creatinine and electrolytes in both groups, the value of UCa/Cr ratio in nephrolithiasis patients was higher than healthy volunteers. We also found 24-h urinary calcium excretion, serum calcium and Ualb/Cr were increased in calcium nephrolithiasis patients as compared to healthy volunteers ($p < 0.001$).

The correlation between fasting UCa/Cr ratio and 24-h urinary calcium excretion was highly significant (Figure 1a). We also found positive correlations between UCa/Cr ratio and 24-h urinary calcium excretion in both groups, being 0.70 ($p < 0.001$) in healthy volunteers and 0.76 ($p < 0.001$) in calcium nephrolithiasis patients.

The ROC curve showed the most appropriate cut-off value of fasting UCa/Cr ratio for the estimation of calcium nephrolithiasis was 0.175 μmol/μmol (sensitivity 85.5%; specificity 71.7%; area under the curve 0.862; 95% CI 0.82–0.91; $p < 0.001$, Figure 1b). Multivariate linear regression analysis showed that 24-h urinary calcium excretion ($p < 0.001$), serum calcium ($p < 0.001$) and gender ($p = 0.002$) influenced the UCa/Cr ratio. Age ($p = 0.710$), BMI ($p = 0.252$), Ualb/Cr ($p = 0.817$), urine pH ($p = 0.498$) and urine density ($p = 0.162$) were unrelated to the UCa/Cr ratio.

⑨ The present study reported the value of fasting UCa/Cr ratio in normal adults and calcium nephrolithiasis patients, 24-h urinary calcium excretion and serum calcium were the influencing factors of UCa/Cr ratio. Our finding that this ratio has correlates with 24-h urinary calcium excretion suggests that it may be used as a predictive index of calcium nephrolithiasis. We emphasise that the subjects involved in this study were free of several factors (e.g. treated with bisphosphonates, calcium, vitamin D, steroids and diuretics) that may influence the UCa/Cr ratio. Although it has been reported that calciuria measured in a fasting second morning void sample is more representative than the 24-h urine collection, we decided to use the first morning void sample, because it has been widely accepted in the literature.[7,8]

Several studies show mixed results of UCa/Cr ratio in different adult populations.[4,9,10] For example, we found the value of UCa/Cr ratio in calcium nephrolithiasis patients to be higher than those reported by Arrabal

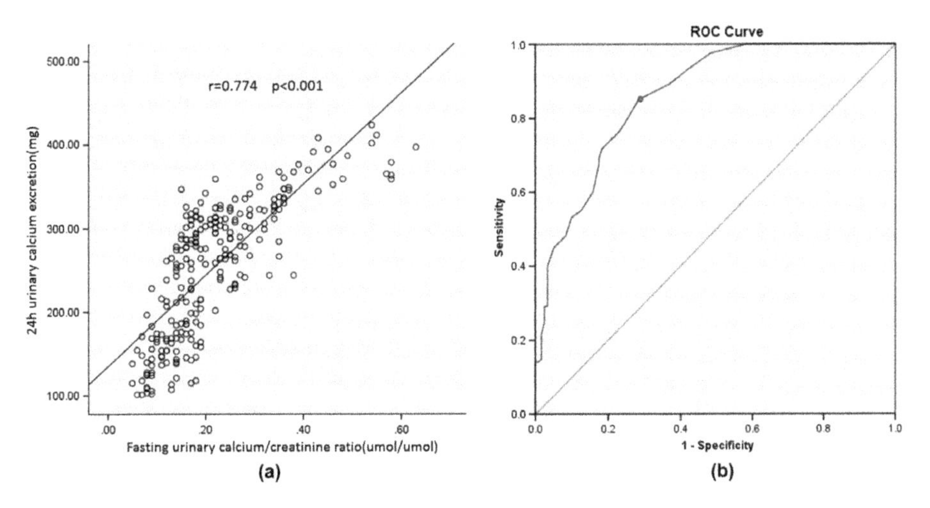

Figure 1. (a). Pearson correlation between fasting urinary calcium/creatinine (UCa/Cr) ratio and 24-h urinary calcium excretion. (b). Receiver operating characteristic (ROC) curve analysis of fasting urinary calcium/creatinine (UCa/Cr) ratio. The cut-off value of UCa/Cr ratio (marked with a black dot on the curve) was defined as 0.175 umol/umol (sensitivity, 85.5%; specificity, 71.7%; area under the curve, 0.862; $p < 0.001$).

FIGURE 9.3 (CONTINUED)

et al. [11]. This, and other differences, may be due to nutrition, age, sex, comorbidity, race and other factors which should be taken into consideration.

The correlation between UCa/Cr ratio and 24-h urinary calcium excretion is controversial. Gokce et al. observed a very strong correlation between Ca/Cr ratio in single-voided urine samples and 24-h total calcium excretion in adults.[12] Other studies showed a strong and positive correlation between the UCa/Cr ratio and 24-h urinary calcium excretion in calcium lithiasis patients.[13] We obtained the same results that the correlation between the UCa/Cr ratio and 24-h urinary calcium excretion was positive in both normal adults and calcium nephrolithiasis patients. Despite this agreement, some studies showed no convincing correlation, while one reported that the early morning spot urine cannot replace the 24-h urine collection in the evaluation of urinary metabolic abnormalities in those who form a renal stone.[14]

There are limitations in our study. First, we did not measure the UCa/Cr ratio on double urine samples, it is possible that samples collected at different times would exhibit different results. Next, was the lack of biochemical parameters that would allow for a full characterisation of calcium metabolism and acid-base status: we have no data on PTH, vitamin D or bicarbonate. However, we believe our study has adequate power to support our conclusions.

(10) ⇨ In conclusion, we recommend that calcium nephrolithiasis patients take a fasting UCa/Cr ratio test, because it is cheaper and simpler than the 24-h urine test and has demonstrated a comparable performance. Our data indicates that the most appropriate cut-off value for the prediction of calcium nephrolithiasis is 0.175 μmol/μmol. This work represents an advance in biomedical science because it shows fasting UCa/Cr ratio could be a predictive index of calcium nephrolithiasis and very simple and convenient for clinical use. ⇦ (11)

Acknowledgments

We thank all our colleagues for their excellent help and the healthy volunteers involved in our study.

Disclosure statement

The authors declare that they have no conflict of interest.

References ⇦ (12)

[1] Møller UK. Effects of adding chymosin to milk on calcium homeostasis: a randomized, double-blind, cross-over study. Calcified Tissue Int. 2015;96:105–112.
[2] Kaneko K, Tsuchiya K, Kawamura R, et al. Low prevalence of hypercalciuria in Japanese children. Nephron. 2002;91:439–443.
[3] Zeng Q, He Y. Age-specific prevalence of kidney stones in Chinese urban inhabitants. Urolithiasis. 2013;41:91–93.
[4] Nordin BE. Assessment of calcium excretion from the urinary calcium/creatinine ratio. Lancet. 1959;274:368–371.
[5] Koyun M, Güven AG, Filiz S, et al. Screening for hypercalciuria in schoolchildren: what should be the criteria for diagnosis? Pediatr Nephrol. 2007;22:1297–1301.
[6] Choi IS, Jung ES, Choi YE, et al. Random urinary calcium/creatinine ratio for screening hypercalciuria in children with hematuria. Ann Lab Med. 2013;33:401–405.
[7] Biyikli NK, Alpay H, Guran T. Hypercalciuria and recurrent urinary tract infections: incidence and symptoms in children over 5 years of age. Pediatr Nephrol. 2005;20:1435–1438.
[8] Madani A, Kermani N, Ataei N, et al. Urinary calcium and uric acid excretion in children with vesicoureteral reflux. Pediatr Nephrol. 2012;27:95–99.
[9] Topal C, Algun E, Sayarlioglu H, et al. diurnal rhythm of urinary calcium excretion in adults. Ren Fail. 2008;30:499–501.
[10] Sargent JD, Stukel TA, Kresel J, et al. Normal values for random urinary calcium to creatinine ratios in infancy. J Pediatr. 1993;123:393–397.
[11] Arrabal-Polo MA, Arrabal-Martin M, Arias-Santiago S. Bone and metabolic markers in women with recurrent calcium stones. Korean J Urol. 2013;54:177–182.
[12] Gokce C, Gokce O, Baydinc C, et al. Use of random urine samples to estimate total urinary calcium and phosphate excretion. Arch Intern Med. 1991;151:1587–1588.
[13] Arrabal-Polo MA, Arias-Santiago S, Girón-Prieto MS, et al. Hypercalciuria, hyperoxaluria, and hypocitraturia screening from random urine samples in patients with calcium lithiasis. Urol Res. 2012;40:511–515.
[14] Hong YH, Dublin N, Razack AH, et al. Twenty-four hour and spot urine metabolic evaluations: correlations versus agreements. Urology. 2010;75:1294–1298.

FIGURE 9.3

An example of an *In Brief* article. Differences from an original article include no abstract, continuous rolling text with no headings, and up to 15 references. Common features with an original article are: (1) the title, (2) the list of authors and their addresses, (3) an introduction that concludes (4) with an original hypothesis, (5) material and methods, that conclude (6) with a statistics section, (7) details of the corresponding author, (8) results, (9) the discussion, which ends (10) with a brief conclusion and (11) a statement of how the work represents an advance in biomedical science, and (12) references.

there will be separate author instructions for these letters. Some editors will also publish correspondence that is critical (or supportive) of previously published papers, in which the criticized authors will be offered a right of reply.

From writing to publication

It is probable that a research or review paper will go through several drafts before it is ready to be submitted. Furthermore, any co-authors will need to read, comment upon, and approve the manuscript. A statement of this is often demanded by the editor of the journal.

It is also worth seeking advice and guidance from colleagues who have writing experience. Furthermore, in the UK, each NHS Trust will have a Department of Research and Development who can offer advice, including the recommendations of a statistician. Certainly, the study of the layout of research and review articles will be invaluable in developing a sense of what is required and the way in which an article is written. But whatever the length (full paper, short report) or nature (research, review) of the manuscript, the most successful papers are those with a well-developed narrative, often bringing the reader along as though telling a story.

Once submitted via the journal's website, the corresponding author will receive an acknowledgement and reference number. The editor will then pass the manuscript to a number of referees, who will judge its quality, and upon whose opinions the editor will make a decision as to whether to publish. It is extremely rare for a paper to be accepted without change. The referees may point out a number of defects and/or make recommendations that must be addressed. Having done so, the author will take account of these recommendations in revision and then re-submit. The editor is then likely to return the manuscript to the referees for their final opinion. It may be that the referees do not consider the changes to have been sufficient in addressing their concerns. If so, it is likely that the editor will reject the paper.

If successful, however, the editor will pass the manuscript to the publisher who will format it for the house style of the journal. The publisher will then pass the formatted manuscript to the author who will scrupulously check these 'proofs' for error, and perhaps address any issues raised by the publisher. Once approved by the author, the proofs will be converted into the final copy to be published. The whole process can easily take six months, and so patience is required, but the wait will be worthwhile.

Plagiarism and dual publication

Plagiarism does not yet have a formal definition, but according to Gipp (2014) it is "the use of ideas, concepts, words, or structures without appropriately acknowledging the source to benefit in a setting where originality is expected". Others may put it more simply, such as the act of passing off someone else's work as one's own, and it is taken very seriously by publishers and in higher education. In its most crude form, this may be simply reproducing entire sections of text, something that can be detected by the use of one of several software packages designed for such a purpose. More subtle forms of plagiarism are more difficult to detect, but are none the less dealt with severely. Persistent offenders are likely to suffer sanctions, such as the refusal by publishers to consider submissions for a set period of time. Many consider plagiarism to be intellectual theft.

What is not plagiarism is the presentation of facts or materials from elsewhere as long as full credit is given to the originators of those facts or materials in the form of references. For example, it could be argued that a paragraph such as:

> One of the major developments in the pathophysiology of cancer is the realization of the importance of angiogenesis, possibly driven by growth factors such as VEGF and angiogenin. However, increased number of endotheliod cells in the blood may also have a role in the pathophysiology of this disease, perhaps reflecting the displacement or destruction of old endothelial cells and their replacement by new bone marrow derived cells (EPCs). Indeed, EPCs may be used as surrogates of the effects of anti-VEGF therapy, as VEGF may partially drive EPCs mobilization from the bone marrow.

is plagiarism, as the manner of presentation gives the impression that the ideas being presented are those of the author. By contrast, exactly the same text with added references does not give the same impression:

> One of the major developments in the pathophysiology of cancer is the realization of the importance of angiogenesis, possibly driven by growth factors such as VEGF and angiogenin[3-5]. However, increased number of endotheliod cells in the blood may also have a role in the pathophysiology of this disease[9-12,14,19], perhaps reflecting the displacement or destruction of old endothelial cells (CECs) and their replacement by new bone marrow derived cells (EPCs)[13,15]. Indeed, EPCs may be used as surrogates of the effects of anti-VEGF therapy[22,23], as VEGF may partially drive EPCs mobilization from the bone marrow[24,25].

The referencing makes it clear that the author is drawing (quite legitimately) on the work of others. In this case, some of the relevant references (adopting the Vancouver style) are:

> 22. Matsusaka, S., Mishima, Y., Suenaga, M., Terui, Y., Kuniyoshi, R., Mizunuma, N., and Hatake, K. Circulating endothelial progenitors and CXCR4-positive circulating endothelial cells are predictive markers for bevacizumab. *Cancer.* 2011;117:4026–32.
>
> 23. Ronzoni, M., Manzoni, M., Mariucci, S., Loupakis, F., Brugnatelli, S., Bencardino, K., Rovati, B., Tinelli, C., Falcone, A., Villa, E., and Danova, M. Circulating endothelial cells and endothelial progenitors as predictive markers of clinical response to bevacizumab-based first-line treatment in advanced colorectal cancer patients. *Ann Oncol.* 2010;21:2382–9.
>
> 24. Li, B., Sharpe, E.E., Maupin, A.B., Teleron, A.A., Pyle, A.L., Carmeliet, P., and Young, P.P. VEGF and PlGF promote adult vasculogenesis by enhancing EPC recruitment and vessel formation at the site of tumor neovascularization. *FASEB J.* 2006;20:1495–7.
>
> 25. Asahara, T., Takahashi, T., Masuda, H., Kalka, C., Chen, D., Iwaguro, H., Inai, Y., Silver, M., and Isner, J.M. VEGF contributes to postnatal neovascularization by mobilizing bone marrow-derived endothelial progenitor cells. *EMBO J.* 1999;18:3964–72.

Textbooks (such as this) don't always have extensive references in the text. They may instead have a bibliography, which lists all the main sources consulted during the preparation of the textbook.

In some cases, authors may directly reproduce the work of others. In the case of the reproduction of entire figures or tables (such as Figure 8.6), the permission of the authors and the publisher is required (as it was in the case of Figure 8.6 from Dr Rothwell), and appropriate credit must be given alongside the reproduced material.

As regards dual publication, data should be published as being original only once—and not merely because two sets of publishers, editors, and referees have been set to work, but also because of copyright issues. Once copyright has been granted to one particular publisher, it can't then be reproduced as new work with a different publisher. Publishing space is highly sought after, and only a fraction of submitted manuscripts are published—in the case of the *British Journal of Biomedical Science*, this is in the region of 15–20%. It follows that dual publication denies another researcher this vital commodity. Dual publication is also grossly mendacious as it can inflate a researcher's apparent output. If discovered, both publishers are likely to remove the citation and bring punitive measures on the culprit, such as refusal to consider other work. However, the publication of the abstract of a poster is not counted as a genuine publication.

Chapter summary

- Good communication is an essential part of good laboratory practice: failure to communicate well in a clinical setting may endanger life, and in other workplaces it may lead to poor experimental or process results.
- Probably the most written communication in the laboratory is the standard operating procedure; others include health and safety.
- Formal presentations of data or updates within and outside the laboratory are generally of posters and oral presentations on PowerPoint.
- All such presentations follow a common format with an introduction, material and methods (that may include details of subjects), results, and a discussion.
- Submission of data to an external meeting often requires an abstract, which will have the same format as above, but focused into perhaps 250 words.
- Should the data be strong enough, it may be submitted for publication, in which case authors must adhere to the instructions to authors.

Suggested reading

- **Systematic critique: the art of scientific reading.** *The Biomedical Scientist:* **2013;659–61.**
- **Gipp, B.** *Citation-based Plagiarism Detection: Detecting Disguised and Cross-language Plagiarism using Citation Pattern Analysis.* **Springer Vieweg, 2014. ISBN 978-3-658-06393-1**

Useful websites

- **The website of the British Journal of Biomedical Science: www.bjbs-online.org**
- **www.thecochranelibrary.com/ A valuable free resource of dozens of reviews.**
- **The 'Harvard' publishing style: http://education.exeter.ac.uk/dll/studyskills/harvard_referencing.htm**
- **The search engine Pub Med: www.ncbi.nlm.nih.gov/pubmed**
- **The 'Vancouver' publishing style: http://www.biomedicaleditor.com/vancouver-style.html**

Questions

9.1 Why is excellence of writing crucial in an SOP?

9.2 How does an original article differ from a short or brief communication?

Answers to end-of-chapter questions

Chapter 1

1.1 The mean is 31 and the median is 30.

1.2 This question is best answered by writing out the values in rank order, and then looking at the difference between individual data points.

Answer (a) is false: set A has a non-normal distribution, set B has a normal distribution.

Answer (b) is true. There are three values smaller and three values larger.

Answer (c) is true. However, as we have established that the data has a normal distribution, then the variation should be expressed as the standard deviation, not as the inter-quartile range.

1.3 Qualitative information can be converted to quantitative data if it is collected in a scale method such as a Likert scale, or the number of times a group of respondents have selected a particular word or words.

1.4 Qualitative: really (got a), terrible (bladder problem), hardly (gets a wink of sleep), most (of the night)

Quantitative: 46 years, 21 years, 5 or 6 times.

Chapter 2

2.1 The atomic weight of calcium is 40, so that 1 mole weighs 40 g. It follows that the average body contains $40 \times 25 = 1000$ g = 1 kg.

2.2 From Table 2.4, the relative molecular mass of sodium hydroxide is 40. Therefore 1 mole weighs 40 g, so for one litre of a 0.25 molar solution we need $40 \times 0.25 = 10$ g. However, since we need only 500 mL of the solution, we need only a quarter of this, that is, 5.0 g.

2.3 The number of protons defines the atomic number, which is therefore 26. The mass number is the sum of the protons and neutrons, so this is 56.

2.4 Potassium has an atomic mass of 39, chlorine of 35.5. So the molecular mass is 74.5, and one mole weighs 74.5 g. Hence 0.1 moles weighs 7.45 g.

2.5 Since pH = $-\log_{10}$ [hydrogen ion concentration], then rearranging the equation gives [hydrogen ion concentration] = $-\text{antilog}_{10}$ pH. This clearly requires some complex mathematics, but in practice pH5 transforms directly to a hydrogen ion concentration of 1×10^{-5} molar.

2.6 Both pH and bicarbonate are above the top of their reference ranges, indicating alkalosis. Diabetes can be a complex disease with various abnormalities. Additional tests are required, one of the most important being blood glucose.

Chapter 3

3.1 The mean is 46.3 and the standard deviation is 2.3. The coefficient of variation is the standard deviation divided by the mean, so is 4.9%.

3.2 Broadly speaking, the relationship is an inverted `U'. The upper and lower horizontal lines mark differences of 5% between the two methods. At low levels of A and B (on the left, perhaps 40 to 54 units/mL), there are eleven points where the difference between A and B is less than 95%. The middle section, where the mean values of A and B are between perhaps 58 and 92 units/mL, almost all data points are within 5% of the line of full agreement (i.e. 100%). The right side of the plot has eight data point were the A/B difference is again less than 95%. So although there is reasonable good agreement between A and B in the range 55 - 95 units/mL, outside this range agreement is poor, most likely because of a methodological discrepancy. Should method A be the existing routine technique, it is most unlikely that it will be replaced by method B.

3.3 The receiver-operating characteristic analysis plots the sensitivity and specificity of a method in its ability to distinguish between two groups. These may be patients with a particular disease versus healthy people, or patients with mild or severe disease. The plot shows us which result gives the best discrimination between groups, and so is a powerful laboratory and clinical tool.

3.4 Probably not. A value of 0.58 lies in the range 0.41–0.6, and so the agreement is moderate (Table 3.4), which for many would be unacceptably low. The agreement between the two methods should be at least good (kappa 0.61–0.80), ideally very good (kappa ≥ 0.81).

Chapter 4

4.1 Skewness refers to distribution. If the set is clustered over to the right or left (Figures 4.2 and 4.3), then it is skewed. The data set in Figure 4.1 is not skewed—the bulk of the data is in the centre.

4.2 A dot plot (such as Figure 4.12) shows each individual data point. A box and whisker plot (such as Figure 4.13) summarizes the data: individual data points rarely appear as outliers.

4.3 The standard deviation (SD) and standard error (SE) are closely related in that the SE is the SD divided by the square root of the sample size (n). The SD is the leading index for assessing the variability of an index. The SE is the leading index to assess the precision of the mean value of a set of data, and is often the basis of the 'error bar' when data is presented as a graph or (erroneously) as a histogram.

4.4 Mean 72.6, SD 26.1, median 63, IQR 55.75–83. From the self-check the data 'seems' to have a non-normal distribution, as most dots are clustered around the lower part of the figure. This is supported by the large difference of 15% or so between mean and median.

However, the SD/mean ratio is about 0.36, so it could have a normal distribution. Best to determine the distribution formally with an Anderson-Darling analysis.

Chapter 5

5.1 The key difference between QA and QC is that QC is done internally, whilst QA is performed in collaboration with a body outside the laboratory, so may be described as external quality assurance (EQA).

5.2 A graphical representation of this data will probably resemble the figure below:

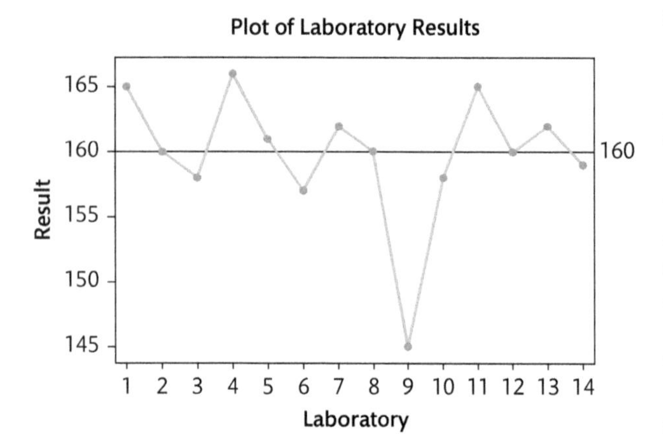

The horizontal bar is the mean of the entire 14 sets. All, except the result from laboratory 9 (whose result is 145), have returned a result between 166 and 157. The outlier is clear: laboratory 9 has returned a result about 10% below the mean of the entire group. If this difference is consistent over time, the senior staff of laboratory 9 may need to check their process.

Incidentally, the coefficient of variation (CV) of this set (the standard deviation divided by the mean) is 3.2%. Removing the outlying result of 145 reduces the CV of the remaining 13 points to 1.8%. This marked improvement of the CV reinforces the view that point 145 is unusual.

5.3 There are no right or wrong answers—each practitioner will have a different view of what they personally feel they are getting from reflective practice.

Chapter 6

6.1 Biomedical research is not done simply for its own sake—it must be directed towards an improvement in the health care of the individual, and so of the nation. This may be in preventing disease from developing, or better ways of treating disease once it has appeared. Because it mostly centres on the clinic, be it in primary care (at a general practice) or secondary care (at hospitals) it is called clinical research. However, research aimed at preventing disease in the general population is also clinical research, despite the fact that these subjects may not present to their GP or hospital for years.

In many cases, prevention and treatment of disease is by drugs, and the development of new drugs demands knowledge of physiology and how the particular disease develops—that being pathophysiology. Accordingly, early stages in new drug development involve testing potential agents to determine their effectiveness and safety in the laboratory before they are tested in people. This is called basic research, and is essential if we are to find new and effective ways of preventing and treating disease.

6.2 A major aspect of experimental work is the avoidance of two types of error: false positives and false negatives. The most likely reason for these appearing is that the number of experiments or people recruited to a study (the sample size) is too small. A formal power calculation is needed to tell us what the minimum sample size must be in order to minimize the risk of these errors being present. The power calculation itself relies on a quantified hypothesis.

6.3 A confidence interval of 22.5–27.5 units means that we are 95% sure that the true result lies somewhere between these numbers. If we want more confidence, such as being 97% sure that the true result is within a certain range, we need more data, such as may be obtained by recruiting more people to the particular study.

6.4 A case-control study compares two groups, one of whom has a particular disease or condition, whilst the other is free of that disease or condition. An intervention study places a group of patients on a particular drug, or has them undergo a procedure such as surgery, and then observes the effects weeks or months later.

6.5 The efficacy of the drug must first be tested in a likely biological model, such as tissue culture, which will provide broad clues as to the effective dose and also any potential cytotoxic effect. This work will inform an animal model, where the true biological effect will be determined. In both models, dose-response experiments will be performed where the dose of the drug is increased. NOTE. This answer is extremely simplified: formal drug development is in fact very complicated, highly regulated, and expensive. The entire process from basic idea to the bedside may take a decade.

Chapter 7

7.1 The t test is determined by the equation,

$$t = \frac{\text{mean of X} - \text{mean of Y}}{\left[(\text{SE X})^2 + (\text{SE Y})^2\right]^{1/2}}.$$

The SE is determined from the SD divided by the square root of the sample size. So for the first set, the SE = 25/4.69, = 5.33. For the second set, the SE = 17/5.2, = 3.27. Therefore,

$$t = \frac{119 - 105}{\left[(5.33)^2 + (3.27)^2\right]^{1/2}}.$$

$$t = \frac{14}{[39.10]^{1/2}}.$$

$$t = 14/6.254 = 2.24$$

Although not part of the question, with a total sample size of 49 and t = 2.21, the probability that the difference is real is <0.05.

7.2 The first step is to rank all the data, perhaps in a table (as shown in Table A), so that the rank of each data point can be determined. The sum of the ranks is then determined, being 34 in one and 44 in the other. The test statistic, W, is taken from the smaller sum, which is 34. From statistical tables, and 12 data points, this gives p=0.471.

TABLE A **Raw and ranked data**

First set	Overall rank	Second set	Overall rank
23	1		
		42	2
44	3		
45	4		
		67	5
		77	6
79	7		
86	8		
		87	9
		92	10
123	11		
		134	12
Sum of the ranks	34		44

Further inspection of the data sets shows that the median of the first set is 62, whilst that of the second is 82, which is 32% higher. This is a large difference, but is not significant because the sample size is so small. This difference would become significant with a sample size of 36 data points in each set.

7.3 There are only two sets of data where the relationship is significant. That between basophils and D-dimers is inverse ($r = -0.49$), so that those with a higher basophil count are likely to have low levels of D-dimers. Of approximately the same strength of relationship with $r = 0.47$ is the relationship between basophils and cholesterol. But this relationship is positive, meaning that someone with high serum cholesterol is likely to have a high basophil count.

Incidentally, since there is a basophil/D-dimer relationship, and a basophil/cholesterol relationship, we may have expected there to be a relationship between D-dimer and cholesterol. However, the lack of significance of these indices ($r = -0.11$, p=0.620) is because the positive relationship between cholesterol and basophils and the inverse relationship between basophils and D-dimers cancel each other out.

7.4 The Anderson-Darling test

7.5

Type of data	Test to be used
Differences between independent groups after the use of an ANOVA	Tukey's
Seeking a relationship between linked pairs of data (e.g. height and weight) where both have a normal distribution	Pearson's correlation
Difference between two independent groups that both have a normal distribution	Student's t
Difference between three or more groups that are linked by a common factor (i.e. are not independent)	Ordered groups
Difference between three or more independent groups where at least one has a non-normal distribution	Kruskal-Wallis
Difference between two sets of paired data (e.g. before/after an intervention) where the difference has a normal distribution	Paired t
Seeking a relationship between linked pairs of data (e.g. height and weight) where at least one has a non-normal distribution	Spearman's correlation
Difference between three or more independent groups that all have a normal distribution	ANOVA
Difference between two sets of paired data (e.g. before/after an intervention) where the difference has a non-normal distribution	Wilcoxon's

Chapter 8

8.1 Two sets of data with a normal distribution (those presented as mean and standard deviation) are compared with a Student's t test. Two data sets, of which one or both have a non-normal distribution (and so are presented as median and inter-quartile range), are compared with the Mann-Whitney U test. Data of a categorical nature (absolute number of people) is analysed by the Chi-squared test.

8.2 Multivariate linear regression analysis can only be used when the dependent variable has a continuously variable distribution. In this analysis, the dependent variable is categorical, and as there are only two categories, then binary logistic regression is used.

8.3 The positivity or negativity of these indices depends entirely upon the designation of an endpoint as 0 or 1—the choice is made by the researcher, often for arbitrary reasons. Similarly, sex may be entered as 1 or 2—whichever of the two sexes is which number does not matter. However, what does matter is to be absolutely sure about the meaning.

In our case an increased C-reactive protein, as a marker of inflammation, is likely to be a predictor of a poor outcome. It follows that the regression analysis shows that low levels are associated with a good outcome, and that a low level brings a reduction of the risk by 39% (that is, 1−0.61 as a percentage).

The disease duration is positive, so the risk will be the reverse of that for C-reactive protein. So a high disease duration indicates an increased risk of suffering an endpoint. This also makes sense—the longer the disease has been present, the more likely it is that it is likely to get worse. However, a high disease duration increases risk by only 8%, and as the 95% confidence interval straddles zero, this increased risk is not significant.

Chapter 9

9.1 Each procedure in the laboratory is important, and so must have a clear set of directions or instructions. Should these be ambiguous or vague, the practitioner will inevitably make errors and the procedure will fail.

9.2 The major difference is the mass of work being presented. A good original article will have a clear original and quantified hypothesis with an appropriate sample size. Complex analyses may include those looking at different aspects, perhaps with a broad range of techniques and/or data. In a clinical setting there may be serial measurements or outcomes. Short communications still present original data, but in a limited form and with few perspectives. This does not necessarily mean that all original articles have a larger sample size than all short reports—it is the depth, quality, and applicability of the data that is important.

Answers to self-check questions

Chapter 1

1.1 Age is measured in years and the distance travelled in units such as miles or kilometres, so this information is quantitative. You might say that their degree of tiredness and hunger is slight, considerable, or marked, and that they would choose cinema, theatre, opera, or another medium of entertainment. These are all qualitative.

1.2 Quantitative data can be continuous and lie on any point within a range (such as height, weight, age in years) or categorical, being in one of a small number of discrete 'boxes', with no intermediate states (such as sex, alive or dead, ABO blood group).

1.3 The mean is 241.7 (being the sum of all the data points, divided by 11), the median is 256 (lies in the middle of the data points when ranked from smallest to largest), and the mode is 267 (it occurs twice, all others occur once).

1.4 Data set 100 (20) has the smallest variation, whereas 75 (30) has the largest variation. Note that the variation (the SD) does not depend on the mean. However, the *relative* variation, being the SD divided by the mean, does vary with the mean.

1.5 We can describe the variation of different types of data in terms of standard deviation and inter-quartile range.

1.6 Focus groups are relatively easy to put together, and can provide a great deal of data. However, this data is strongly dependent on the nature of the focus group, and it may be difficult to extrapolate their conclusions to other situations.

Chapter 2

2.1 There are many, many possible results. For example, the weight of an adult female is approximately 75 kg, whilst her height may be 1.5 m. In haematology, a normal prothrombin time would be 12 seconds, whilst the freezing point of water is 273 K. In biochemistry, an acceptable level of potassium in the blood is 4.5 mmol/L.

2.2 Derived units of relevance to biomedical science are frequency (such as number of breaths per minute), pressure (partial pressure of blood gases), temperature (of the body), and radioactivity (the decay of a radioisotope).

2.3 0.000005 units has six zeros (that is, a million) in front of the number five. The prefix for one millionth is micro- (μ), so that 0.000005 units is also 5 μU.

2.4 450 000 converts first to $4.5 \times 100\,000$, and so to 4.5×10^5.

2.5 The atomic number defines the number of protons, the atomic mass the sum of the number of protons and neutrons. Hence phosphorus has $31 - 15 = 16$ neutrons.

2.6 The relative molecular mass of potassium carbonate is the sum of the relative atomic masses of potassium (39×2), carbon (12) and oxygen (16×3), being 138 daltons.

2.7 The relative molecular mass of calcium carbonate is the sum of the relative atomic mass of calcium (40), carbon (12), and oxygen (16×3), being 100. Hence:

$$\text{Number of moles} = 40\text{ g}/100\text{ g mol}^{-1}$$

$$\text{Amount of substance (mol)} = \frac{40\text{g}}{100\text{g mol}^{-1}} = 0.4\text{mol}$$

2.8 The relative molecular mass of magnesium chloride is the sum of the relative atomic masses of magnesium (24) and chlorine (35.5×2), being 95. So as 1 mole weighs 95 g, then 9.5 grams is 0.1 moles. This mass dissolved in a litre of water would produce a solution with a molarity of 0.1, but since the volume of water is only a quarter of a litre (250 mL) then the molarity is correspondingly higher, that is, 0.4 molar.

2.9 Since $\text{pH} = -\log[\text{H}^+]$, and the concentration of hydrogen ions is 10^{-8} mol L^{-1}, then $\text{pH} = -\log_{10}[10^{-8}]$, which is pH 8.

2.10 The sodium result is above the top of the reference range, whilst the potassium result is low. The patient may have a degree of renal disease, and additional tests may be necessary.

Chapter 3

3.1 Method A gives a mean value of 126.6 (to one decimal place), with a range of 122 to 133. Method B gives a result of 125.8, with a range of 119 to 134. Method C gives a mean value of 128.5, with a range of 127 to 130.

Although Method B gives the closest result to that of the existing method (125), it has the largest range ($134-119 = 15$). So, although the most accurate, it is also the most imprecise. Method C is the most precise, with a range of values of only 3, but the mean value is far from the target (128.5 versus 125), so it is the least accurate method. Method A is between the two, with an accuracy better than that of C, but less than that of B.

The question of which is best is answered by balancing precision versus accuracy. Unfortunately, the most accurate method is also the least precise (Method B). Although Method C is most precise, it is least accurate. This therefore leaves Method A, as the best compromise, with the best combination of accuracy and precision.

3.2 The CV is the standard deviation (SD) divided by the mean. The SD of each method is 3.59, 5.35, and 1.08 respectively. Hence:

$$\text{Method A: CV} = 3.61/127.2 \times 100 = 2.84\%$$

$$\text{Method B: CV} = 4.60/126.7 \times 100 = 3.63\%$$

$$\text{Method C: CV} = 1.08/128.5 \times 100 = 0.08\%$$

Therefore Method C is the most reproducible, with the smallest CV, and Method B is the least reproducible, as it has the largest CV.

3.3 This question is answered by constructing a table as follows:

	Scientist A positive	Scientist A negative	Total
Scientist B positive	32	8	40
Scientist B negative	11	54	65
Total	43	62	105

The kappa statistic is derived by the following mathematics:

1. There is total agreement in 32 + 54 of the 105 samples, giving a proportion of 0.82
2. The total proportion of negative results is $40 \times 43/(105)^2 = 0.156$
3. The total proportion of positive results is $65 \times 62/(105)^2 = 0.365$
4. Add these two proportions: $0.156 + 0.365 = 0.521$

$$\text{Kappa} = \frac{0.82 - 0.52}{1 - 0.52} = \frac{0.3}{0.48} = 0.62$$

According to Table 3.4, a kappa value of 0.62 suggests that the level of agreement between the two scientists is 'good'. However, note that they agreed about 82% of the samples, failing to agree on 18%. The analysis makes no judgement about right or wrong, or of the implications of the comparison. This is for senior staff to make.

3.4 There are 95 true positives (TPs), 13 false positives (FPs), 11 false negatives (FNs), and 120 true negatives (TNs). These give:

$$\text{Sensitivity} = \text{TP}/(\text{TP} + \text{FN})$$
$$= 95/(95 + 11) = 95/106 = 90\% \text{ or } 0.90$$

$$\text{Specificity} = \text{TN}/(\text{TN} + \text{FP}) = 120/(120 + 13)$$
$$= 120/133 = 90\% \text{ or } 0.90$$

$$\text{PPV} = \text{TP}/(\text{TP} + \text{FP}) = 95/(95 + 13) = 88\% = 0.88$$

$$\text{NPV} = \text{TN}/(\text{FN} + \text{TN}) = 120/(11 + 120) = 92\% = 0.92$$

3.5 The positive likelihood ratio is given by the formula positive LR = sensitivity/(1−specificity), hence 0.9/1−0.9 = 0.9/0.1 = 9. The negative likelihood ratio is similarly given by the equation negative LR = 1−sensitivity/specificity, hence 1−0.9/0.9 = 0.1/0.9 = 0.11

Chapter 4

4.1 Plotting the X and Y values is likely to give a graph and line of best fit such as that shown in Figure A.

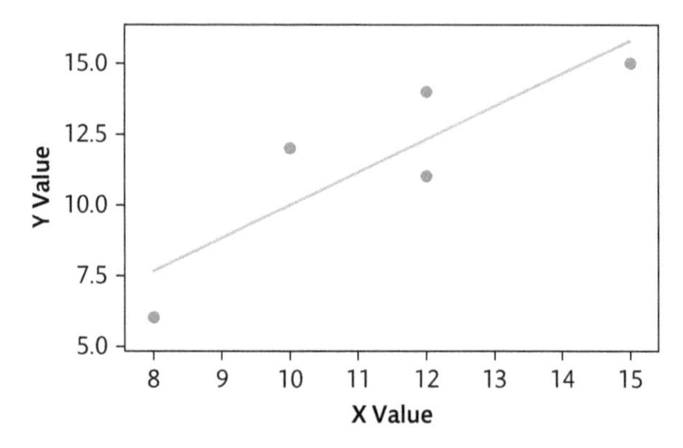

FIGURE A Scatterplot of Y value versus X value

4.2 Plotting these data points should give a figure such as that shown in Figure B.

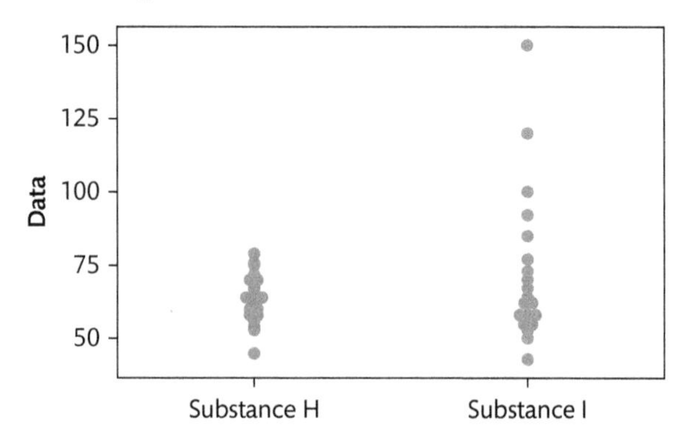

FIGURE B Individual value plot of Substance H and Substance I

Note that the Substance H data is clustered quite tightly together, between about 48 and 77, and accordingly is likely to be normally distributed. Conversely, the Substance I data is clustered towards the bottom-there are two high outliers (data points 120 and 150). Accordingly, the data is likely to have a non-normal distribution.

4.3 Plotting the temperature against month should produce a figure such as that shown in Figure C.

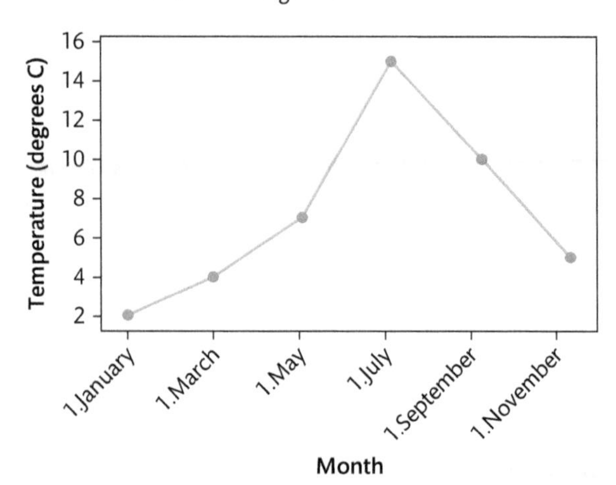

FIGURE C Scatterplot of temperature (degrees Celsius) versus month

The temperature is highest in July, strongly suggesting the data arises from a geographical location in the northern hemisphere.

Chapter 5

5.1 The International Organization for Standardization (the ISO) publishes guidelines and recommendations regarding audit. Its website is http://www.iso.org.

5.2 The four fundamental steps in audit are (a) Plan: prepare an audit checklist, (b) Do: perform the audit, (c) Respond: analyse the results, (d) Act: initiate corrective action (if required).

5.3 There are several possible examples of horizontal audit that can be found in a laboratory. One is in ensuring that all items with an electrical power supply are regularly checked by an independent electrician. Another might be the distribution of various grades of staff at different times of the day and week.

Chapter 6

6.1 This is a false positive in that it finds that men live longer than women. The reverse is the case–women live longer than men.

6.2 A null hypothesis may be that men and women have the same length of hair. An alternative hypothesis is that women have longer hair than men. A quantified hypothesis is that women's hair is 6 cm longer than men's hair. Probably the major confounder would again be age, as men tend to lose hair naturally when older.

6.3

(a) There are four queens and 52 cards, so the probability is 4 in 52, which is 0.077.

(b) The probability that one card will be a spade is 1 in 4, which is 0.25. If this is repeated twice more the probability will be $0.25 \times 0.25 \times 0.25 = 0.016$ (to two significant figures).

6.4 By applying the data into the equation, we get,

$$70.9 \pm 1.96 \times (9.0 / \text{square root of } 36)$$
$$= 70.9 \pm 1.96 \times (9/6)$$
$$= 70.9 \pm 1.96 \times (1.5)$$
$$= 70.9 \pm 2.9$$

Therefore the 95% CI is 68.0 to 73.8 kg.

Chapter 7

7.1 The first step is to construct a two-by-two table, placing all those with rheumatoid arthritis (RA) or osteoarthritis (OA) in different boxes, as in Table 7.4. This is shown in Table A. According to the null hypothesis, the proportion of those with RA free of an event (50%) is the same as those with OA (50%). So of 145 people free of an event, there should be 72.5 in each category–this is the expected number. However, the actual (observed) number is 65 and 80 respectively. So the equation $(O-E)^2/E$ becomes $(80-72.5)^2/72.5$ (= 0.776) and $(65-72.5)^2/72.5$ (= 0.776) for the RA and OA patients respectively.

TABLE A **A two-by-two table for a Chi-squared analysis**

	Free of an event	Suffered an event	Total
Rheumatoid arthritis	65	35	100
Osteoarthritis	80	20	100
Total	145	55	200

Similarly, the null hypothesis states that there is an equal proportion of RA and OA people suffering an event, and since there are 55 events, we should expect 27.5 events per group. However, we have observed 35 and 20 events per respective group, giving $(O-E)^2/E$ values of $(35-27.5)^2/27.5$ (= 2.045) and $(20-27.5)^2/27.5$ (= 2.045) respectively.

The sum of these four equations (0.776 + 0.776 + 2.045 + 2.045) is the Chi-squared statistic, which is 5.643. Applying this statistic to an appropriate table will give p=0.018, and that the null hypothesis has been rejected. So we can be 98.2% sure that the alternative hypothesis is correct, and that there *is* a significant difference, which is genuine.

This tells us that in rheumatoid arthritis the greater risk of suffering an event (35%) is statistically significant when compared with the risk in osteoarthritis (20%).

7.2 A preliminary check of Figure 7.4 is to look at the histogram graphic, which implies the data has a non-normal distribution, and this is supported by the Anderson–Darling value of 1.44, giving $p<0.005$. The SD is about 25% of the mean, suggesting a normal distribution, but this is countered by a skewness greater than 1.

The mean is only 6.4% higher than the median (suggesting normality), but the median (20) is not in the middle of the 1st and 3rd quartiles (18–24), nor the minimum and maximum (14–35). Therefore, as most of the indicators suggest a non-normal distribution, then it should be compared using non-normal tests.

7.3 The key is Table 7.10, which summarizes different methods for assessing distribution. There are no graphics for Groups G, H, and I, so we rely on the descriptive statistics. In each case the SD is less than 12.5% of the mean, and the difference between the mean and the median is less than 3.5%. Each median value is relatively central between the IQR and minimum/maximum. All of these are evidence of a normal distribution.

7.4 Methods for determining distribution without a formal test such as that of Anderson–Darling include the relative size of the standard deviation in terms of the mean, the difference between the mean and the median, and the degree to which the median is roughly in the middle of the inter-quartile range or the full range. These are summarized in Table 7.10.

7.5 A paired t test is used when the distribution of the difference between two data sets has a normal distribution. A Wilcoxon's test is used when the distribution of the difference has a non-normal distribution.

Chapter 8

8.1 The equation is $y = mx + c$. The y component is the index on the vertical axis, and is called the dependent variable. The x component is the index on the horizontal axis, and is called the independent variable. The m is the gradient of the straight line, which in multivariate analysis is called the coefficient. The c component is the point where the straight line crosses the y axis.

8.2 In linear regression analysis, the coefficient is an equivalent of the gradient of the straight line (the 'm' in $y = mx + c$). But as with almost all statistical indices, there is variability in this index. The standard error effectively tells us how confident we can be that the coefficient is reliable. The ratio between the coefficient and the standard error (the t value) goes one step further in this search for assurance. The value and the sample size give us the probability that these indices from which they are derived are genuinely related, and that any such relationship is not due to chance.

8.3 An odds ratio of 1.16 translates to 16%. This may mean a 16% increased risk of being in a particular group, such as suffering or being free of an endpoint. However, the 95% confidence interval of 0.95–1.46 (which includes 1) translates to a -5% to 46% increase in risk, and as this straddles zero, this 16% increase is not significant. Similarly, an odds ratio of 0.85 translates to a 15% reduction in the risk of an event compared with the reference range. As the 95% confidence interval of 0.71–0.96 (which does not include 1) translates to a 4%–29% risk reduction, which does not include zero, this improvement is significant.

8.4 Factors that are significantly different in those with diabetes compared with those free of diabetes are age ($p<0.001$), splenectomy ($p=0.001$), hypogonadism ($p<0.001$), being positive for hepatitis C virus RNA ($p=0.01$), and the frequency of myocardial iron overload ($p=0.026$).

Chapter 9

9.1 Poor communication may well lead to error, misunderstanding, and so failure to correctly process a sample. This is turn could have serious clinical consequences.

9.2 Introduction, Materials and Methods, Results, Discussion.

9.3 There is no specific answer—just those described by colleagues.

9.4 As described in Box 9.4: Major aspects are that the subject matter must be appropriate for the journal and original. The sample size must be sufficiently large so the hypotheses are robustly tested, and the manuscript must be formatted according to the instructions to authors.

Index

Note: Tables, figures, and boxes are indicated by an italic *t*, *f*, and *b* following the page number.